W0042038

CALCIUM BINDING PROTEINS IN NORMAL AND TRANSFORMED CELLS

ADVANCES IN EXPERIMENTAL MEDICINE AND BIOLOGY

Editorial Board:

NATHAN BACK, *State University of New York at Buffalo*

IRUN R. COHEN, *The Weizmann Institute of Science*

DAVID KRITCHEVSKY, *Wistar Institute*

ABEL LAJTHA, *N. S. Kline Institute for Psychiatric Research*

RODOLFO PAOLETTI, *University of Milan*

Recent Volumes in this Series

A Continuation Order Plan is available for this series. A continuation order will bring delivery of each new volume immediately upon publication. Volumes are billed only upon actual shipment. For further information please contact the publisher.

CALCIUM BINDING PROTEINS IN NORMAL AND TRANSFORMED CELLS

Edited by

Roland Pochet
Free University of Brussels
Brussels, Belgium

D. Eric M. Lawson
AFRC Institute of Animal Physiology and Genetics Research
Cambridge, United Kingdom

and

Claus W. Heizmann
Division of Clinical Chemistry, Department of Pediatrics
University of Zurich
Zurich, Switzerland

PLENUM PRESS • NEW YORK AND LONDON

Library of Congress Cataloging in Publication Data

European Symposium on Calcium Binding Proteins in Normal and Transformed Cells
(1st: 1989: Brussels, Belgium)
  Calcium binding proteins in normal and transformed cells / edited by Roland
Pochet, D. Eric M. Lawson, and Claus W. Heizmann.
     p.   cm. — (Advances in experimental medicine and biology; v. 269)
  "Proceedings of the First European Symposium on Calcium Binding Proteins in Nor-
mal and Transformed Cells" — T.p. verso.
  Includes bibliographical references.
  ISBN-13: 978-1-4684-5756-8   e-ISBN-13: 978-1-4684-5754-4
  DOI: 10.1007/ 978-1-4684-5754-4
  1. Calcium-binding proteins — Congresses. 2. Calcium-binding protein genes —
Expression — Congresses. 3. Cancer cells — Congresses. I. Pochet, Roland. II. Lawson,
D. Eric M. III. Heizmann, Claus W. IV. Title. V. Series.
QP552.C24E93   1989   90-6704
591.19'245 — dc20   CIP

Proceedings of the First European Symposium on
Calcium Binding Proteins in Normal and Transformed Cells,
held April 20–22, 1989, in Brussels, Belgium

© 1990 Plenum Press, New York
Softcover reprint of the hardcover 1st edition 1990
A Division of Plenum Publishing Corporation
233 Spring Street, New York, N.Y. 10013

All rights reserved

No part of this book may be reproduced, stored in a retrieval system, or transmitted
in any form or by any means, electronic, mechanical, photocopying, microfilming,
recording, or otherwise, without written permission from the Publisher

PREFACE

The First European Symposium on Calcium-Binding Proteins in Normal and Transformed Cells was held at the Faculty of Medicine of the "Université Libre de Bruxelles" in Brussels, Belgium, April 20-22, 1989. Delegates from seventeen countries attended. This Symposium was initiated through an EEC Stimulation Program.

The formal program included forty verbal presentations by invited speakers and sixty miniposter presentations, and was formulated by the Organizing and Scientific Committee : E. Carmeliet (Leuven), J.P. Collin (Poitiers), S. Forsen (Lund), C.W. Heizmann (Zürich), D.E.M. Lawson (Cambridge), P. Miroir (Brussels), J.L. Pasteels (Brussels) and R. Pochet (Brussels).

This volume contains the papers prepared by the invited speakers. The contributions are grouped according to their general subject matter : Genes of Calcium-Binding Protein Family, Structure/Function Relationships, The Cytoskeleton and Calcium-Binding Proteins, Calcium-Binding Proteins in Transformed Cells, Calcium/Lipid-Binding Proteins, Calcium-Binding Proteins Substrates and Immunohistochemistry of Calbindin and Calretinin.

The highlights of the symposium are numerous. Among the items to be noted are the growing number of abundant proteins which interact with calcium and sometimes with other second messenger systems; specifically p11 associated with the tyrosine kinase calpactin, calcyclin, p9Ka induced by growth factors, MRP-8 and MRP-14 (also called cystic fibrosis antigen, L1 or calgranulins) forming half the soluble protein of granulocytes. New structure/function relationships on calbindin D9K and calmodulin have emerged from nuclear magnetic resonance and site-directed mutagenesis studies. Questions of function were raised for many calcium-binding protein families and there is an increasing need to investigate their physiological roles.

Calcium-binding proteins is a field of scientific endeavour which will continue to have great implications on modern biology.

The Organizing Committee would like to acknowledge the financial support of Amersham Belgium, Applitek, Banque Bruxelles Lambert, Banque Degroof, Banque Nationale de Belgique, Bayer Belgium, Beecham Pharmaceuticals, Boehringer Pharma, The British Council, Duphar, European Economic Communities, Fonds National de la Recherche Scientifique, H.V.L., I.B.M., Jeol (Europe), Labaz-Sanofi, Loterie Nationale, Ministère de la Région Bruxelloise, Northern Shipping Company, Pfizer, Pharmacia-LKB, Sandoz, Schweizerische Naturforschende Gesellschaft, Swissair, Teijin Ltd (Japan), Travex Voyages and Van Hopplynus. Without this support, the Symposium would never have taken place.

Roland Pochet
Eric Lawson
Claus W. Heizmann

PREFACE

CONTENTS

4

CALCIUM AND CELL STEADY STATES

R.J.P. Williams

Inorganic Chemistry Laboratory, University of Oxford, South Parks Road, Oxford OX1 3QR, UK

INTRODUCTION

Since the realisation that calcium was a major trigger of activity both in extracellular and intracellular systems a change has come about in our knowledge of calcium binding in different biological zones. We know now that there is a wide diversity of calcium controlled systems. Triggering is not just a matter of a simple switch in calcium ion levels associated with a cell followed by simple binding to a target. Each eukaryotic cell has a complex shape and a complicated set of compartments, reticula and organelles. Within different zones of the cell cytoplasmic membrane and of the different compartment membranes there are differentially placed transport and receptor systems which use calcium. Again within the compartment and cytoplasmic aqueous solutions there is a variety of response systems, binding units, which act in rather different ways to the calcium. This is not just a matter of the calcium binding constants but is reflected in the effect of calcium on the organisation of protein components in the cell. The extent of these organised reactions is not known but includes contraction elements, themselves related to cell shape changes, exocytosis and perhaps cell division. The extent of the findings, see Table I, leads me to suggest that calcium is more than a trigger : it is also an internal coordinator of the cell integrating its compartmentalised activities through differential, kinetically controlled, i.e. energy-linked, homeostatic devices while the cell is at rest. Rest state homeostasis needs a re-examination.

An appropriate starting point for the discussion of calcium levels and their variations in cellular systems is a list of the sources of calcium within the whole organism. There are three major zones which will be subdivided later.

(1) The external body fluids often holding calcium concentration at a fixed level and in equilibrium or close to equilibrium with bone, see later. The calcium concentration is here $> 10^{-3}$M.

(2) The cell cytoplasm keeping calcium concentration below 10^{-7}M by the expenditure of energy which is used to pump calcium outwards in opposition to minor leaks inward.

(3) Intracellular vesicular compartments, reticula, etc..., into which much calcium is pumped to give a level around 10^{-3}M.

Given this calcium distribution it can be used in two major triggering modes.

(a) Organic chemicals (proteins), which have not seen high calcium concentrations since synthesis in the cytoplasm, can be released to the extracellular fluids.

(b) External or vesicular calcium can be triggered so as to enter the cell cytoplasm when calcium is a messenger.

Now if such a complex system is to have a reliable trigger response function the different compartments must be held at rest with constant calcium concentrations. In the case of the bone supply to body fluids the buffering is managed by the huge bone store. The only problem is the slow release of calcium

Table 1 . Calcium controlled events in cells

Activity	Control
Photosynthesis	Dioxygen Release
Oxidative Phosphorylation	Dehydrogenases
Receptor responses	(a) Nerve Synapse
	(b) IP_3-linked reactions
Contractile Devices	(a) Muscle triggering (actomyosin)
	(b) Cell filament controls (tubulins)
Digestion	Activation of hydrolases
Adhesion and Cell Association	Surface Glycoproteins
Immune reaction and Clotting System	Complement Reactions and Gla-proteins
Membrane/filament organisation	Calpactin-like proteins provide tension
Cell Division	S-100 proteins(?)

from bones. To this end there are two sets of cells which make and dissolve bone respectively under hormonal control. The simplest device for breakdown is the local adjustment of pH by the osteoclast together with release of enzymes to break-down the protein matrix of bone. Associated with the control mechanism and limiting bone dissolution is the inhibitory protein osteocalcin to which I shall return and there are several other calcium-binding proteins in bones which prevent fast dissolution. The second device is in the cells which deposit bone i.e. the osteoblasts. The sensitivity of the system, solution/precipitation, of bone is such that bone growth responds not only to genetic controls but also mirrors stresses on itself (fig. 1). The bone/solution "equilibrium" is seen to be really a steady state under kinetic control and may be supersaturated though not far from equilibrium, see below. The distance from the equilibrium is very valuable and at the same time full of risks.

When we examine the problem of the bone matrix more closely we find that there are additional problems. The solubility product (fig. 1) for small particles is a function of their surface energy. This surface energy is however not just a problem of the crystal chemistry of apatite nor even of the particle size, but also concerns the control over the surface energy by the binding matrix as well as by physical stresses. The binding matrix is almost always maintained in mechanical tension and so the crystals are in compression. The apatite of biology has a range of solubility products and quite possibly for small particles the observed product of $[Ca]^2[OH][PO_4]$ *only exceeds the solubility product* of bound, unstressed, small crystals at pH = 7.

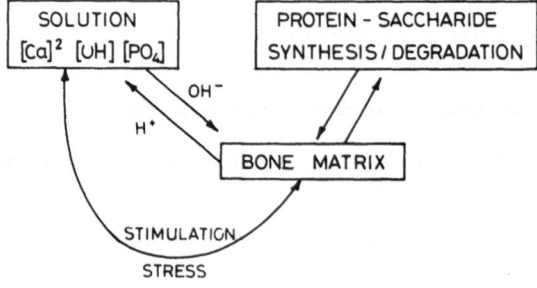

Fig. 1 . The complex controls on the precipitation of calcium phosphate apart from the availability of calcium and phosphate include the synthesis of matrices, the pH, the effect of cellular activities and the application of stress.

Table 2 . Extracellular Calcium Proteins

Protein	Binding Centre
EGF-Domains of Proteins C, S, Z and VII, IX and X	Hydroxy Aspartate plus Two Aspartates
Protein C Osteocalcin Prothrombin	γ-Glutamate plus Aspartate or Glutamate
Complement C_{1r} and C_{1s}	EGF-like?

The ionic product may not even be sufficient to nucleate apatite in the absence of a matrix. The removal of matrix, collagen, the lowering of pH, or the application of stress then cause bone to dissolve. The levels of calcium in external fluids to cells is a critical matter for the cells as we shall see. The first point is then that bone is part of a curious kinetically controlled local homeostasis not really related to thermodynamic buffering. We shall want to know if there is feed back to collagen synthesis from the levels of calcium outside the cell.

The need for the homeostasis of external calcium as a kinetic phenomenon can be seen too in the balance between clotting of blood and clot dissolution which are taking place continuously. The controls of the two rates rest on the actions of calcium on the thrombin cascade (and its regulation by the thrombomodulin system) and the plasmin activation series.

The "resting" state is a fixed binding level of calcium to the multiple calcium sites of the various proteins in all three parts of the system using especially -carboxyglutamate (Gla) and hydroxyasparate centres (Table 2). The sensitivity to calcium is then extreme. The resting flow state of the whole system is perturbed by the release of more calcium locally by platelets following injury. It is therefore essential that bone calcium and extracellular calcium concentration should be well regulated - once again by Gla-proteins. The stress in this article is not the activation of the cascade systems by calcium after injury but is on the resting state of the cascades which are in constant fluxes. Calcium controls the flux rates at all times so that the actions of fibrin clot formation and dissolution are not switched on and off from a zero level but their relative rates are manipulated. This is a well-known feature of rapid-response control mechanisms and we shall see that a parallel circumstance holds in the cells. In the blood stream it is thought that platelets can increase calcium levels *locally* after injury and if this is so it is a third triggering mechanism to be added to the two above. Platelets must have a very high stored-calcium.

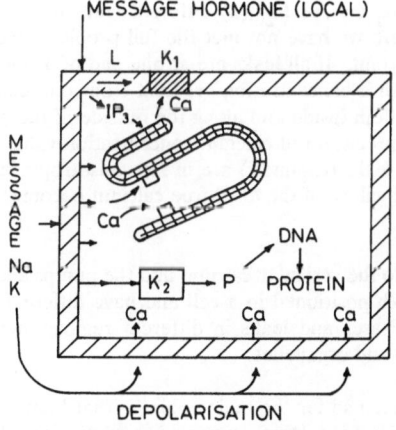

Fig. 2 . The major modes of calcium triggering (1) by depolarisation of the whole cell or a region (synapse) (2) by receptor communication to the inner stores of reticula (shown hatched) by calcium or IP_3. K represents a ATP utilising process e.g. kinases or pumps. The integration of calcium and phosphate activities is clear.

Turning to the other large stores of calcium there are the different vesicular reticula. The best known is the sacroplasmic reticulum but it seems that all complex cells have calcium stores in somewhat similar reticula vesicles. The calcium may be stored with the help of a protein such as calciquestrin. Again I shall return to this protein. Unfortunately the exact level of calcium in the different reticula is not known but it is maintained by an outward (from the cytoplasm) calcium ATP-ase pump. In animal cells the mitochondria also take up calcium from the cytoplasm and there is little doubt that calcium levels are raised in organelles such as the Golgi (Fig. 2).

There is a big difference between the relationship of bone to the buffering of body fluid calcium and the relationship of the reticula stores to the vesicular and cytoplasmic fluid concentrations. Homeostasis based on a crystalline solid is virtually invariant provided the kinetics of solid/solution exchange are rapid since the solubility product is fixed for reasonably large crystals (but note the caveats above)

$$[Ca]_o^2[OH][PO_4] = K_{sp}$$

Here calcium concentration is invariant if pH and phosphate concentrations are fixed. The level of calcium in the reticula is related to the pumping ability from the cytoplasm. In steady state

$$[Ca]_c.pumping = K_1[Ca]_c[ATP]_c = [Ca]_R.leaks$$

where c is the cytoplasm, K_1 is a combination of kinetic and thermodynamic constants related to pump activity, and R the reticular solution and leaks are back to the cytoplasm. By itself this could allow considerable variation in calcium concentrations in vesicles and in the cytoplasm but we know that there are further considerations in that calcium is also pumped to the extracellular fluids by a similar ATP-ase

$$[Ca]_c(pumping) = K_2[Ca]_c[ATP]_c = [Ca]_o leaks$$

where o represents the outside solution and leaks are again to the cytoplasm. K_2 is defined as for K_1 above. These two equations imply that calcium concentrations in different vesicles are not just dependent on the number and character of the pumps and leaks in the vesicle and cytoplasmic membranes but also are dependent on the extracellular calcium. In some vesicles the calcium concentration is close to that in the intracellular fluid, as these equations might lead one to expect. Care must be taken not to exceed this value since it could cause calcium precipitation, but notice the absence of the collagen matrix. [Some vesicles in some organs or cells of particular species do have precipitates of calcium salts in their vesicles]. There has to be a manipulation of the terms K_1 and K_2 to be sure of the calcium levels. Before we look more closely at the homeostatis in the cytoplasmic and vesicle fluids we must examine more carefully the concentration levels which are being discussed.

Looking first at the cytoplasm the most intriguing problem arises as to the free calcium level. Is it the same in all cells or does it vary from say $10^{-7}M$ to $10^{-8}M$? Such variation would require either (1) different pumps i.e. differences in K_1 and/or K_2, (2) different leaks or (3) different numbers of leaks and pumps in different cells. Even here we have not met the full problem since the disposition of pumps and leaks in any one cell will be important. If all leaks are at one end of a cell and all pumps at the other then calcium flows through the cell with a concentration gradient from one end of the cytoplasm to the other. This generates a calcium current both inside and along the outside of the cell. There is then only homeostasis of calcium gradients and no precise fixed calcium concentration value in or around cells. Increasingly there is evidence that pumps and leaks (channels) are in fact anisotropic in their positions in cells and currents are present. Now the actual values of the *local* free calcium becomes important since there are many anisotropic features of cells.

Applying the same logic to the vesicular calcium and the pumps and leaks of the vesicle membranes and given that the vesicles are both positioned in a cell and have different curvature in different zones we may suppose that there are ATP-ases and leaks in different regions and there can be calcium currents around the surfaces of the vesicles and organelles.

To summarise the discussion so far the major point is that homeostasis, the resting condition, is a dynamic energy-dependent condition for calcium currents in a particular cell and not a thermodynamic condition such as is "buffering" in chemistry. The rate constants controlling homeostasis, i.e. the steady state balance, can be dependent upon ATP levels (energy), numbers of leaks and pumps, and the efficiency of the pumps, see K in the above equations, the calcium concentrations on the opposite side of a membrane, and the control over especially pumps and channels by other chemicals, hormones etc..., and by calcium itself in a feed-back control. We imply too that even in cells of uniform calcium external concentration/internal concentration ratio this latter concentration may vary greatly from cell to cell. We must then remember that if there are activities of cells at rest which depend on $[Ca^{2+}]^n$, where n is a large number, some cells will behave very differently from others although they have similar cytoplasmic protein complements. In an anisotropic homeostasis where there is a fixed *gradient* of calcium ions in the cell these remarks apply to *local* regions of a given cell, e.g. the synapse of a nerve cell, which can therefore differ in its calcium activities from the bulk of the cell. It is probable that at rest there is a constant flow of calcium in a variety of aqueous media in and around cells.

Putting these views of the connections between $[Ca]_o$, $[Ca]_c$ and $[Ca]_R$ together with the discussion of the bone matrix there is a sense in which it is all one homeostatic mechanism at rest which can be understood in terms of exchange monitoring from one compartment to another. In other words there are in the resting state an assembly of calcium pools, each pool talking to others, and this set of pools is deeply involved in the whole management of the organism. Calcium *energised resting levels* are then an overall organising device and changes in calcium resting levels can generate a change in the whole organism. This is quite a different problem from triggering and is in fact related to regulation.

CALCIUM AND PHOSPHATE

Before we continue the analysis of the resting state, which is a state of activity, we must note the equal involvement of phosphate. This is obvious in bone. There may not be an interesting concentration gradient of phosphate in a resting cell but there must be fixed ratios of differently bound forms of phosphate through which phosphate flows continuously. Some of these forms are nucleotide phosphates (NTP, NDP, NMP, c-NMP, etc...), some are substrate phosphates (glucose phosphate), and some are protein phosphates. The inter-relationship between this group of compounds containing phosphates in different energised conditions and bound in different signalling modes is in parallel with the differently energised calcium compartments and bound calcium signalling modes *at rest*. Quite apart from triggering signals the cell must have these maintenance signals. Now the calcium and the phosphate systems are not independent but speak to one another and separately to different parts of themselves by feed-back mechanisms e.g. ATP dependent calcium channels and calcium dependent phosphate metabolism. There is a Ca/P active network in the resting state, i.e. a joint homeostasis (fig. 3).

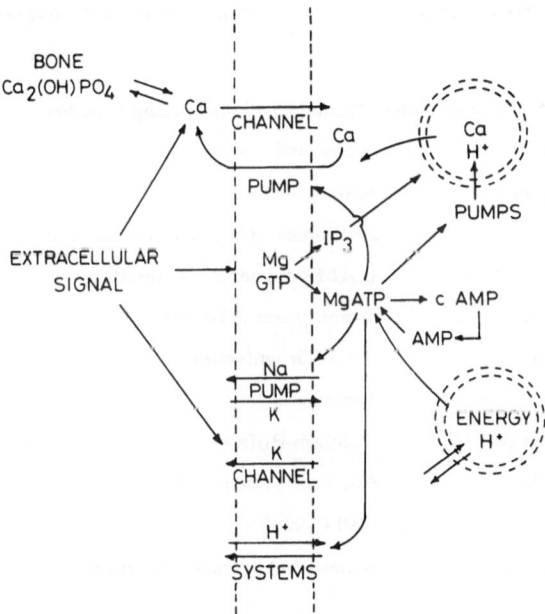

Fig 3. The connection between extra- and intra-cellular networks including protons (energy), NTP (energy), and ionic gradients.

Table 3 . Some extracellular response proteins

Protein	Response
Osteocalcin	Unfolded → Folded
Prothrombin[*]	Gla-peptide → Folded
Trypsinogen	Terminus unfolded for hydrolysis
Phospholipase A.2	Ca-binding generates active site

N.B. All these proteins are -S-S- cross linked and/or have considerable β-structure.

[*]Prothrombin is just one example of many calcium-dependent proteins in the blood-clotting cascade.

Now many of the calcium-binding systems bind to several calcium ions (Tables 3 and 4) and it is known that the calcium pumps themselves are regulated by proteins, calmodulins, which bind more than one calcium. The sensitivity of the system, dynamic homeostasis, $[Ca]/T_{Ca}$ is a function $[Ca]^n$ locally. The advantage is obvious in that a small change in free calcium now gives a switch from say 10% to 90% activity. (For $n=4$ the change in $[Ca]$ required is only 3-fold which contrasts with $n=1$ which requires a 100-fold change to cover the same range) (see fig. 4).

Let us consider what are the conditions of re-setting in an anisotropic cell which makes contact say with a surface (another cell perhaps) with which it "wishes" to cooperate. Calcium entry may be caused *locally* at the point of contact and local phosphate reactions follow. If such entry stimulates responses internally it is understandable that these can be such that they generate growth of filaments in the activated *local* region which then helps to secure the attachment by generating a pseudo-pod. This mechanism of local "growth" of a cell in a specific direction at the expense of loss of cell material elsewhere can arise from many forms of stimulation. For example it may well be that the stimulation is of a nerve cell by a message input so that the cell extends making new contacts very locally. Such a device produces a cell shape or network of cell shapes representing external events which represents a new resting state once they come into balance. It

Table 4 . Some internal Calcium-Binding Proteins

Protein	Response
Calmodulin	Kinase Activation
	Phosphase (Calcineurin) Activation
	c-AMP diesterase Activation
Troponin-C	Actomyosin ATP-ase
Gelsolin	Actin Organisation
Calpain	Protease
Parvalbumin	Calcium Buffer?
Calbindin	Calcium Transport?
S-100	Cell Cycle?
Annexin	Kinase (membrane) activation

N.B. All these proteins are largely helical and undergo helix/helix motions.

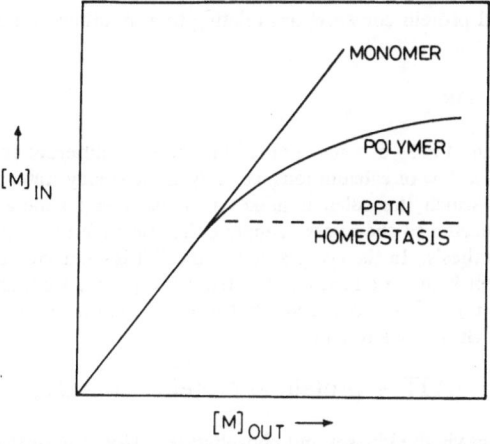

Fig. 4 . The homeostatic controls over Ca(M) inside a cell. The linear plot represents a simple buffer, the curve shows buffering by a cluster or a polymer with many binding sites, while precipitation gives the broken line. The two axes are not on the same scale.

is a growth dependent memory and would benefit by reinforcement or repetition to consolidate the growth. We consider this to be through a phosphorylation cascade (fig. 5 and see below). Another example would be the regrowth of cells to one another to form new contacts after injury.

All such responses are associated with filamental growth which have large numbers of units so that the response is both local and precipitous, large n, i.e. extremely sensitive. It is initially reversible but finally irreversible due to the establishment of a new cell pattern. This is a change of rest state regulation. Every different shape has a different calcium flow pattern at rest.

      In summary at rest a complex system of cells can be in a homeostatic steady state each type of cell differentiated and each organelle differentiated. All of the homeostasis can be linked to an external material such as bone or any other salt or matrix which has little dependence for its own thermodynamic activity upon the free calcium e.g. a condensed gel. Given such a generalised energised gradient of an ion specific messages are sent to local targets, that is local cells or parts of cells, by manipulation, of the energised calcium gradients differentially. The action generated depnds on the concentration, the value of n, the level of calcium at rest, and the proteins which receive the calcium. If the system responds by growth it reaches a new

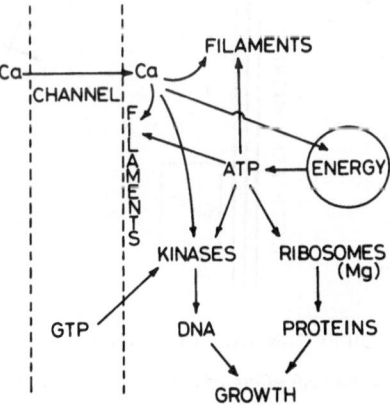

Fig. 5 . A much simplified view of the way in which calcium levels can influence the whole cell via phosphorylation.

resting state which has novel protein constructions relating to a modified homeostatic level of calcium and phosphate metabolism.

THE CALCIUM INTEGRATOR

The picture (fig. 2 and fig. 3) of a series of calcium pools of different concentrations in different cell compartments connected by a flow of calcium ions, i.e. a dynamic steady state, is highly suggestive when we remember the connections which the calcium makes in these compartments. In the mitochondria or chloroplasts calcium is connected to the energy supply either through the dehydrogenases or through the dioxygen release of photosynthesis. In the cytoplasm of the cell it is connected to the tension of the filamentous constructions of different kinds in an extremely intricate way since we know that it is not just the triggering of tension (troponin C) which is connected but also the maintenance of tension. In particular the level of phosphorylation, which is a steady state

$$ATP + protein \rightarrow protein.P + ADP,$$

is controlled partly by calcium via the kinases and phosphatases. Note the further connection to polymerisation of filaments

$$\underline{n}protein\ P \rightleftharpoons [protein\ P]_n$$

Since the organelle energy generation is of ATP there is now an integrated link via calcium communication between the metabolism of the cell and the mechanical devices and even the synthesis of fibres. We do not know yet the extent of the stress fibres but quite probably they link the whole of the cell i.e. membrane, cytoplasm, nucleus and vesicular systems (fig 6). The integration is via mechanical/calcium-controlled and phosphate controlled devices. An exactly parallel treatment can be used to describe the cell/cell interactions.

SUMMARY OF INTEGRATING FUNCTION OF CALCIUM

Before asking how calcium acts at the molecular level through special sets of proteins it may be useful to give a brief conclusion to which this analysis of calcium functions appears to lead.

Calcium is the (universal) connector between changes of electrical or chemical concentrations in ex-

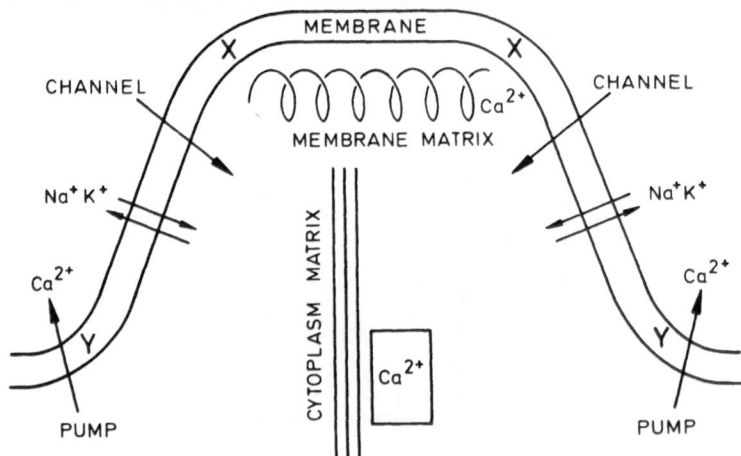

X,Y DIFFERENTIAL CURVATURE

Fig. 6 . A representation of the mechanical effects of calcium in a cell at rest or after triggering. Calcium inputs (channel) are far from the pumps. Different enzymes and proteins flow into regions of different curvature X and Y. The combination of adjustments we presume to be related to steady state activity of the cell.

ternal fluids or body fluids and mechanical activity inside and outside cells. The mechanical interactions are the major constructs of cells and cell/cell systems. In order to serve this function the calcium ion is held at rest in a state of balanced flow throughout the organism. Only against the background of such a strict kinetic homeostasis could external changes produce precise responses including growth. A metal ion with no redox properties and in fast exchange is peculiarly suitable for this overall integrating control function since it is not metabolised. Calcium ions have the peculiar advantages of combining charge, modest binding, no redox states and fast kinetics. However while the calcium ion can be linked to these many activities of cells it is not directly connected to the energy supply, the synthetic mechanisms, and to DNA. All are essential for a complete description of homeostasis and all three are connected to phosphate metabolism which itself is intimately linked to calcium (fig. 3). The power of phosphate chemistry in biology rests in three features

(a)   the carrying of negative charge;
(b)   the formation of energised condensation products as the first chemical synthetic step, e.g. pyrophosphate (ATP);
(c)   the absence of redox chemistry - no free radicals.

While (a) and (c) are parallels with calcium chemistry (b) gives the connection to a variety of sustained kinetic controls over DNA, enzymes and so on, as well as with triggering via second messengers such as c-AMP, GTP hydrolysis and so forth. However neither the concentration nor the cross-membrane distribution of phosphate itself seem to be so critical.

The present article is concerned with calcium but everywhere one turns one finds that calcium chemistry is linked to phosphate chemistry inside cells although phosphate usually has a lower profile outside cells. Here I include filament growth and structure, metabolism, signalling, generation of tension (cell division). The conclusion is almost inevitable that the dynamic homeostatic control of cells rests with some product function of calcium and phosphate. An immediate suggestive connection is of course the solubility product of bone, $[Ca]^2[PO_4][OH]$.

[A parallel point could be made with regard to organisms which do not have bone but use $CaCO_3$ as their mineral deposits. The control of CO_2 is of course pH and the atmosphere which then links the external environment to the organism via calcium controls].

While suggesting the primary standing of phosphate metabolism and calcium with respect to the maintenance of a given cellular homeostasis we do not neglect the known effects of the vast range of hormones which are connected to cell activities and balances. Many of course act directly through calcium and/or phosphate metabolism that is to say that they adjust the homeostasis locally at the organ level. Here we return to the idea that the base levels for calcium and phosphate metabolism may well differ from cell type to cell type. Other hormones act directly at the level of DNA. Here one could say that the adjustment is to the protein complement of the cell representing a switch in cell function. The question then arises as to the effect of such a switch on the calcium and phosahte homeostasis.

A change in the all-embracing Ca^{2+}/P system of homeostasis might lead to failure. The failure could lead to cell death of course but it could generate a whole system of changes and a failure of many of the control mechanisms before leading to a new steady state. In this state the form of the cell, the surface responses and the internal metabolic system could be adjusted. Does this represent a metamorphosis or it could be a cancerous state ? The link of these changes to calcium is often suggested and the link to kinases and phosphorylations is known. In other words because the homeostatis or integrating role of calcium/phosphate-controls is so complete there is a knock-on effect from any change in it to a wide range of functions both in the cell and here we include the transcription of the DNA and the surface of the cell. We do not see just new gene products due to two or three mutations but a switch in the on/off state of many genes, a change in cell shape and form, i.e. a changed mechanical character of the cell, and an adjusted metabolism all of which are now in a new homeostatic relationship of the Ca/P system. One major obvious connection is to the variety of kinases.

The next section describes the interaction of the calcium ion with the special evolved proteins which match its properties. There is a similar set of proteins opposite phosphate signalling.

THE PROTEIN INVOLVED

Assuming a dynamic (flow) homeostasis using calcium and phosphate metabolism there must be a corresponding dynamic set of proteins to respond to changes. We know that there is a set of calcium sites in proteins which are unoccupied in the range of calcium concentration 10^{-7} to 10^{-5}M and which respond in 10^{-3} sec. This is the calcium trigger function and it depends on the dynamics of proteins such as calmodulins. I am now stating that there is also another set of calcium sites in proteins with binding constants in the range $> 10^7$. These sites belong to sites of a very similar kind, say EF-hands, but they respond more slowly. They monitor the dynamic rest condition not triggering. Both series of calcium binding proteins speak to phosphate metabolism through bound phosphates e.g. kinases and phosphatases. These kinases are also dynamic proteins which can be activated and lead on phosphorylation changes of proteins and so adjust the nature of the whole cell. There is one set for phosphate trigger responses which reverse quickly to the rest state and another set for the control of the resting state which are more or less set with the calcium resting levels. They need not respond so rapidly and their reversal will be relatively slow. Prolonged changes of calcium resting steady states therefore mean a change in the nature of the cell. Since changes in the phosphate metabolism have the same effect and there is a constant feedback between calcium levels and phosphate metabolism via dynamic protein structure much of resting dynamic homeostasis is built around these two ingredients and their corresponding dynamic proteins. A cell is therefore characterised by its temporary homeostasis, where temporary has a time scale build in by the nature of the organism. Changing the inputs to calcium levels of a long-term kind generate changes in the cell e.g. of shape (i.e. growth of nerves), of constant inhibition (tumour cells) or of growth potential (fertilisation) or differentiation. Representative proteins are in Table 1 compare the dynamic proteins of triggering Table 4.

If this hypothesis is to be sustained there must be a corresponding set of dynamic homeostatic proteins in the extracellular fluids which look after the extracellular matrix. These proteins will be very like those which are ejected from cells in order to trigger extracellular activity but now the binding constant range of the ejected proteins can be greater than the permanent proteins for extracellular homeostasis. Some proteins of the appropriate kind are listed in Tables 2 and 3.

TRIGGERS AND SWITCHES

There is a great difference between (a) triggering plus relaxation and (b) switching of resting states. A trigger/relaxation sequence lasts some milliseconds; a switch is sustained for virtually infinite time, i.e. it is regulatory. Both effects are under kinetic control. In the first a calcium pulse is removed before a long-term effect on phosphorylation cascades is initiated. The calcium level returns to the pre-trigger state. In a switch the effect of the change calcium input is sustained either by rapid repetition of the input message and/or by a sufficiently changed phosphorylation sequence. For example the trigger could be just an effect on channels, tensions and pumps, while the switch would be on a kinase, say kinase C, which causes a cascade of effects including synthesis and altering calcium level changes, the whole being irreversible. The constantly running steady state is illustrated in fig. 7.

ACKNOWLEDGEMENTS

A review article such as this needs no references. It depends on the work of hundreds of people reported in this and other volumes on calcium biochemistry and is intended as a possible stimulation of thinking.

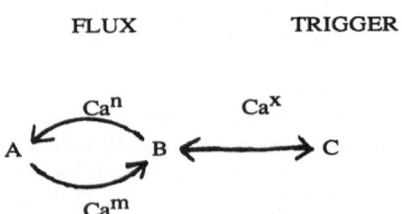

Fig. 7 . A representation of the dynamic states, A and B, of calcium involvement in cellular activity. The left hand side shows the constantly energised, i.e. energy consuming, calcium flow in and around all cells at rest. Any change of the level of free calcium affects the whole balance of the resting state through the different power levels, $\underline{n}$ and $\underline{m}$, to which the calcium concentration is raised on the input and output arrows. This metabolic rest state can be triggered (and reversed) by a sudden single calcium pulse, which has a different power dependence, $\underline{x}$, to a state C. Repetitive pulsing will change the nature of the cell.

EVOLUTION OF THE "EF-HAND" FAMILY OF CALCIUM-BINDING PROTEINS

Christine Perret, Nour-eddine Lomri and Monique Thomasset

INSERM, U 120, 44 Chemin de Ronde, 78110 Le Vésinet, France

The family of the intracellular calciproteins that bind calcium with an affinity constant Ka $\simeq 10^4$ to $10^8 M^{-1}$ includes the two vitamin D-induced calcium-binding proteins (CaBP) now termed calbindins D. In this family of proteins each calcium-binding domain is organized as a consensus structure, "EF-hand" domain (1). The "EF-hand" calcium-binding family can be divided into two groups :
- group 1 comprises proteins having four or less "EF-hand" domain such as calmodulin (Cd), myosin light chain (MLC_1, MLC_3, MLC_2), parvalbumin, calpain, CaBP 9K;
- group 2 is made up of proteins with six "EF-hand".

Several authors have proposed that the proteins of the first group are all derived from a four-domain ancestor resulting from two sucessive duplications of a single "EF-hand" domain (see ref. 2 for review). Indications of gene duplications should be recognizable in the structure of the genes. We have therefore analyzed the position of the introns in relation to the "EF-hand" domain and the phases of the intron to detect if any conservation revealed any symmetry that might exists.

The structure of the genes is shown schematically in fig. 1. The positions and the phases of certain introns are highly conserved, confirming their common evolutionary origin. But such a scheme does not shows the two original duplications. However, if we carefully examine the sequence, we find that two amino acids are highly conserved :
- the first is the glycine residue in the interdomain region just after site I and site II;
- the second is the first amino acid of the F helix, which is very often a leucine residue (L) in site I and site II and a phenylalanine residue (F) in sites II and IV. This observation supports the concept of two successive duplications in the formation of the four domain ancestral gene. When an intron interrupts the conserved glycine codon, it is always of phase 1 (fig. 1). MLC_2 possesses the same intron between sites II and III and shows the duplication event, LFLF. We propose that this intron is the vestigial intron which would have permitted the duplications (fig. 2). This hypothesis is in full agreement with the theory of Gilbert (4) on the role of introns. The distinct branches of this family would be derived from the four domain ancestral gene (PA_4, fig. 2) by remodeling the structure of this ancestral gene, principally by loss and insertion of introns (see ref. 2 for more details). We have made a sliding alignement of the calpain genes (fig. 1) to be in agreement with the alternative LFLF event and to align the conserved glycine codon. Site I of the calpain is aligned with site II of the ancestral progenitor. Dot matrix analysis with Staden's diagon of the calpain versus calmodulin, which is the protein closest to the ancestral progenitor (PA_4), is an argument in favor of this sliding (2).

Analysis of the structure of the genes coding for the CaBPs with two domains, similar to S100, is outlined in fig. 3. The structures of all these genes are similar, with a phase 0 intron in the interdomain region between each site. The genes could be derived from the two domain primordial ancestor (PA_2) or from PA_1 by loss of two domains. According to Goodman et al (5), and to Baba et al (6), they probably arose from PA_4. The only proteins of the second group whose complete structure is known is the chicken CaBP28K gene. It is outlined in fig. 4. Analysis indicates that there is no alternating LFLF sequence. The first residue of the F helix is always a leucine residue. The conserved glycine codon, which was highly conserved in the others members of this family, is also not present.

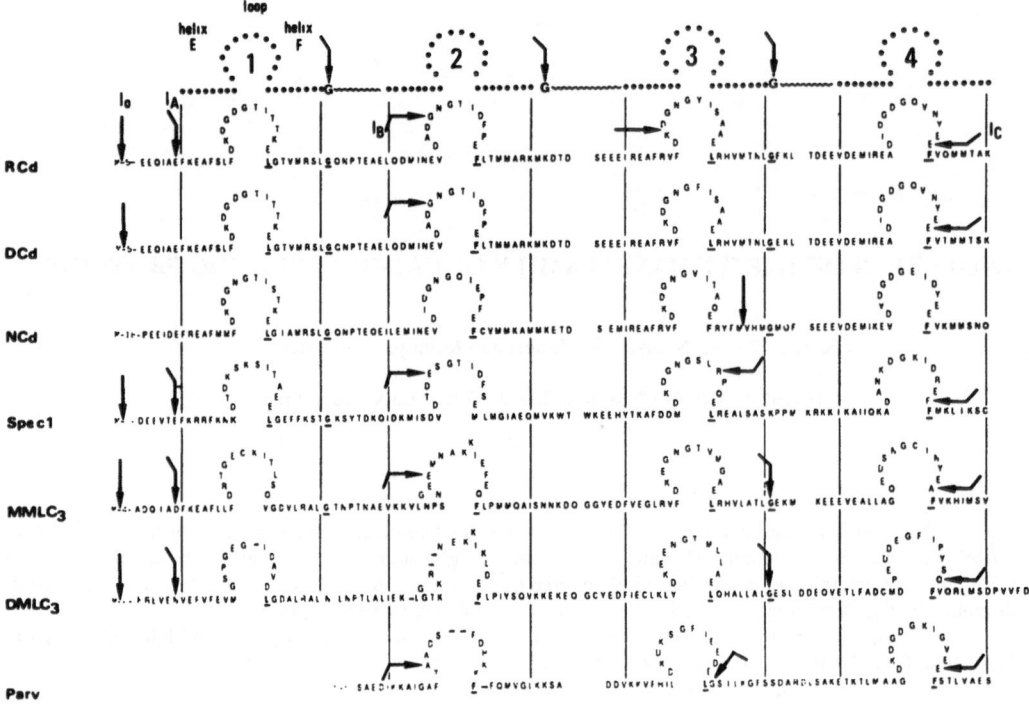

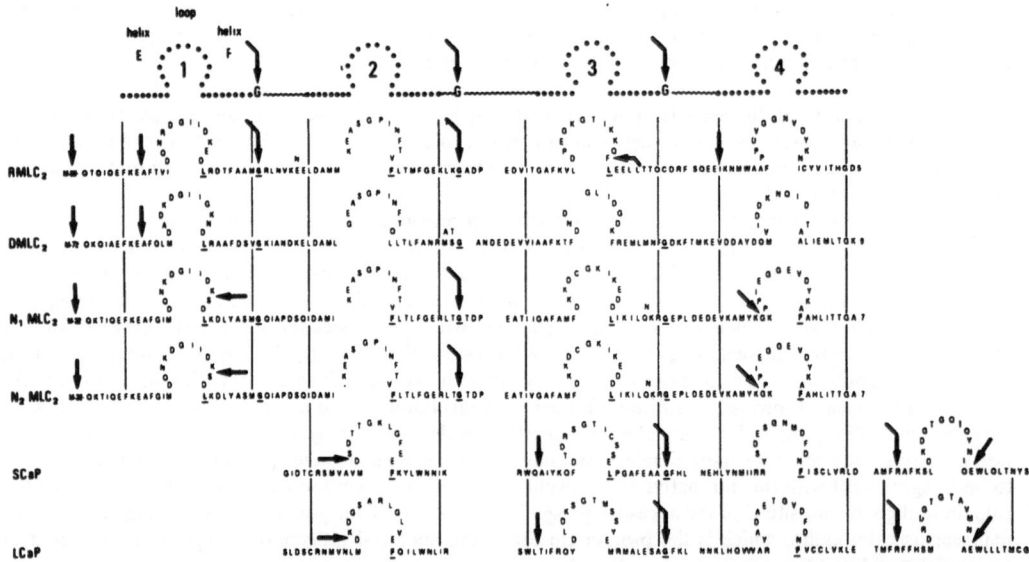

Fig. 1 . Analysis of intron position in relation to the "EF-hand" domain : Each "EF-hand" domain is shown by a sequence of amino-acid representing the loop and the two helices E and F (1, 2). The conserved glycine residue, in the interdomain and the alternating LFLF sequence are underlined. The arrows indicate the positions of the intron and the shape of the arrows indicates the phase of the intron. RCd : Rat Cd; DCd : Drosophila Cd; NCd : Caenorhabditis elegans Cd; Spec 1 : sea Urchin pec 1; $MMLC_3$: mouse MLC_3; $DMLC_3$: Drosophila MLC_3; Parv : rat parvalbumin; $RMLC_2$: rat MLC_2; $DMLC_2$: Drosophila MLC_2; N_1MLC_2 and N_2MLC_2 : Caenorhabditis elegans MLC_2; SCaP and LCaP : small and large calpains. For references, see ref 2 and ref 3 for Caenorhabditis elegans MLC genes.

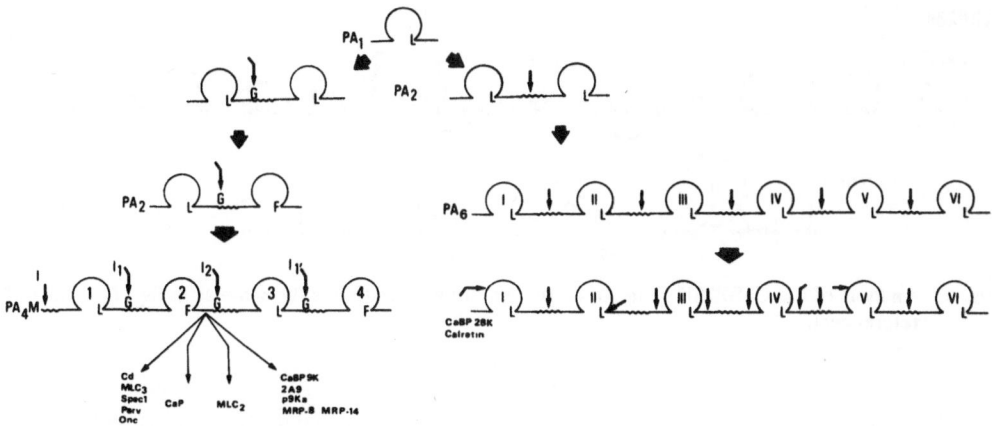

Fig. 2 . Schematic outline of our hypothesis for the evolution of different lineages of the "EF-hand" calcium-binding protein family. Each calcium-binding site is represented by a loop flanked by two straight lines for the two helices E and F. Large dotted arrows indicate the duplication and triplication events. Roman numerals indicate the number of the sites in the present day proteins, the arabic numerals indicate those of the proposed ancestral proteins - PA_1 : one-site primordial ancestor - PA_2 : two-site primordial ancestor, PA_4 : four-site primordial ancestor.

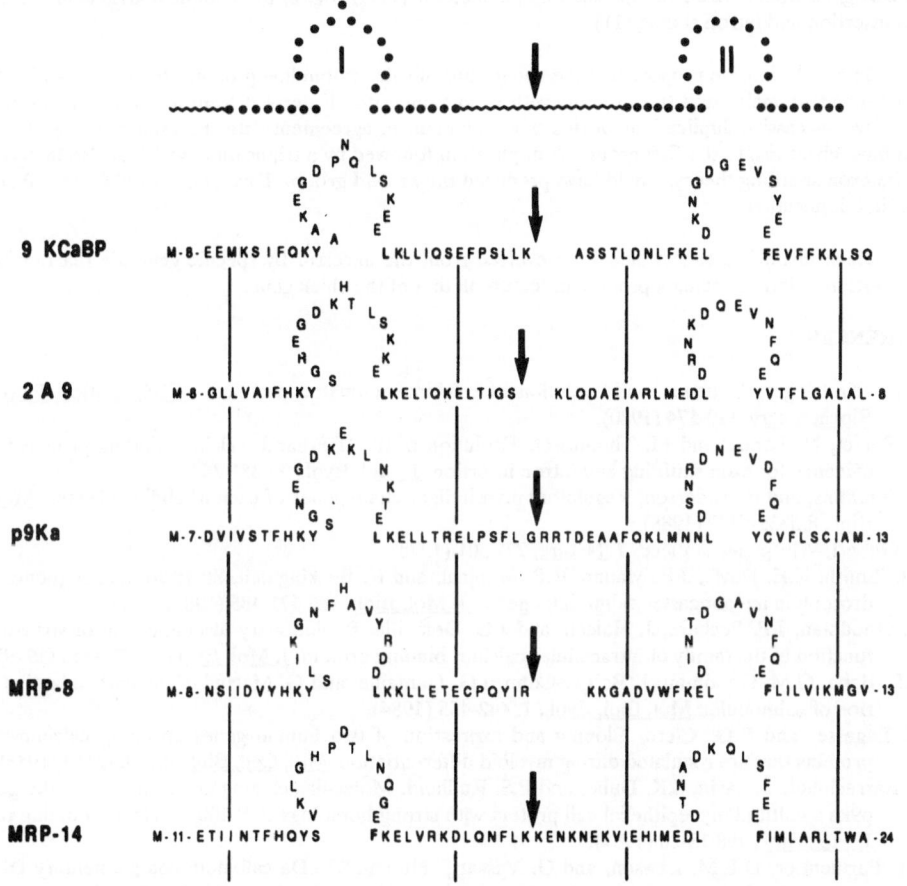

Fig. 3 . Analysis of the structure of the genes similar to the S100 protein. The representation is the same as in fig. 1. CaBP9K : calbindin-D9K. 2A9: calcyclin (see ref 2 for references) - p9Ka (ref 9) MRP-8 and MRP-14 (ref 8).

19

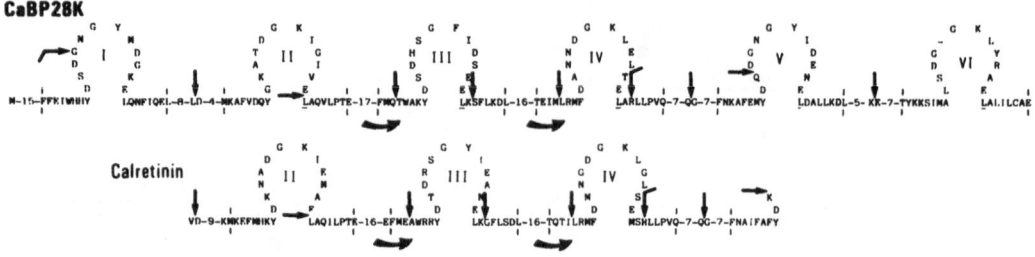

CaBP28K

Fig. 4. Analysis of the CaBP28K gene structure. The representation is the same as in fig. (see ref 11 for references).

It seems therefore that the evolutionary origin of CaBP28K is the most divergent from the rest of the calcium-binding protein. Parmentier et al (8) have proposed that CaBP28K would arise from a duplication followed by a triplication, of a single "EF-hand domain". Analysis of the structure of the gene (fig. 4, ref 10) shows that a phase 0 intron is present in the interdomain region after sites I, IV and V. The same intron is present in the "E-helix" of domain II and III and could have been moved from the interdomain region by a sliding mechanism as the sequences in between are not highly conserved. We propose that this intron is the vestigial intron that would have mediated the duplication/triplication events. The present structure of the CaBP28K gene would arise from the six-domain ancestor (PA_6) (fig. 2) by genomic rearrangements, such as intron insertion and intron sliding (11).

In conclusion, we propose that the two groups of calcium-binding proteins, the proteins with four or less "EF-hand" domain, and the proteins with 6 domains, are all derived from a signle domain ancestral gene. Two successive duplications of this single domain, in agreement with the exon shuffling theory (4), would have given rise to the first group. A duplication followed by a triplication, which is also in agreement with the exon shuffling theory, would have produced the second group. Divergence would have taken place at the first duplication.

Each distinct lineage would have evolved from the ancestor by specific genomic rearrangements, with insertion of introns being a prominent feature in that of the chick gene.

REFERENCES

1. R.H. Kretsinger, Structure and evolution of calcium-modulated proteins. <u>CRC Critical Reviews in Biochemistry</u>, 119-174 (1980).
2. C. Perret, N. Lomri, and M. Thomasset, Evolution of the "EF-hand" calcium-binding protein family - evidence for exon shuffling and intron insertion, <u>J. Mol. Evol.</u>, 27:351-364.
3. C. Cummins, and P. Anderson, Regulatory myosin light-chain genes of caenorhabditis elegans, <u>Mol. Cell. Biol.</u>, 8:5339-5349 (1988).
4. W. Gilbert, Why genes in pieces ? <u>Nature</u>, 271:501 (1978).
5. V.L. Smith, K.E. Doyle, J.F. Maune, R.P. Munjaal, and K. Beckingham, Structure and sequence of the drosophila melanogaster calmodulin gene, <u>J. Mol. Biol.</u>, 196:471-485 (1987).
6. M. Goodman, J.F. Pechere, J. Haiech, and J.G. Demaille, Evolutionary diversification of structure and function in the family of intracellular calcium-binding protein, <u>J. Mol. Evol.</u>, 13:331-352 (1979).
7. M.L. Baba, G.M. Goodman, J. Berger-Cohn, J.G. Demaille, and G. Matsuda, The early adaptive evolution of calmodulin, <u>Mol. Biol. Evol.</u>, 1: 442-455 (1984).
8. E. Lagasse, and R.G. Clerc, Cloning and expression of two human genes encoding calcium-binding proteins that are regulated during myeloid differentiation, <u>Mol. Cell. Biol.</u>, 8:2402-2410 (1988).
9. R. Barraclough, J. Savin, S.K. Dube, and P.S. Rudland, Molecular cloning and sequence of the gene for p9ka a culturel myoepithelial cell protein with strong homology to S-100, a calcium-binding protein, <u>J. Mol. Biol.</u>, 198:13-20 (1987).
10. M. Parmentier, D.E.M. Lawson, and G. Vassart, Human 27-kDa calbindin complementary DNA sequence. Evolutionary and functional implications, <u>Eur. J. Biochem.</u>, 170:207-215 (1987).
11. N. Lomri, C. Perret, N. Gouhier, and M. Thomasset, Cloning and analysis of calbindin-D28K cDNA and its expression in the central nervous system, <u>Gene</u>, 80:87-98 (1989).

FUNCTIONAL ANALYSIS OF THE PROMOTER REGION OF THE GENE ENCODING CHICKEN CALBINDIN D28K

Stefano Ferrari,[1] Renata Battini,[1] and Wesley J. Pike[2]

[1]Istituto di Chimica Biologica, Universita di Modena, Via Campi 287, 41100 Modena, Italy

[2]Baylor College of Medicine, Department of Pediatrics, Endocrinology and Metabolism Section, 8080 N. Stadium Drive, Houston, TX 77054, USA

INTRODUCTION

Vitamin D controls the expression of many genes, among which are those encoding osteocalcin (Price and Baukol, 1980), collagen type I (Kream et al., 1986), matrix Gla protein (Fraser et al., 1988), parathyroid hormone (Okazaki et al., 1988), c myc (Simpson et al., 1987), and a family of calcium-binding proteins termed calbindins. At least two types of calbindins exist: a M_r 9,000 protein (Calbindin D9K), present in mammalian intestine (Kallfelz et al., 1967), placenta (Bruns et al., 1978), yolk sac (Bruns et al., 1986), and a larger M_r 28,000 protein (calbindin D28K) identified in avian intestine (Wasserman and Taylor, 1966) and many other tissues (Christakos and Norman, 1980). Subsequent studies have shown that mammalian kidney (Pansini et al., 1984), brain (Jande et al., 1981) and other tissues (Norman et al., 1982) also express a M_r 28,000 protein which is immunologically similar but not identical to the avian Calbindin D28K. 1,25 dihydroxycholecalciferol [$1,25(OH)_2D_3$], the hormonally active form of vitamin D, has been shown to regulate the levels of avian intestinal calbindin D28K from undetectable in vitamin D deficient chickens, to up to 13% of the cytoplasmic protein of the intestinal cell (Christakos et al., 1979). This hormonal induction by 1,25(OH)2D3 is mediated through a high affinity soluble receptor protein (Pike et al., 1987) in a manner analogous to other steroid hormone systems. It is therefore hypothesized that following hormone receptor binding, this complex interacts with a cis acting element in the promoter region of the calbindin D28K gene and regulates transcription of the corresponding mRNA.

In this study we have analyzed the activity of the promoter of the gene encoding chicken calbindin D28K, by transient transfection of several cell lines with plasmids containing variable lengths of the 5'-flanking region linked to the bacterial chloramphenicol acetyltransferase (CAT) gene. While sequence domains were identified, that have the property of affecting the activity of the promoter, it was not possible to detect any sequence element responsive to $1,25(OH)_2D_3$.

MATERIALS AND METHODS

Plasmid DNA constructs, cell transfection and CAT assays. The recombinant chicken genomic clone CBC1 contains a portion of the gene encoding calbindin D28K and sequences which flank the 5'-end by 4.4 kb (Ferrari et al., 1988a, Ferrari et al., 1988b). The plasmid pCB-2200 was obtained by inserting the -221EcoRI/SacII fragment of CBC1 (coordinates 2200/+46) in the polylinker of the CAT vector BlueCAT (Ferrari, unpublished). pCB-451 and pCB-182 were obtained by 5'deletions of pCB 2200, using appropriate restriction sites. The chimeric plasmid -568/-413 osteo/-182/+46 CaBP was obtained by inserting the vitamin D responsive element of the human osteocalcin gene in the SacII site of pCB-182. 10 to 20 μg of recombinant plasmid DNA were tranfected in mouse 3T6, dog MDCK1, rat ROS 17/2.8, human HL60, human MCF7 or chicken kidney primary cells (Craviso et al., 1987), by either the calcium phosphate

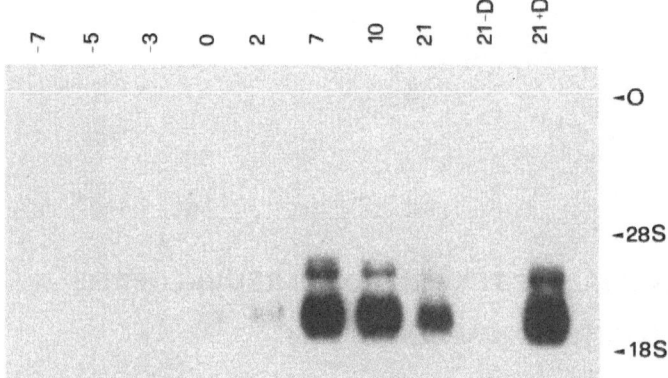

Fig. 1 . Effect of development and 1,25(OH)₂D₃ on the concentration of Calbindin D28K mRNAs in the chicken intestine. Negative numbers refer to days before hatching, which is taken as day O. 21-D and 21+D respectively indicate intestinal RNA extracted from 21 days old vitamin D deficient chickens and vitamin D deficient chickens injected with 1nM of 1,25(OH)₂D₃. The origin and the position of the 28S and 18S rRNA subunits are indicated at right.

Gorman et al., 1982), the DEAE dextran (Glass et al., 1987), or the lipofection (Felgner et al., 1987) methods. CAT assays were usually performed 36 hours after transfection, as described earlier (Gorman et al., 1982).

RNA extraction and Northern blot analysis. Total RNA was extracted, by the guanidinium hydrochloride method, from the small intestine of chickens at various stages of development (from 7 days before to 21 days after hatching) or from 21 days old chickens either maintained at a vitamin D-deficient diet until sacrifice or injected intracardially with a single dose of 1,25(OH)2D3 6 hours before sacrifice (Ferrari et al. 1984). 10 μg of RNA were separated by agarose gel electrophoresis in denaturing conditions and transferred to Hybond membranes (Amersham) as previously described (Ferrari et al., 1984). Hybridizations were to the ^{32}P-labelled fragment of the chicken calbindin D28K gene encompassing the region with coordinates -451/+2100. Hybridization, washing conditions and exposure to Hyperfilm MP (Amersham) were as described (Ferrari et al.,1984).

RESULTS

Northern blot analysis of Calbindin mRNA in the chicken intestine

Figure 1 shows that the concentration of calbindin D28K mRNA in the intestine is affected by both the stage of development and the presence of the hormonally active metabolite 1,25(OH)₂D₃. Calbindin D28K mRNA is not detectable in the embryonic intestine, it appears at hatching (O) and steeply increases to reach a maximum at day 10. As already reported in the literature (Hunziker et al., 1986; Mangelsdorf et al., 1987; Clemens et al., 1988) Calbindin D28K mRNA consists of three discrete species of 2.1, 2.8 and 3.1 kb, which are equally dependent on 1,25(OH)₂D₃. In fact no hybridization signal can be evidentiated in intestinal RNA extracted from 21 days old animals raised at a vitamin D-deficent diet (21-D), while the signal is restored in vitamin D-deficient chickens injected intracardially with 1,25(OH)₂D₃ and sacrificed 6 hours later (21+D).

Functional analysis of the DNA region encompassing the 5'end of the Calbindin gene

As it is shown in figure 2, a fragment of the calbindin D28K gene with coordinates -2200/+46 (-2200) is capable of promoting the expression of CAT in transiently transfected MDCK1 cells. However no enhancement of CAT expression could be observed upon incubation of transfected cells with 10⁷M 1,25(OH)₂D₃. The same results were obtained when other cell lines (3T6, ROS 17/2.8, MCF 7, HL 60) or chicken kidney primary cells were used (not shown). Transfection procedures other than calcium phosphate coprecipitation (DEAE dextrane, lipofection) resulted in variations of the transfection efficiency but did not affect the lack of induction by 1,25(OH)₂D₃ (not shown). Furthermore figure 2 shows that the deletion

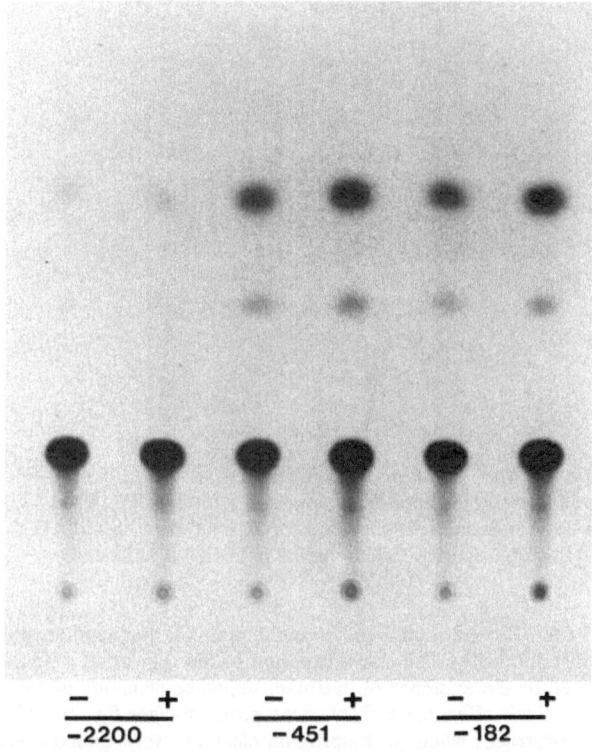

$$\underline{\quad - \quad + \quad} \quad \underline{\quad - \quad + \quad} \quad \underline{\quad - \quad + \quad}$$
$$-2200 \qquad\qquad -451 \qquad\qquad -182$$

Fig. 2 . Promoter activity of the chicken Calbindin D28K gene. The autoradiogram shows the CAT activity of MDCK 1 cells transfected with pCB-2200/+46 (-2200), pCB-451/+46 (451) and pCB-182/+46 (-182). - and + respectively indicate whether $1,25(OH)_2D_3$ was not added or added to the medium.

mutants pCB451 and pCB182 are considerably more efficient than pCB2200 in promoting CAT activity, however independently on $1,25(OH)_2D_3$.

The possibility exists that the lack of induction by $1,25(OH)_2D_3$ might be related to either technical reasons or be an inherent property of the promoter. To check this possibility ROS 17/2.8 cells were transfected with a chimeric construct where the vitamin D-responsive element of the human osteocalcin gene was cloned ahead of pCB 182. As figure 3 shows a 5 fold induction of CAT is achieved upon incubation of transfected cells with $1,25(OH)_2D_3$.

DISCUSSION

It has been shown in the past that the mechanism of action of steroid hormones is based on the ability of the hormone receptor to bind specific DNA sequences (HRE) with enhancer properties. The search for a similar element (VDRE) in the chicken calbindin D28K gene, which is capable of mediating the action of the vitamin D receptor, has not led to conclusive results. We have previously shown that no VDRE's can be demonstrated in the region of the chicken Calbindin D28K gene between nt -451 and +46 (Ferrari et al., 1988). The investigation of a much larger region of the gene (not shown here are also experiments with DNA fragments starting from nt -4300 up to the entire first intron) similarly has not allowed the identification of any VDRE. The insertion of the VDRE of the human osteocalcin gene upstream of nt -182 in the promoter of the Calbindin D28K gene significantly enhances its activity upon the addition of $1,25(OH)_2D_3$ to cells transfected with this construct. It therefore appears that either VDRE's are not present in the chicken Calbindin D28K gene (at least in the region that we investigated) or that a combination of vitamin D receptor and other transcriptional, possibly tissue specific, factors are necessary for inducible expression.

Furthermore, recent data appeared in the literature (Clemens et al., 1988) indicate that the rapid increase of calbindin D28K mRNA in the intestine (but not in the kidney) of vitamin D-deficient

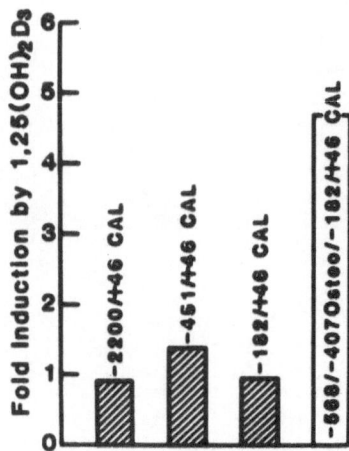

Fig. 3 . Relative activity of the chicken Calbindin D28K promoter and of the chimeric Osteocalcin/Calbindin promoter upon induction with $1,25(OH)_2D_3$. ROS 17/2.8 cells were transfected with the indicated plasmids, treated with or without 10^7M $1,25(OH)_2D_3$ for 48 hours, harvested and assayed for CAT activity. Fold hormone induction is indicated.

chickens injected with $1,25(OH)_2D_3$, is not inhibited by actinomycin D, thus suggesting that the hormone, at least initially, acts post-transcriptionally. The complex nature of the regulation of chicken calbindin D28K synthesis is further confirmed by the existence of upstream sequence elements in the 5'-flanking region of the gene, which seem to negatively affect the activity of the promoter (see fig. 2). Preliminary transfection experiments with plasmids constructs, where such upstream elements were cloned ahead of a heterologous promoter, apparently confirm this observation.

ACKNOWLEDGMENTS

Work supported by the "Progetto Finalizzato Biotecnologia" of the CNR.

REFERENCES

Bruns, M.E.H., Fausto, A., and Avioli, L.V., 1978, Placental calcium-binding protein in rats, J. Biol. Chem., 253:3186.

Bruns, M.E., Kleeman, E., and Bruns, D.E., 1986, Vitamin D-dependent Calcium-binding Protein of mouse yolk sac, J. Biol. Chem., 261:7485.

Christakos, S., Friedlander, E.J., Frandsen, B.R., and Norman A.W., 1979, Studies on the mode of action of Calciferol. XIII. Development of a radioimmunoassay for vitamin D dependent chick intestinal Calcium-Binding Protein and tissue distribution, Endocrinology, 104:1495.

Christakos, S., and Norman, A.W., 1980, Vitamin D-dependent Calcium-binding Protein synthesis by chick kidney and duodenal polysomes, Arch. Biochem. Biophys., 203:809.

Clemens, T.L., McGlade, S.A., Garret, K.P., Horiuchi, N., and Hendy, G.N., 1988, Tissue specific regulation of avian vitamin D dependent Calcium-binding Protein 28kDa mRNA by 1,25dihydroxyvitamin D3, J. Biol. Chem., 263:13112.

Craviso, G.L., Garrett, K.P., and Clemens, T.L., 1987, 1,25 dihydroxyvitamin D3 induces the synthesis of vitamin D dependent Calcium-binding Protein in cultured chick kidney cells, Endocrinology, 120:894.

Felgner, P.L., Gadek, T.R., Holm, M., Roman, R., Chan, H.W., Wenz, M., Northrop, J.P., Ringold, G.M., and Danielsen, M., 1987, Lipofection: a highly efficient, lipid mediated DNA transfection procedure, Proc. Natl. Acad. Sci. USA, 84:7413.

Ferrari, S., Battini, R., Leone, A., Ferrari, S., Torelli, G., and Barbiroli, B., 1984, Isolation of a cDNA clone containing a sequence complementary to the intestinal Calcium-binding Protein of the chick, Gene, 30:233.

Ferrari, S., Drusiani, E., Battini, R., and Fregni, M., 1988a, Nucleotide sequence of the promoter region of the gene encoding chicken Calbindin D28K, Nucl. Acids Res., 16:353.

Ferrari, S., Battini, R., Drusiani, E., and Fregni, M., 1988b, Functional analysis of the promoter region of the gene encoding chicken calbindin D28K, in: "Vitamin D. Molecular, Cellular and Clinical Endocrinology," A.W. Norman, K. Schaefer, H.G. Grigoleit, and D. v. Herrath eds., Walter de Gruyter and Co., Berlin New York.

Fraser, J., Otawara, Y., and Price, D.A., 1988, 1,25 dihydroxyvitamin D3 stimulates the synthesis of matrix gammacarboxyglutamic acid protein by osteosarcoma cells, J. Biol. Chem., 263:911.

Glass, C.K., Franco, R., Weinberger, C., Albert, V., Evans, R.M., and Rosenfeld, M.G., 1987, A erbA binding site in the rat growth hormone gene mediates transactivation by thyroid hormone, Nature, 329:738.

Gorman, C.M., Moffat, L.F., and Howard, B.H., 1982, Recombinant genomes which express chloramphenicol acetyltransferase in mammalian cells, Mol. Cell. Biol., 2:1044.

Hunziker, W., 1986, The 28kDa vitamin D-dependent calcium binding protein has a six domain structure, Proc. Natl. Acad. Sci. USA, 83:7578.

Jande, S. S., Maler, L., and Lawson, D. E. M., 1981, Immunohistochemical mapping of vitamin D dependent calcium binding protein in brain, Nature, 294:765.

Kallfelz, F.A., Taylor, A.N., and Wasserman, R.H., 1967, Vitamin D3 induced calcium binding factor in rat intestinal mucosa, Proc. Soc. Exp. Biol. Med., 125:54.

Kream, B.E., Rowe, D., Smith, M.D., Maher, V., and Majeska, R.J., 1986, Hormonal regulation of collagen synthesis in a clonal rat osteosarcoma cell line, Endocrinology, 119:1922.

Mangelsdorf,D. J., Komm, B.S., McDonnell, D.P., Pike, J. W., and Haussler, M. R., 1987, Immunoselection of cDNAs to avian intestinal calcium binding protein 28K and a novel calmodulin-like protein: assessment of mRNA regulation by the vitamin D hormone, Biochemistry, 26:8332.

Norman, A.W., Roth, J., and Orci, L., 1982, The vitamin D endocrine system: steroid metabolism, hormone receptors, and biological response, Endocrine Rev., 3:331.

Okazaki, T., Igarashi, T, and Kronenberg, H.M., 1988, 5'-flanking region of the parathyroid hormone gene mediates negative regulation by 1,25(OH)2 vitamin D3, J. Biol. Chem., 263:2203.

Pansini, A.R., and Christakos, S., 1984, Vitamin D dependent calcium-binding protein in rat kidney, J. Biol. Chem., 259:9735.

Pike, J.W., Sleator, N.M., and Haussler, M.R., 1987, Chicken intestinal receptor for 1,25 dihydroxyvitamin D3, J. Biol. Chem., 262:1305.

Price, P.A., and Baukol, S.A., 1980, 1,25 dihydroxyvitamin D3 increases synthesis of the vitamin K dependent bone protein by osteosarcoma, J. Biol. Chem., 255:11660.

Simpson, R.U., Hsu, T., Beagley, D.A., Mitchell, B.S. and Alizadeh, B.N., 1987, Transcriptional regulation of the c-myc protooncogene by 1,25 dihydroxyvitamin D3 in HL60 promyelocytic leukemia cells, J. Biol. Chem., 262:4104.

Wasserman, R.H., and Taylor, A.N., 1966, Vitamin D3 induced calcium-binding protein in chick intestinal mucosa, Science, 152:791.

Forrest, S., Bartlett, R., and Pregar, M., 1988a. Sinusoidal synthetic. A-O-a prompter report of the gene encoding chicken Calbindin D28K. *Natl. Acad. Sci.* 46224.

Forrest, S., Bartlett, R., Douglas, G., 1988b, pp. 174, 1988b. Functional analysis of the promoter region of the gene encoding chicken Calbindin D28K, in Vitamin D: Molecular, Cellular and Clinical Endocrinology, A. W. Norman, K. Schaefer, H.G. Grigoleit and D. v. Herrath eds., Walter de Gruyter and Co., Berlin, New York.

Franceschi, R. and Gina, B. A. 1986, 1,25-dihydroxyvitamin D₃ stimulates the synthesis of calcium-binding proteins in cultured rat intestinal epithelium, *J. Biol. Chem.* 2, 5011.

Gaier, C.H.J. Stein, R., Whitecaps, G., Albert V., Orwell, B.M., and Rasmussen, M.G., 1978. A hormonal in vivo for the cell growth-coupled gene induced neutralization by thyroid hormone, *Nature* 1, 1228.

Gurcan, C.M., DeMull, D., and Haussler, M.R., 1982, Intracellular proteins which require electronic physical conformations for its interactions with Mol. Cell. Biol., 70, 74.

Haussler, et al, 1984, The 25-its vitamin D concerning linkage binding in Hep-3 a cell, in short structure, *J. Nut. Acad. Sci.* 22, 5.00/00.

Inaba, S. and Lawson, D.E.M., 1984. Immunohistochemical mapping of vitamin D-dependent calcium binding protein in rat, *Biochem. Biophys. J.* 1, 11.

Jande, S., and Legault-Demare, F., 1984, Immunohistochemical binding calcium-binding protein in rat brain, *Proc. Natl. Acad. Sci.* 1028.

Jones, K.L., and Waterman, S.H., 1980, 1,25-dihydroxyvitamin D-induced calcium-binding factor, *Proc. Natl. Acad. Sci.* 1020.

Krisinger, J., et al, Sontre, N.M., Baillie, C., and Morley, W.C. 1984. Homeostatic regulation of intestinal calcium-binding protein in the laying hen occurs with a molecular binding accompanied by a decrease in the levels of its mRNA, in Vitamin D: Molecular, Cellular and Clinical Endocrinology, A.W. Norman, K. Schaefer, D. v. Herrath eds., Walter de Gruyter, Berlin, New York.

MacManus, J., Whitfield, J.F., 1972, The heat-D-induced synthesis of a calcium-binding protein, induced analogous response, *Endocrinology* Res. 4681.

McNeill, P.E., Haussler, M., and Grandison, L., 1983. Vitamin D and its metabolites in the neonatal rat and mouse, *Endocrinology* 1020, Vitamin D, 1. *Endocrinology Suppl.*

Norman, A.W., et al, 1984. Vitamin D: Molecular and Cellular Endocrinology, Walter de Gruyter, Berlin, New York.

Price, D.W., Waterman-Hazuda, J.C., 1984, 1,25-Dihydroxyvitamin D₃ sterol receptor-mediated actions, *Biol. Chem.* 58, 5030.

Pike, J.W., Haussler, M.R., 1979, Purification of chicken intestinal receptor for 1,25-dihydroxyvitamin D, *Proc. Natl. Acad. Sci.* 76, 5485.

Price, D.R., and Richards, C.M., 1979, The dihydroxyvitamin D binding in vertebrate in the vitamin D-dependent calcium binding, *J. Biol. Chem.* 254, 1260.

Simpson, R.U., Hsu, T., Wendt, K., and Simpson, D.C., 1984, 1,25-dihydroxyvitamin D receptors in cultured brain cells and 1,25-dihydroxyvitamin D₃ and 1,24,25-trihydroxyvitamin D₃ biosynthesis, *J. Biol. Chem.* 53, 4700.

Simpson, R.U., 1986, Vitamin D₃ and cell calcium-binding protein interaction, *Proc. Soc.* 2009.

STRUCTURE OF THE HUMAN cDNAS AND GENES CODING FOR

CALBINDIN D28K AND CALRETININ

Marc Parmentier

I.R.I.B.H.N., ULB Campus ERASME, 808 route de Lennik, 1070 Bruxelles, Belgium

INTRODUCTION

Calbindin D28K was first described by Wasserman and Taylor (1966) and is now considered as the main direct molecular effect of vitamin D derived hormones on gut epithelial cells. It is thought to play a major role in the absorption of calcium from the intestinal lumen and from the distal convoluted tubule of the kidney. The protein is postulated to act as a calcium ferry, facilitating the transport of calcium ions through the cell and keeping the actual free calcium below its toxic level (Jande et al., 1981, McBurney and Neering, 1987). Other cell types containing high amounts of calbindin are the neurons of the central and peripheral nervous systems (Roth et al. 1981), as well as the connected sensory organs (Verstappen et al. 1986), and the alpha and beta cells of the pancreatic islets (Pochet et al. 1987).

CALBINDIN IN THE BRAIN

We described previously that the brain of the rat contains two proteins reacting immunologically with antibodies directed against chick intestinal calbindin D28K (Pochet et al. 1985). Additionally to the calbindin D28K band (27 kDa in rat), a second band of 29 kDa was present. These two proteins were shown to derive from different messenger RNA populations. Both proteins were binding calcium, after partial purification by affinity chromatography. It was postulated that the second protein was related to calbindin. It was unknown however if their respective mRNA were derived from a single gene by a brain specific alternative splicing, or if a second gene was specifically expressed in the nervous system. In a phylogenetic study of calbindin in lower vertebrates (Parmentier et al., 1987a), we showed that only one band was present in the fish brain while two bands were detected in the brain of amphibians, reptiles, birds and mammals. Moreover, calbindin was not detected in the intestine and kidney of fish species. It was concluded that from an evolutionary point of view, calbindin was primarily a neuronal protein. When amphibians ancestors became terrestrial, the need for a calcium-binding protein somehow appeared in cells where Ca^{2+} exchange with the environment was taking place, namely the digestive tract epithelium and the kidney tubules. One of the existing brain calcium-binding protein was selected for this purpose. It is interesting to note that in mammals, a smaller and completely different protein (calbindin D9k, a protein related to S100 subunits) has functionally replaced calbindin D28K in the intestine. The second conclusion of the phylogenetic study was that, additionally to the well characterized calbindin D28K, a second protein (named calbindin 29 kDa) of higher molecular weight and specific to the brain, was present in all vertebrates from amphibians to mammals. In order to determine the structure of both brain proteins and their relationship, we decided to clone their cDNAs and genes.

STRUCTURE OF HUMAN CALBINDIN D28K AND CALRETININ

Antibodies raised against chicken intestinal calbindin were used to screen two complementary DNA libraries (human brain stem and human cerebellum) constructed in gt11 by EcoRI linker addition to double stranded cDNA (Huynh et al. 1985). 1.1 10^6 recombinant clones were screened at high density (10^5 pfu/15

```
                              EL--LL--L*-*-*G-L*---*L--LL--L
      hCalbin           MAESHLQSSLITAS QFFEIWLHFDADGSGYLEGKELQNLIQEL QQARKKA
                                    ■ ■■■ ■■ ■■■ ■■■■■■ ■■ ■■■■■ ■ ■■■   ■■■
      hCalret  MAGPQQQPPYLHLAELTAS QFLEIWKHFDADGNGYIEGKELENFFQEL EKARKGS

      hCalbin          GLE------LSP EMKTFVDQYGQRDDGKIGIVELAHVLPTE ENFLLLF
                                  ■    ■■ ■   ■    ■■■   ■■■ ■■■■ ■■■■■ ■
      hCalret       GMMSKSDNFGE KMKEFMQKYDKNSDGKIEMAELAQILPTE ENFLLCF

      hCalbin            RCQQLKSCE EFMKTWRKYDTDHSGFIETEELKNFLKDL LEKANKT
                             ■ ■   ■    ■■■ ■■■■■■■ ■■ ■■ ■■■ ■■ ■■ ■ ■■■
      hCalret          R-QHVGSSA EFMEAWRKYDTDRSGYIEANELKGFLSDL LKKANRP

      hCalbin           VDDTKLAE YTDLMLKLFDSNNDGKLELTEMARLLPVQ ENFLL
                             ■ ■■ ■ ■■    ■ ■■ ■ ■■■■ ■ ■■ ■■■■■ ■■■■■
      hCalret           YDEPKLQE YTQTILRMFDLNGDGKLGLSEMSRLLPVQ ENFLL

      hCalbin         KFQGIKMCGK EFNKAFELYDQDGNGYIDENELDALLKDL CEKNKQDL
                          ■■■■ ■     ■■■ ■ ■■ ■ ■■■■■ ■■■■■■■■■ ■■■■
      hCalret         KFQGMKLTSE EFNAIFTFYDKDRSGYIDEHELDALLKDL YEKNKKEI

      hCalbin             DINN ITTYKKNIMALSDGGKLYRTDLALILCAG DN
                           ■   ■■■ ■ ■■     ■■■■■ ■■  ■■
      hCalret             NIQQ LTNYRKSVMSLAEAGKLYRKDLEIVLCSE PPM
```

Fig. 1. Aminoacid sequence of human calbindin D28K (hCalbin) and calretinin (hCalret) as deduced from their cDNA sequences. The consensus of the EF-hand domain is represented on top (E=glutamine, L = hydrophobic residue, G=glycine, * = oxygen-containing residue). The sequences are arbitrarily cut in the interdomain regions. Residues identical in both sequences are indicated by ■.

cm Petri dish) (Benton and Davis, 1977). 105 clones were positive in the first screening, 44 clones were purified to homogeneity by two successive cycles of screening at low density, and their insert size determined. From restriction maps and cross-hybridization experiments, it appeared clearly that the isolated clones could be separated into two distinct populations. The longest clones from each population were subcloned in M13mp derivatives and sequenced using the dideoxy-nucleotides chain termination reaction of Sanger et al. (1977). A combination of forced cloning, exonuclease III deletions (Henikoff, 1984) and sequence specific primers was used. Additional screening were necessary to obtain the full coding sequence for both proteins. Open reading frames of 261 and 271 amino acids were found, corresponding to proteins of 30,300 and 31,760 daltons respectively. Despite the difference with the apparent molecular weights deduced from SDS gel electrophoresis, it was logically postulated that the 31,8 and 30,3 kDa proteins corresponded respectively to the 29 and 27 kDa proteins on Western blots. As expected, the 27 kDa protein turned out to be human calbindin D28K. It was found to be 79% identical to chicken calbindin (Wilson et al. 1985), and 98% identical to bovine calbindin (Takagi et al. 1986). The second protein, homologous to calbindin, was compared to the partial sequence of the chicken retinal protein described by Rogers (1987) and named calretinin. The overlapping sequence was found 88% identical. Additional evidence of the orthologous relation between calretinin and calbindin 29 kDa recently came from in situ hybridization preliminary data. The distribution of calbindin 29 kda in the rat nervous system, as determined by in situ hybridization using a human cRNA probe, coincides with the distribution of calretinin, as determined by immunohistochemistry, using antibodies against the chicken recombinant protein.

Analysis of the sequences showed that the two proteins shared 58% identical residues. They both contain six EF-hand domains as described by Kretsinger (1980). Up to 14 consecutive amino acids are identical in both proteins (between domains 4 and 5). It is therefore easy to understand that at least some antibodies directed against one protein would recognize the other. Several studies dealing with the distribution of calbindin in the nervous system of chick and mammalian species have been published in the past 10 years (Roth et al., 1981, Jande et al. 1981, Baimbridge et al., 1982, Legrand et al., 1983, Feldman and Christakos, 1986). Since calretinin was not known, the specificity of the antibodies for calbindin could not be tested in these studies. It is therefore possible that some cell populations that were described as containing calbindin D28K do actually contain calretinin.

The higher molecular weight of calretinin corresponds mainly to five additional amino acids at the NH$_2$ terminus, and to an insertion of five residues between domains 1 and 2. There is also a single amino acid deletion between domains 2 and 3, and an additional residue at the C-terminus. It is known that calbindin only binds 4 calcium ions and it is postulated that the inactive domains are the second and the sixth,

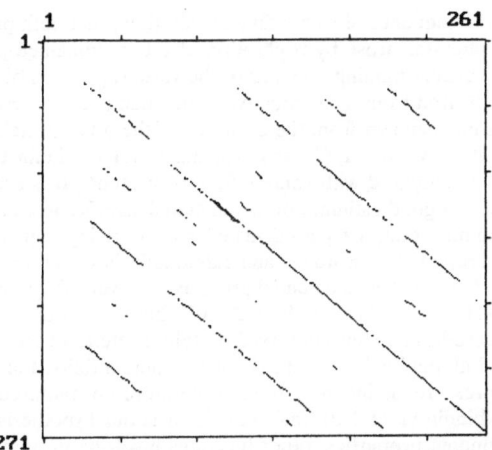

Fig. 2 . Dot matrix showing the homologies between human calbindin D28K (horizontally), and human calretinin (vertically), as well as internal repeats. Each dot indicates a minimum of 10 matches over 30 residues, using the Prosis software (Hitachi, Japan). It is clear that the proteins are composed of three repeats, each comprising two calcium-binding domains. The homologies between even and odd domains is less obvious.

because of their divergence from the EF-hand consensus. From the sequence of calretinin, it is speculated that the first five domains of the protein are active in binding calcium. The sixth domain is probably inactive, since it diverges strongly from the consensus.

```
 -840 ggttgcttgaaacagaaagaaggagacgtgtgtgttaaggcgactgaatggggctgtggt
 -780 ggtagcactaacgtacctacttatgaacctggtcagatccttcctttctctttgttttag
 -720 tttactcactccttctaaactgaaaggcagtaaactctagtatctgtaagatgacttcct
 -660 tatctaagattttaaggtttitataattaactgtattactttccaattataattctactt
 -600 catgctgaaaagaaatccataatctagcttttttcgtgttggcaaggttagttcatagaa
 -540 accatcttaagaaacattgacatgatggcacaaaatagtgtgattattgatagcaatcaa
 -480 ctcattgtatacattaatcaaattattctaatgcgtgccaaggtttcacatgctttttt
 -420 ttttcaaacagatactgaaaagtcttggagaaatcacacaagaatataataatcatttat
 -360 gtgattttacccgtcatcagctttcccctitctatctgaacatttaaacctctctagataa
 -300 caagagtttitactataatattcaggaataagattcttcctggactaggaggatgtaggg
 -240 gaaatgagcctctgtttctcagcatccaatctctgagcaaaggatctcccaccacctgctg
 -180 cttccaacaagcccaaactctcgctccagctagctttcctgggaacagagcagaaaactg
                                          SP1
 -120 gaagggaggggaagaaggctggtgagagcaagaggcggggagtgaggaatgggaggctgg
  -60 aggaggcgtggcccggcttggggccgtcgggATAAATActgagaactgggtgcggggtgt
    1 AGGGAGAGAACTCTGGAGGAACGCTGAGCTGAGCAGCACCGAGGACAGCGCCCGGCAGCG
   61 CCCGCGCCCAGGTCTCCCTCCGCAGCCCTGACTCGCGCACACGCTGAGCTTTTGCTCACT
  121 CCCCTTCGCGCGGACACAGACACACGCATATTCACACACCCAGACACACACCCCGCTGTA
  181 CAATGGCAGAATCCCACCTGCAGTCATCCCTCATCACAGCCTCACAGTTTTTTCGAGATCT
       MetAlaGluSerHisLeuGlnSerSerLeuIleThrAlaSerGlnPhePheGluIle
  241 GGCTCCATTTCGACGCTGACGgtgattatcttctcttttaactgtttcaagacccgttt
       TrpLeuHisPheAspAlaAsp
```

Fig. 3 . Sequence of the promoter region of the human calbindin D28K gene showing the first exon (upper case), the start codon and the TATA box (bold), the putative cap site (ª), the SP1 binding site consensus sequence, and several imperfect palindromes (over- and underlined). Additional sequence information can be found in Parmentier et al, 1989.

It is obvious, from the sequence of calbindin and calretinin, that both proteins evolved from a common ancestor and that this ancestor arose by triplication of a two domained protein (fig. 2). From comparisons between individual calcium-binding domains of the various proteins of the troponin C superfamily, it appeared that all proteins derived from a common two-domained ancestor. However, the pathway leading to calbindin and calretinin is independent from the evolution of the other proteins of the family (Parmentier et al. 1987). Evolutionary rates (Kimura 1983) were calculated for calbindin D28K, calretinin, and other calcium-binding proteins, and compared with some reference proteins. It is accepted that the evolutionary rate of amino acid sequences is a good estimate of its functional significance. Fibrinopeptides, the sequence of which is physiologically unimportant, are paradigms of freely evolving sequences, with a rate of $8.30 \ 10^{-9}$ mutations per aa per year. Proteins like histones and calmodulin have the lowest evolutionary rates (0.01 and $0.04 \ 10^{-9}$ aa^{-1} $year^{-1}$), and are known as essential proteins with extensive interactions with DNA or other proteins. Calbindin D28K ($0.30 \ 10^{-9}$) and calretinin ($0.27 \ 10^{-9}$) have evolutionary rates similar to cytochrome C ($0.30 \ 10^{-9}$) which is considered as a highly conserved protein. Proteins of the troponin C superfamily have very divergent sequences and all bind calcium with about the same affinity. For the sole purpose of binding calcium, a strong selective pressure against mutations is therefore not required. This is exemplified by the high evolutionary rate of calbindin 9k ($1.8 \ 10^{-9}$ aa^{-1} $year^{-1}$). It is our hypothesis that calbindin would have, additionally to its calcium-binding properties, other functions involving protein-protein interactions. These functions could be restricted to the primary localization of calbindin, the nervous system.

STRUCTURE OF THE HUMAN D28K GENE PROMOTER

A clone (HG29M19) containing the promoter region of the human calbindin gene was selected by screening a genomic library constructed in lambda Charon 4A with the 350 bp EcoRI fragment of the calbindin cDNA clone HBSC21. The insert size was 14 kb long. Overlapping fragments of the 5' region, comprising 1800 bp of upstream sequence and the first two exons, were subcloned and sequenced. The 3' region of the gene was cloned and sequenced as described in Parmentier et al. (1989). The position of the mapped introns were at identical position as those in the chicken gene (Wilson et al. 1988). An ATAAATA sequence was found 30 nucleotides ahead of the first residue obtained on cDNA clones (fig. 3). This sequence satisfies the TATA box consensus. No "CAAT" box could be found at reasonable proximity of the TATA box. The

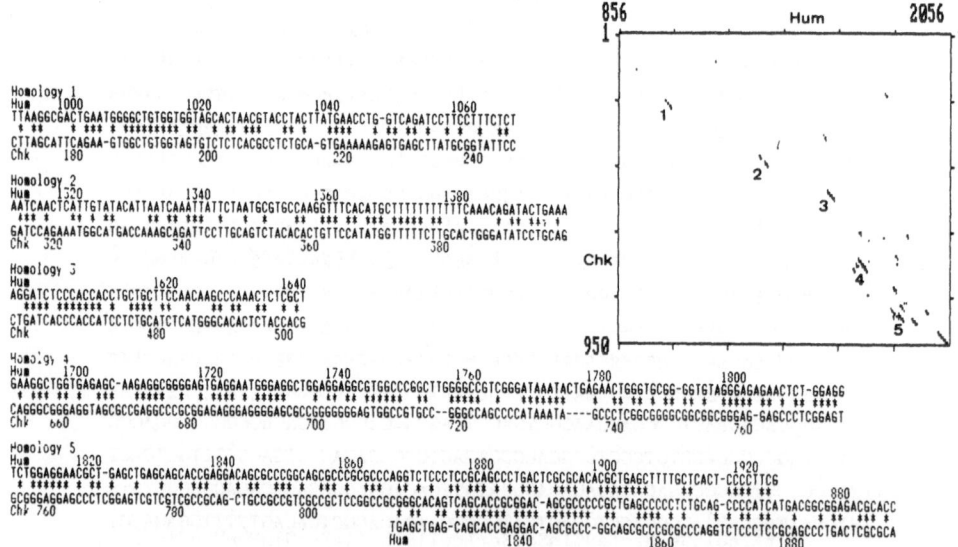

Fig. 4. Harr plot showing similarities between human and chicken calbindin D28K gene promoter regions. Each dot indicates that 22 out of 40 residues are identical. The coding regions are in the lower right corner. The alignments corresponding to the 5 regions showing similarity on the Harr plot are displayed. For the fifth one, two possible alignment between human and chicken sequences are displayed. The significance of such similarities between two GC rich sequences appears particularly low. The human and chicken sequences are numbered according to Wilson et al. (1987) and Parmentier et al. (1989) respectively.

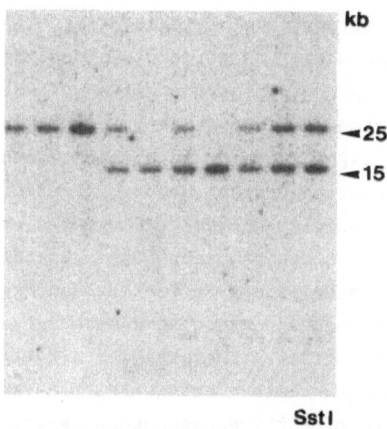

kb

◄25

◄15

SstI

Fig. 5. Southern blot of DNA from 10 unrelated Caucasian individuals digested with SstI, run on a 0.6% agarose gel, and hybridized with the 1.8 kb EcoRI fragment of the calbindin D28K cDNA clone HBSC21. A 2 allele restriction fragment length polymorphism (25 and 15 kb bands) separates individuals in three different phenotypes.

closest one (CCAATCT) was located 187 nucleotides upstream of the ATAAATA motif. The region surrounding the TATA box is highly GC rich, with a typical clustering of CpG dinucleotides. A potential SP1 binding site was found 58 bp ahead of the TATA box. The promoter was searched for homologies between human and chicken sequences by dot matrix analysis (fig. 4). No strongly conserved features were found. The best alignment was located for the region immediately surrounding the TATA boxes. This lack of conserved boxes between the human and chicken promoters can be explained by the evolutionary distance between the two species, and by the fact that calbindin D28K does not have the same distribution pattern in birds and mammals (intestinal localization). As a consequence, differences in regulatory regions and/or trans acting factors are to be expected. Several imperfect palindromes were found in the promoter region (fig. 3). A homopurine-homopyrimidine box (-911) and a alternating purine-pyrimidine box (+1012) were found. These two elements are reported to be regularly associated with the promoter regions of eukaryotic genes. Homopurine-homopyrimidine boxes are responsible for the S1 nuclease hypersensitivity often described within regulatory regions (Nickol and Felsenberg 1983, Evans et al. 1984), probably because of their unusual DNA structure (Wells, 1988). It has been suggested that it could play an important role in the regulation of gene expression. Alternating (A-T)n sequences have also been found often in eukaryotic promoters (McClellan et al. 1986). These sequences exhibit a high flexibility which is thought to be related to a biological role. The functional significance of these elements is purely speculative, but they should be considered in future expression studies.

The chromosomal location of the human calbindin gene was determined using panels of rodent-human cell hybrids. DNA isolated from the cell lines were digested with BamHI, EcoRI or HindIII and the presence of the human calbindin D28K locus was determined on Southern blots, using the 1800 bp EcoRI fragment of the cDNA clone HBSC21 as a probe. The calbindin gene could unambiguously be assigned to the human chromosome 8. Using the same probe, restriction fragment length polymorphisms (RFLP) were obtained after digestion with BglII, HindIII, RsaI and SstI. The HindIII and SstI polymorphisms were fully characterized. The A1 (4.7 kb) and A2 (4.3 kb) alleles of the HindIII RFLP were estimated to represent 72 and 28% of the Caucasian alleles. The frequencies of the B1 (25 kb) and B2 (15 kb) of the SstI RFLP were estimated to 65 and 35% respectively fig. 5). The frequencies of the A1B1, A1B2, A2B1 and A2B2 gametes were estimated to be 65, 7, 0 and 28% respectively. No polymorphism was detected in five unrelated individuals after ApaI, BanII, BstEII, BstXI, EcoRI, EcoRV, HaeIII, HindII, KpnI, MspI, PstI, PvuII, TaqI and XbaI.

STRUCTURE OF THE HUMAN CALRETININ GENE PROMOTER

The human genomic library was screened using a 1300 bp cDNA clone as a probe. Three independent clones were purified to homogeneity. They were shown to contain exons 2, 3 and 4 and other exons that

```
-960 tcctgtctgctgccctcaggatgtgcgctcagtagctgcgtctattttctctgagaccag
-900 ctcagaacatccccaagacagagttggacgttgttctctgtccactggagcaggcacatt
-840 cccacgatgtccctaggtgggcgtggttaagacctggtacgggttgatctagttctgcca
-780 ccccctggccaaatgtctaaaagccccccagagcatggccagcgtgaggcaggtaccagg
-720 gtgaagggaggctccgaggggacggatatacgaagacccaaacagacagtggaagccccc
-660 caccccacccacaccacttccatcggaatcctcccggggcactgctgattccagctgc
-600 tccccactaaagccttgagaactcttggctgctctgcaagactgagccccatgaaggagc
-540 cacgtgcggcgtggaaagagtgctgagttcaaattgtagccctgccactaatttgctggg
-480 ccagtcacttaatcatctgaagtcacaagtacctcatcagaaaagtggtcccagctcttc
-420 ctgctgtggaaggatcagaagagaggaggcacgacagagacctagtgaactccgaagccc
-360 gagtgctaaatatttgtcaagtttgtgttagtattactattagtgttgttactgctgtta
-300 ttattattgctactgccagcaataataagtggtggatgtactcaagacggtcgggaggga
-240 aggcaagggcagcctctccctcattttcaccgaaaatcctccgggtgtccctggccccgc
-180 gccgaggggtctcagcgcagaggtaagggccctctaggagtccgggccgagcctctcgcg
-120 ccgccgcccccgccgcgccgcgccccggtcggattccctgagcgcgcgcgcccccttctg
-60 gcggccgggcgcaggcgcaggctccagagcgTATATAAgggcagcgtggcgcacaacccc
1 agcgcgagtgccagagcccagccggcgcggagcgggagcggtgcaggctgaggtctccga
61 gcggctcgccATGGCTGGCCCGCAGCAGCAGCCCCCTTACCTGCACCTGGCCGAGCTGAC
   MetAlaGlyProGlnGlnGlnProProTyrLeuHisLeuAlaGluLeuThr
121 GGCGTCCCAGTTCCTGGAAATATGGAAGCACTTTGACGCAGACGgtcagtaaagctccca
   AlaSerGlnPheLeuGluIleTrpLysHisPheAspAlaAspGly
181 acttctgtgcccattggcacccaggggacctgcggtgaagggggcaggggcggcgctga
241 atgcgggcaggtgtacgtttgccctcaggaggtggtctggtcgcagagcctgctggggta
301 gtggtcccgggtttgggagacatccggctgctctgggcacaactgtcagtaactttcttc
```

Fig. 6 . Partial sequence of the promoter region of the human calretinin gene. showing the first exon (upper case), the ATG codon and the TATA box (bold), and three putative SP1 binding sites (underlined). Note the high GC content of the promoter and the frequency of CpG dinucleotides around the TATA box.

were not precisely mapped. An additional clone (HG29BP) was obtained by screening the library with two synthetic 30-mer from the 5' end of the cDNA. The intronic junctions of the first four exons were sequenced showing that the location of these introns in calretinin and calbindin genes are identical. The clone HG29BP contained about 4.7 kb of sequence ahead of the ATG codon . Most of this upstream region was sequenced, as well as the first exon and 900 bp of the first intron which occupies the remaining part of this 18 kb clone. A part of this sequence is presented in fig. 6. A TATATAA sequence was found 57 bp ahead of the 5' end of our cDNA sequence. The precise transcriptional start was not determined. No typical CAAT box was found. Several putative SP1 binding sites are present in this region, especially 88 and 793 bp ahead of the TATA box, and 255 bp downstream, in the beginning of the first intron. A classical Alu repeat is located 1900 bp ahead of the TATA box, in a reversed orientation, and flanked by 7 bp repeats.

A panel of 13 rat-human and mouse-human hybrid cell lines were screened for the presence of calretinin locus, after digestion with *Hind*III and hybridization with a 1300 bp *Acc*I-*Hind*III genomic probe containing exon 2. The gene was assigned to the human chromosome 16. It is obvious the calbindin and calretinin genes derive from a common ancestor by gene duplication (sequence homology, introns position). From the present assignments, it appears that the two genes were separated on different chromosomes during the genome evolution leading to man.

ACKNOWLEDGEMENTS

This work was partially supported by the Belgian Fonds de la Recherche Scientifique Médicale, the Ministère de la Politique Scientifique (Sciences de la Vie), the European Economic Community (Stimulation program), the British Council, Solvay and the Association Recherche Biomédicale et Diagnostic. We thank

D.E.M. Lawson for the gift of anti-calbindin serum, E. Muir for her participation in the sequencing work on the calbindin 27 kDa gene, J. De Vijlder, A. Geurts van Kessel, C. and J. Szpirer, G. Levan and M.Q. Islam for their collaboration in the chromosomal assignments, V. Pohl for the preliminary in situ hybridization study, and G. Vassart and R. Pochet for helpful discussions. M.P. was Chargé de Recherche of the Fonds National de la Recherche Scientifique of Belgium.

REFERENCES

Baimbridge, K.G., and Miller, J.J., 1982, Immunohistochemical localization of calcium-binding protein in the cerebellum, hippocampal formation and olfactory bulb of the rat. Brain Res., 245: 223-229.

Benton, W.D., and Davis, R.W., 1977, Screening lgt recombinant clones by hybridization to single plaques in situ. Science, 196: 180-182.

Bucher, P., and Trifonov, E.N., 1986, Compilation and analysis of eukaryotic Pol II promoter sequences. Nucl. Acids Res., 14: 10009-10026.

Evans, T., Schon, E., Gora-Maslak, G., Patterson, J., and Efstratiadis, A., 1984, S1-hypersensitive sites in eukaryotic promoter regions. Nucl. Acids Res., 12: 8043-8058.

Feldman, S.C., and Christakos, S., 1983, Vitamin D-dependent calcium-binding protein in rat brain: biochemical and immunocytochemical characterization. Endocrinology, 112: 290-302.

Fournet, N., Garcia-Segura, L.M., Norman, A.W., and Orci, L., 1986, Selective localization of calcium-binding protein in human brain stem, cerebellum and spinal cord. Brain Res., 399: 310-316.

Henikoff, S., 1984, Unidirectional digestion with exonuclease III creates breakpoints for DNA sequencing. Gene, 28: 351-359.

Huynh, T.V., Young, R.A., and Davis, R.W., 1985, Constructing and screening cDNA libraries in gt10 and gt11. in: "DNA cloning techniques", D.M. Glover, ed., pp. 49-78, IRL Press, Oxford.

Jande, S.S., Maler, L., and Lawson, D.E.M., 1981, Immunohistochemical mapping of vitamin D-dependent calcium-binding protein in brain. Nature (London), 294: 765-767.

Kallfelz, F.A., Taylor, A.N., and Wasserman, R.H., 1967, Vitamin D-induced calcium-binding factor in rat intestinal mucosa. Proc. Soc. Exp. Biol. Med., 125: 54-58.

Kimura, M., 1983, "The Neutral Theory of Evolution" Cambridge University Press, Cambridge.

Kretsinger, R.H., 1980, Structure and evolution of calcium-modulated proteins. C.R.C. Crit. Rev. Biochem., 8: 119-174.

Legrand, Ch., Thomasset, M., Parkes, C.O., Clavel, M.C., and RabiÅ, A., 1983, Calcium-binding protein in the developing rat cerebellum. Cell Tissue Res., 233: 389-402.

McBurney, R.N., and Neering, I.R., 1987, Neuronal calcium homeostasis. Trends Neurosc., 10: 164-169.

McClellan, J.A., Palacek, E., and Lilley, D.M.J., 1986, (A-T)$_n$ tracts embedded in random sequence DNA - Formation of a structure which is chemically reactive and torsionally deformable. Nucl. Acids Res., 14: 9291-9309.

Nickol, J.M., and Felsenfeld, G., 1983, DNA conformation at the 5' end of the chicken adult Þ-globin gene. Cell, 35: 467-477.

Nussinov, R., Owens, J., and Maizel, J.V., 1986, Sequence signals in eukaryotic upstream regions. Bioch. Bioph. Acta, 866: 109-119.

Parmentier, M., Lawson, D.E.M., and Vassart, G., 1987a, Human 27-kDa calbindin complementary DNA sequence: Evolutionary and functional implications. Eur. J. Biochem., 170: 207-215.

Parmentier, M., Ghysels, M., Rypens, F., Lawson, D.E.M., Pasteels, J.L., and Pochet, R., 1987b, Calbindin in vertebrate classes: Immunohistochemical localization and Western blot analysis. Gen. Comp. Endocr., 65: 399-407.

Parmentier, M., and Vassart, G., 1988, HindIII RFLP on chromosome 8 detected with a Calbindin 27kDa cDNA probe, HBSC21. Nucl. Acids Res., 16: 9373.

Parmentier, M., De Vijlder, J.J.M., Muir, E., Szpirer, C., Islam, M.Q.,Geurts van Kessel, A., Lawson, D.E.M., and Vassart, G., 1989, The human calbindin 27 kDa gene: structural organization of the 5' and 3' regions, chromosomal assignment, and restriction fragment length polymorphism. Genomics, 4: 309-319.

Pochet, R., Parmentier, M., Lawson, D.E.M., and Pasteels, J.L., 1985, Rat brain synthesizes two 'Vitamin D-dependent' calcium binding proteins. Brain Res., 345: 251-256.

Pochet, R., Pipeleers, D.G., and Malaisse, W.J., 1987, Calbindin D27kDa: preferential localization in non-B cells of the rat pancreas. Biol Cell, 61: 155-161.

Rogers, J., 1987, Calretinin: a gene for a novel calcium-binding protein expressed principally in neurons. J. Cell Biol., 105: 1343-1353.

Roth, J., Baetens, D., Norman, A.W., and Garcia-Segura, L.-M., 1981, Specific neurons in chick central nervous system stain with an antibody against chick intestinal vitamin D-dependent calcium-binding protein. Brain Res., 222: 452-457.

Sanger, F., Nicklen, S., and Coulson, A.R., 1977, DNA sequencing with chain-terminating inhibitors. Proc. Natl. Acad. Sci. USA, 74: 5463-5467.

Takagi, T., Nojiri, M., Konishi, K., Maruyama, K., and Nonomura, Y., 1986, Amino acid sequence of vitamin D -dependent calcium-binding protein from bovine brain. FEBS Lett., 201: 41-45.

Verstappen, A., Parmentier, M., Chirnoaga, M., Lawson, D.E.M., Pasteels, J.L., and Pochet, R., 1988, Vitamin D-dependent calcium-binding protein immunoreactivity in human retina. Ophtalmic Res., 18: 209-214.

Wasserman, R.H., and Taylor, A.N., 1966, Vitamin D3-induced calcium-binding protein in chick intestinal mucosa. Science 152: 791-793.

Wells, R.D., 1988, Unusual DNA structures. J. Biol. Chem., 263: 1095-1098.

Wilson, P.W., Harding, M., and Lawson, D.E.M., 1985, Putative amino acid sequence of chick calcium-binding protein deduced from a complementary DNA sequence. Nucl. Acids Res., 13: 8867-8881.

Wilson, P.W., Rogers, J., Harding, M., Pohl, V., Pattyn, G., and Lawson, D.E.M., 1988, Structure of chick chromosomal genes for calbindin and calretinin. J. Mol. Biol., 200: 615-625.

CALBINDIN-D9K (CaBP9K) GENE : A MODEL FOR STUDYING THE

GENOMIC ACTIONS OF CACITRIOL AND CALCIUM IN MAMMALS

Monique Thomasset, Jean-Marc Dupret, Arlette Brehier and Christine Perret

INSERM, U120, Alliée CNRS, 44 Chemin de Ronde, 78110 Le Vésinet, France

INTRODUCTION

The 9KDa vitamin D-dependent calcium-binding protein (CaBP9K) now termed calbindin-D9K (1), is characteristic of the rat and other mammals and was discovered by Wasserman's group in 1967 (2). This acidic protein has 2 calcium-binding sites organized as "EF-hands" (3, 4). CaBP9K was first located in the absorptive cells of the duodenum but is also found in the epithelial cells of the placenta and uterus, as demonstrated by in situ hybridization (5, 6) using a specific cloned CaBP cDNA (7). Northern analysis shows a single 0.5 kb long transcript from rat duodenum, fetal and maternal placenta and uterus (5, 6). CaBP9K gene expression in rat duodenum is hormonally controlled by $1,25(OH)_2D_3$ (calcitriol), the active form of vitamin D. $1,25(OH)_2D_3$ receptors have been identified and localized in the duodenum (8) and duodenal concentration of CaBP9K is dependent upon the vitamin D status (9).

MATERIALS AND METHODS

A specific cloned CaBP9K cDNA (7) and antibodies (9) have been used to study *in vivo* and *in vitro* regulation of CaBP9K gene expression by calcitriol and calcium in rat duodenum.

RESULTS AND DISCUSSION

A single injection of $1,25(OH)_2D_3$ stimulates duodenal CaBP biosynthesis in vitamin D-deficient rat. CaBP9K mRNA level increases significantly 1 h after injection and three fold in 3 hours. CaBP9K level increases later (10).

We have also used mucosal cells from fetal rat duodena, isolated at 18 days of gestation and maintained in culture for 10 days in a serum-free medium, to examine the regulation of duodenal CaBP9K biosynthesis. Under these experimental conditions cells differentiated into enterocytes. In this *in vitro* system CaBP9K mRNA level was significantly higher as early as 1 hour after calcitriol injection, whereas CaBP9K itself measured by radioimmunoassay increased after 24 hours. Actinomycin (0.1 and 1 μM), a drug blocking transcription, inhibited the increase in CaBP9K mRNA induced by $1,25(OH)_2D_3$. This is consistent with a regulation of CaBP9K gene expression by $1,25(OH)_2D_3$. This transcriptional effect of $1,25(OH)_2D_3$ was demonstrated by run-on assay, i.e. *in vitro* elongation of CaBP9K gene transcript initiated *in vivo* from isolated duodenal nuclei. Synthesis of CaBP9K mRNA in the vitamin D deficient rats increased 15 min after a single $1,25(OH)_2D_3$ injection. Therefore $1,25(OH)_2D_3$ has a rapid effect on CaBP9K gene transcription (11). Maximum rate of CaBP9K mRNA synthesis (run on) was at 1 hour. The CaBP9K mRNA accumulation increased gradually from 1 to 16 h after treatment as measured by dot-blot assay. From this experiment we cannot exclude a post-transcriptional effect of $1,25(OH)_2D_3$, as for other steroid hormones.

The CaBP9K gene was isolated to facilitate further study of the mechanism of $1,25(OH)_2D_3$ action at the genomic level (4). This gene contains 3 exons interrupted by 2 introns. Each calcium site is encoded by a separate exon with a helix-loop-helix structure. One phase 0 intron is present in the interdomain between each Ca-binding site.

We have analyzed the promoter region. Transcription is initiated at a single site as demonstrated by S_1 nuclease mapping. The promoter region contains the usual regulatory elements (one TATA box and two CAAT boxes). The 2 kb upstream the cap site, contains sequences similar to those of the hormone regulatory elements (hre) characteristic of other steroid hormones, and other cis-acting elements (AP_1, SP_1, heatshock) recognized by other trans-acting factors. These results indicate that $1,25(OH)_2D_3$ effects CaBP gene transcription, as demonstrated in organ culture and by run-on assay *in vivo*. The vitamin D-responsive element is presently being investigated.

To determine whether $1,25(OH)_2D_3$ is the only effector we have investigated the role of calcium in CaBP9K gene expression :
 - in vitamin D-deficient rats by altering dietary Ca and
 - in organ culture by modifying extracellular [Ca].

In vivo, when vitamin D-deficient rats diet is changed from a low-Ca diet to a high-Ca diet there is an increase in CaBP9K mRNA and CaBP9K protein. This increase in CaBP9K mRNA and protein itself also occurs after $1,25(OH)_2D_3$ injection. We then tested the effect of changes in calcium concentration in the medium used for fetal duodenal cells maintained in culture.

Increase of calcium concentration in the medium from 0 to 1.2 mM increased CaBP9K mRNA level both in absence and presence of $1,25(OH)_2D_3$. Actinomycin blocked this increase in CaBP9K mRNA.

In conclusion these results indicate that not only $1,25(OH)_2D_3$ has a rapid effect on CaBP9K gene transcription but also Calcium seems able to modulate CaBP9K gene expression .

ACKNOWLEDGEMENTS

We are grateful to N. Gouhier and M. Eb for their technical assistance.

REFERENCES

1. R.H. Wasserman, 1967, Nomenclature of the vitamin D-induced calcium-binding proteins in Vitamin D, in: "Chemical, biochemical and clinical update", A.W. Norman et al., eds, p. 321, De Cruyter, Berlin.
2. F.A. Kallfelz, A.N. Taylor, and R.H. Wasserman, Vitamin D-induced calcium-binding factor in rat intestinal mucosa, Proc. Soc. Biol. Med., 125:54-58 (1967).
3. D.M.E. Szebenyi, and K. Moffat, The refined structure of vitamin D-dependent calcium-binding protein from bovine intestine : molecular details, ion binding, and implications for the structure of other calcium-binding proteins, J. Biol. Chem., 261:8761-8777 (1986).
4. C. Perret, N. Lomri, N. Gouhier, C. Auffray, and M. Thomasset, The rat vitamin D-dependent calcium-binding protein (9-kDa CaBP) gene. Complete nucleotide sequence and structural organization, Eur. J. Biochem., 172:43-51 (1988).
5. M. Warembourg, C. Perret, and M. Thomasset, Distribution of vitamin D-dependent calcium-binding protein messenger ribonucleic acid in rat placenta and duodenum, Endocrinology, 119:176-184 (1986).
6. M. Warembourg, C. Perret, and M. Thomasset, Analysis and in situ detection of cholecalcin messenger RNA (9000 Mr CaBP) in the uterus of the pregnant rat, Cell Tissue Res., 247:51-57 (1987).
7. C. Desplan, O. Heidmann, J.W. Lillie, C. Auffray, and M. Thomasset, Sequence of rat intestinal vitamin D-dependent calcium-binding protein derived from a cDNA clone, J. Biol. Chem., 258:13502-13505 (1983).
8. T.L. Clemens, K.P. Garrett, X.Y. Zhou, J.W. Pike, M.R. Haussler, and D.W. Dempster, Immunocytochemical localization of the 1,25-dihydroxyvitamin D_3 receptor in target cells, Endocrinology, 122:1224-1230 (1988).
9. M. Thomasset, O. Parkes, and P. Cuisinier-Gleizes, Rat calcium-binding proteins : distribution, development and vitamin D-dependence, Am. J. Physiol., E483-E488 (1982).
10. C. Perret, C. Desplan, and M. Thomasset, Cholecalcin (a 9-kDa cholecalciferol-induced calcium-binding protein) messenger RNA : distribution and induction by calcitriol in the rat digestive tract, Eur. J. Biochem., 150:211-217 (1985).
11. J.M. Dupret, P. Brun, C. Perret, N. Lomri, M. Thomasset, and P. Cuisinier-Gleizes, Transcriptional and post-transcriptional regulation of vitamin D-dependent calcium-binding protein gene expression in the rat duodenum by 1,25-dihydroxycholecalciferol, J. Biol. Chem., 262:16553-16557 (1987).

PROTEIN ENGINEERING AND STRUCTURE/FUNCTION RELATIONS IN BOVINE CALBINDIN D_{9k}

S. Forsén, T. Drakenberg, C. Johansson, S. Linse, E. Thulin, J. Kördel

Physical Chemistry 2, Chemical Centre, University of Lund, P.O. Box 124, S-221 00 Lund Sweden

INTRODUCTION

In an attempt to elucidate the relationships between structure, dynamics and function in the calmodulin superfamily of Ca^{2+}-binding intracellular proteins we have undertaken a detailed study of bovine calbindin D_{9k}. This protein has a size ($M_r \approx 8,500$) and tertiary structure similar to that of the globular domains of calmodulin and troponin C and binds two Ca^{2+}-ions strongly ($K \approx 10^7 - 10^8 M^{-1}$, depending on the ionic strength). The schematic structure of the molecule is shown in figure 1.

Genes encoding wild-type bovine calbindin D_{9k}, as well as some 30 mutants with single, double and triple amino acid substitutions and/or deletions, have been synthesized and expressed in E. coli (Brodin, 1986; Linse, 1987). The combination of site-specific mutations and biophysical measurements of dynamic and equilibrium properties of the mutant proteins has proved to be very informative. Results from studies of a number of mutants - mostly confined to the N-terminal Ca^{2+}-site (site I) - have been reported elsewhere (Linse, 1987; Linse, 1988; Forsén, 1988; Wendt, 1988). In addition we have initiated studies of the solution structure of wild-type calbindin D_{9k} in the calcium-loaded and calcium-free (apo) forms using 2D ^1H NMR spectroscopy. During the course of these studies we have made observations that prove the existence in solution off different iso-forms of the protein due to cis-trans isomerization around a proline amide bond (Pro43) (Chazin, 1989a). Also the occurence of an iso-aspartyl linkage formed upon deamidation of calbindin (at Asn56) has been conclusively demonstrated by 2D ^1H NMR (Chazin, 1989b).

SALT EFFECTS ON EQUILIBRIUM Ca^{2+}-BINDING CONSTANTS

We have recently finished an investigation of the effect of added KCl on the equilibrium Ca^{2+}-binding properties of the wild-type calbindin and seven single, double and triple mutants involving replacements of Glu17, Asp19 and Glu26 by the corresponding side chain amides, see Figure 1 for their location. The apparent macroscopic Ca^{2+}-binding constants, K_1 and K_2, were determined in the presence of 0, 0.05, 0.10 and 0.15 M KCl and either Quin2 or $5,5'$-Br_2BAPTA as described (Linse, 1987; Linse, 1988). The analysis was based entirely on concentration. The large spread in Ca^{2+}-affinity for the proteins made it necessary to use both chelators, whose Ca^{2+}-affinity differ by a factor of 20. K_1 and K_2 were used to calculate the apparent free energy of binding of two Ca^{2+}-ions, $\Delta G_{tot} = -RTln(K_1.K_2)$, and the lower limit of the free energy of interaction between the two Ca^{2+}-sites, $\Delta\Delta G_{\eta=1} = -RTln(4.K_2/K_1)$ (derived for the case of equally strong sites, see Linse, 1988).

The results are presented in Figure 2A and B. Note that for clarity the error bars are left out, but the uncertainties are quoted in the figure text. The precision of the titration method used here is very high and thus effects of individual mutations can be compared. It is clear that the contributions to ΔG_{tot} from the negative side chains of Glu17, Asp19 and Glu26 are not equally large. Being situated further away from the Ca^{2+}-ions (cf. Figure 1) Glu26 contribute only half as much to the free energy of binding as do Glu17 and Asp19.

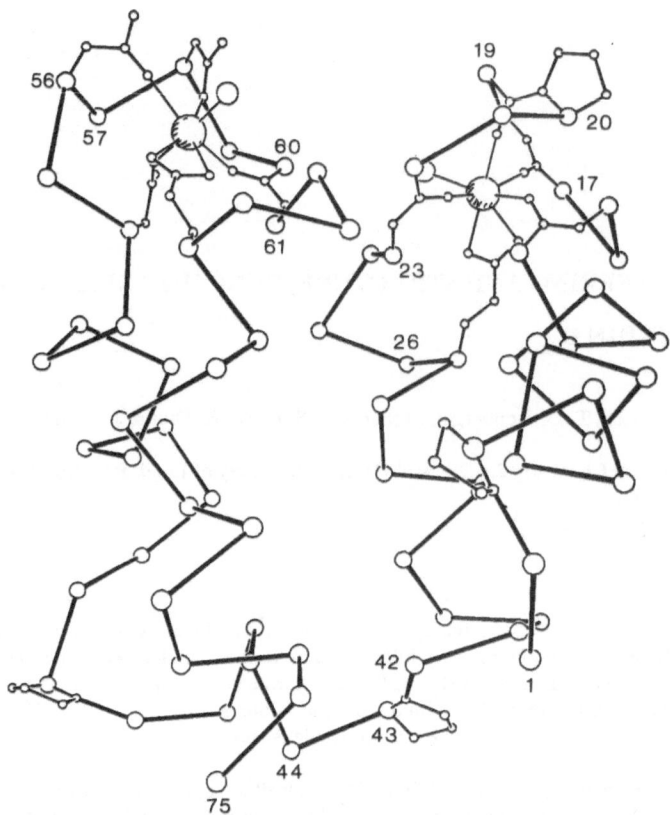

Fig. 1 .   Schematic structure of Ca^{2+}-loaded bovine calbindin D_{9k} (adapted after Szebenyi, 1986).

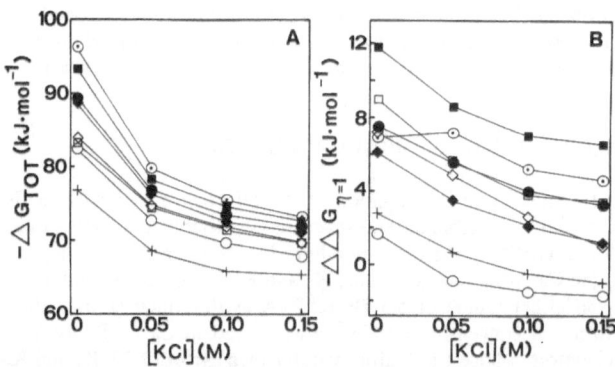

Fig. 2 .   Apparent free energy of binding of two Ca^{2+}-ions, ΔG_{tot}, and the lower limit of the free energy of interaction between the two Ca^{2+}-sites, $-\Delta\Delta G_{\eta=1}$, as a function of KCl concentration for (○) the wild-type protein, (●) E17Q, (♦) D19N, (■) E26Q, (○) (E17Q, D19N), (□) (E17Q, E26Q), (◊) (D19N, E26Q), (+) (E17Q, D19N, E26Q). ΔG_{tot} is highly reproducible between individual titrations of the same protein at the same value of [KCl]. The uncertainties in ΔG_{tot} are ±0.5kJ/mol. For $\Delta G_{\eta=1}$ the uncertainties are ±0.5 kJ/mol in the range $-1.5 < -\Delta\Delta G_{\eta=1} < 1.5$, but become succesively larger as $-\Delta\Delta G_{\eta=1}$ increases - ±1 KJ/mol at 5 kJ/mol and ±2 kJ/mol at 10 kJ/mol. This is due to the inherent physical properties of cooperativity. The binding curve becomes less sensitive to the same absolute change in $\Delta\Delta G_{\eta=1}$ as infinitive cooperativity is approached (Weber, 1975).

It is also evident that the differences among the individual proteins are most pronounced at low ionic strength. Extrapolation of the curves leads to the conclusion that ΔG_{tot} of the wild-type is reduced by approximately one third at high salt concentration. This might be entirely an ionic strength effect (screening of electrostatic interactions) but there might also be contributions from specific interactions between K^+-ions and the Ca^{2+}-sites (competitive binding). This ambiguity could probably be resolved through a ^{39}K NMR study of the binding of potassium ions to calbindin. The cooperativity of Ca^{2+}-binding can be characterized in energetic terms by the free energy of interaction between the two sites, $\Delta\Delta G$. This is equal to the difference in free energy of binding of Ca^{2+} to site I as a result of Ca^{2+}-binding to site II (and vice versa).

$$\Delta\Delta G = \Delta G_{I,II} - \Delta G_I = \Delta G_{II,I} - \Delta G_{II}$$

But, since the titration method used does not sense the distribution of Ca^{2+} between the two sites of the protein, we can only obtain a lower limit of $-\Delta\Delta G$ (denoted $-\Delta\Delta G_{\eta=1}$) which is equal to $-\Delta\Delta G$ if the sites are equally strong, i.e. $\Delta G_I = \Delta G_{II}$. Although $-\Delta\Delta G_{\eta=1}$ is not the actual free energy of interaction between the two sites in the protein molecule, it still has some relevance to use the quantity $-\Delta\Delta G_{\eta=1}$ in comparisons of individual proteins, since the actual free Ca^{2+} concentration in the presence of a 2-site Ca^{2+}-binding protein depends on ΔG_{tot} and $-\Delta\Delta G_{\eta=1}$ (not $-\Delta\Delta G$). Thus while $-\Delta\Delta G$ pertains to the extent of interaction between the Ca^{2+}-sites $-\Delta\Delta G_{\eta=1}$ describes the action on the surrounding solution. As can be seen in Figure 2b $-\Delta\Delta G_{\eta=1}$ decreases with increasing salt concentration, and there are differences among the individual proteins. Important to note is that there is still a fair amount of cooperativity in the wild-type at physiological ionic strength $-\Delta\Delta G_{\eta=1} = 5$ kJ/mol corresponds to an at least 8-fold increase in the Ca^{2+}-affinity for site I as Ca^{2+} binds to site II (and vice versa). If we compare the effects of removing one charge at Glu17, Asp19 or Glu26, the negative sidechain of Asp19 appears to contribute to positive cooperativity whereas the negative charge at Glu26 seems to counteract it. This is not an unexpected result since the cooperative Ca^{2+}-binding observed in EF-hand proteins probably results from a delicate balance of several intramolecular interactions. At high ionic strength $-\Delta\Delta G_{\eta=1}$ of the triple mutant falls below zero. This could mean that the positive cooperativity is lost, or that the Ca^{2+}-affinity for one of the two sites has decreased more than the other, leading to sequential Ca^{2+}-binding and thus "invisible" cooperativity.

CALCIUM OFF-RATES

From the temperature dependence of the line shape of ^{43}Ca NMR signal the off-rates (k_{off}) of the Ca^{2+} ion from its binding sites in the protein may be determined (Drakenberg, 1983; Tsai, 1987). For the native calbindin k_{off} for both site I and site II is approximately 3 s^{-1} (Linse, 1987). In a set of mutant proteins where various amino acid residues in site I have been altered and/or deleted the value of k_{off} pertinent to site II remains unaffected. There are, however, dramatic changes in k_{off} for site I in the same series (Linse, 1987; Johansson, 1989). In figure 3 the temperature dependence of the ^{43}Ca lineshape for the mutant proteins A14 Δ and P20G,A14Δ is shown. For the mutant protein A14 Δ the value of $k_{off}(I)$ has been

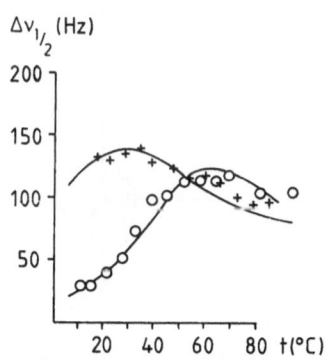

Fig. 3 . The temperature dependence of the ^{43}Ca line width for the mutant proteins (o):A14Δ and (+):P20G,A14 Δ. The continuous line shows the best theoretical fit to the experimental points. The dynamic parameters thus obtained for the intermediate exchanging site are for P20G,A14Δ: $\Delta H^{\ddagger} = 35$ kJ/mol, $\Delta S^{\ddagger} = -60$ J/mol/K, $\chi = 1$ MHz and $\Delta G^{\ddagger} = 25$ kJ/mol and for A14Δ: $\Delta H^{\ddagger} = 39$ kJ/mol, $\Delta S^{\ddagger} = -62$ J/K/mol, $\chi = 1.05$ MHz and $\Delta G^+ = 25.7$ kJ/mol. χ is the quadrupole coupling constant of the ^{43}Ca nucleus in the binding sites.

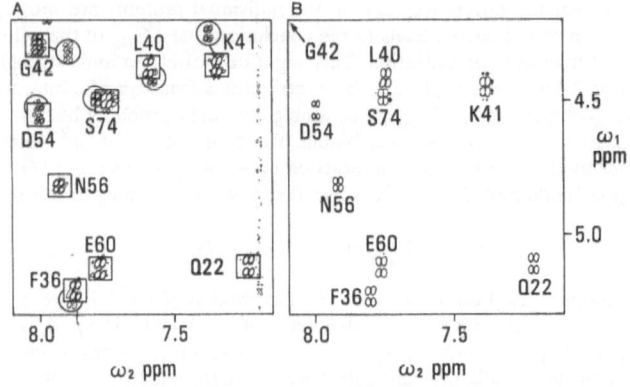

Fig. 4 . Expanded COSY spectrum of A) native calbindin D_{9k} and B) the Pro43 →Gly mutant. Major resonances are boxed and, when distinguishable, minor resonances are circled.

increased approximately 100 fold to 500 ± 100 s^{-1} and upon additional deletion of P20 $k_{off}(I)$ is further increased about 10 fold to 7500 ± 1000 s^{-1}. A similar increase in $k_{off}(I)$ is seen between the mutant proteins P20G ($k_{off}(I) = 430 \pm 100$ s^{-1}) and P20G,A14$_\Delta$ ($k_{off}(I) = 3400 \pm 1000$ s^{-1}).If, however, the only modification is a deletion of P20 the value of $k_{off}(I)$ is immediately raised to 5000 ± 1000 s^{-1} and further deletion of A14 has no additional effect on $k_{off}(I)$. The same dramatic increase in $k_{off}(I)$ is seen for P20G,N21 Δ were $k_{off}(I) = 5000 \pm 1000$ s^{-1} and again deletion of A14 to P20G,N21 Δ,A14Δ leaves koff virtually unchanged. Thus the effect of the mutations are not additive and the order in which the mutations are made is important. It is also interesting to note the pronounced stabilizing effect that P20 has on the calcium binding of site I.

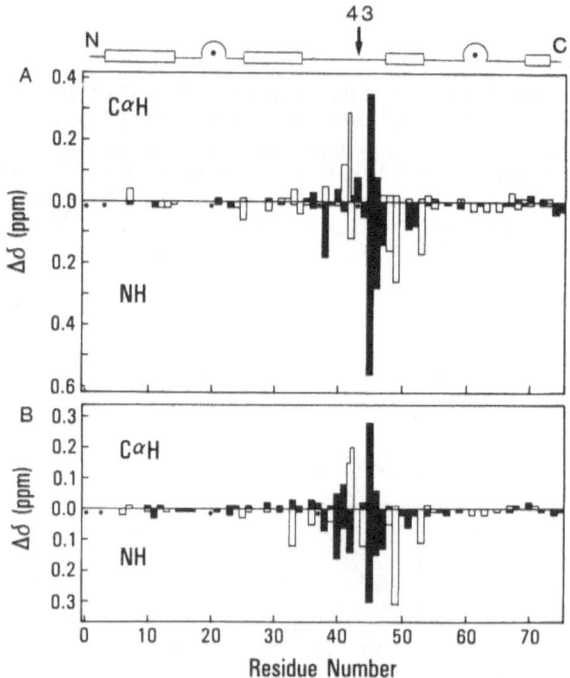

Fig. 5 . Chemical shift differences as a function of sequence for the backbone protons. A) *trans*-Pro43 - cis-Pro43 calbindin. B) *trans*-Pro43 - Pro43 →Gly calbindin. Filled bars represent a negative chemical shift difference and open bars a positive. The (helix-loop-helix)$_2$ secondary structure of calbindin as a function of the sequence is outlined at the top of the figure.

All ^{1}H NMR spectra of calbindin show that approximately half of the proton resonances are resolved into major and minor resonances in a ratio of 3:1. Figure 4A shows a small region of the backbone fingerprint in a two-dimensional (2D) scalar correlated spectroscopy (COSY) spectrum of calbindin D_{9k}. The resolution of major and minor resonances is indicated and it is clear that the chemical shift difference between the two forms is variable. The two forms are in equilibrium and the interconversion rate can be estimated to 10^{-1} - 10^{-2} s^{-1} at room temperature (Chazin, 1989a).

To gain further insight into the origin of this conformational heterogeneity the complete sequence specific assignments are needed (Kördel, 1989a; Kördel, 1989b). In summary, these were acquired using a strategy integrating relayed coherence transfer and multiple quantum techniques to identify the spin system of each amino acid (Chazin, 1988). These amino acid spin systems were thereafter assigned to their location in the sequence by use of the sequential assignment procedure (Billeter, 1982). For a general review see Wüthrich (1986).

In figure 5A the chemical shift differences between the backbone protons of the two forms as a function of sequence are depicted. The global structure is essentially identical for the two forms and the structural differences, as deduced from chemical shifts, are highly localized to the linker region between the two domains of the protein. The exchange rate and the sequential location of the perturbation indicates that *cis/trans* isomerism at Pro43 is the cause of the multiple conformations. This conclusion is strengthened by the fact that in all regions of ^{1}H NMR spectra of the Pro43→Gly mutant only one set of resonances appear (cf. fig 4B).

As figure 6 shows, it is possible to distinguish *cis* from *trans* Gly-Pro peptide bonds from the through-space proton pair connectivities. While the proline C^{δ} protons, but not the C^{α} proton, are within 5 Å of the glycine C^{α} protons in the *trans* form (figure 6A) the opposite is true for the *cis* form where only the C^{α} protons of glycine and that of proline are within 5 Å of each other (figure 6B). Using 2D nuclear Overhauser spectroscopy (NOESY) experiments adjusted to detect only proton pairs closer than 5 Å, the major form of calbindin could be shown to correspond to *trans*-Pro43 calbindin and the minor form to the

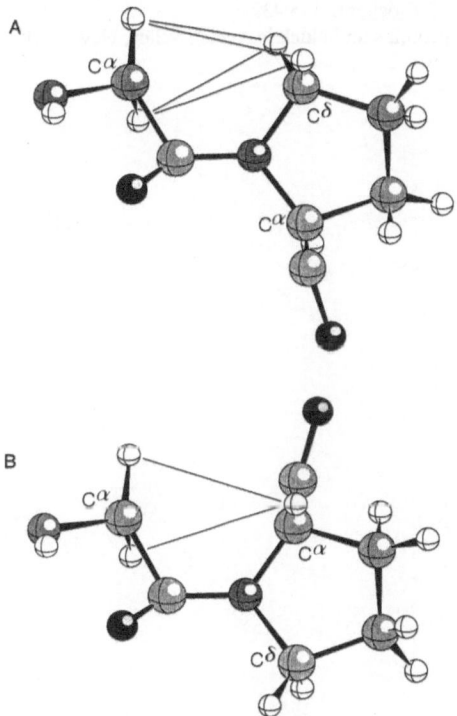

Fig. 6 . A Gly-Pro dipeptide in A) *trans* conformation and B) in *cis* conformation. Nuclear Overhauser enhancements possible to observe in NOESY experiments are indicated with solid lines.

cis-Pro43 isoform (Chazin, 1989a). This conformational heterogeneity is also present in porcine calbindin (Drakenberg, 1989) as well as in deamidation products of calbindin (Chazin, 1989b). The structural heterogeneity can be eradicated by substituting Pro43 with a glycine residue. As can be seen in figure 5B, 2D ^{1}H NMR analysis of this mutant demonstrates that the global conformation is very similar to that of the two wild-type forms while there is no conformational heterogeneity as confirmed in fig 4B.

REFERENCES

Billeter, M., Braun, W., and Wüthrich, K., 1982, J. Mol. Biol., 155:321.

Brodin, P., Grundström, T., Hofmann, T., Drakenberg, T., Thulin, E., and Forsén, S., 1986, Biochemistry, 25:5371.

Chazin, W.J., Rance, M., and Wright, P., 1988, J. Mol. Biol., 202:603.

Chazin, W.J., Kördel, J., Thulin, E., Drakenberg, T., Brodin, P., Grundström, T., and Forsén, S., 1989a, Proc. Natl. Acad. Sci. USA, 86:2195.

Chazin, W.J., Kördel, J., Thulin, E., Hofmann, T., Drakenberg, T., and Forsén, S, 1989b, Biochemistry, in press.

Drakenberg, T., Forsén, S., and Lilja, H., 1983, J. Magn. Reson., 53:412.

Drakenberg, T., Hofmann, T., and Chazin, W.J., 1989, Biochemistry, 28:5946.

Forsén, S., Linse, S., Thulin, E., Lindegard, B., Martin, S.R., Bayley, P.B., Brodin, P., and Grundström, T., 1988, Eur. J. Biochem., 177:47.

Johansson, C., Thulin, E., Brodin, P., Grundström, T., Forsén, S., and Drakenberg, T., 1989, submitted.

Kördel, J., Forsén, S., and Chazin, W.J., 1989a, Biochemistry, 28:7065.

Kördel, J., Forsén, S., Drakenberg, T., and Chazin, W.J., 1989b, submitted.

Linse, S., Brodin, P., Drakenberg, T., Thulin, E., Sellers, P., Elmdén, K., Grundström, T., Forsén, S., 1987, Biochemistry, 26:6723.

Linse, S., Brodin, P., Johansson, C., Thulin, E., Grundström, T., and Forsén, S., 1988, Nature, 335:651.

Szebenyi, D., and Moffat, K., 1986, J. Biol. Chem., 261:8761.

Tsai, M-D., Drakenberg, T. Thulin, E., and Forsén, S., 1987, Biochemistry, 26:3635.

Weber, G., 1975, Adv. Protein Chem., 29:1.

Wendt, B., Hofmann, T., Martin, S.R., Bayley, P.B. Brodin, P., Grundström, T., Thulin, E., Linse, S., and Forsén, S., 1988, Eur. J. Biochem., 175:439.

Wüthrich, K., 1986, "NMR of proteins and Nucleic Acids", Wiley, New York.

MUTANT ANALYSIS APPROACHES TO UNDERSTANDING CALCIUM SIGNAL TRANSDUCTION THROUGH CALMODULIN AND CALMODULIN REGULATED ENZYMES

J. Haiech,[1] M.-C. Kilhoffer,[2] T.A. Craig,[3] T.J. Lukas,[3] E. Wilson,[3] L. Guerra-Santos,[3] and D.M. Watterson[3]

[1]LCB, CNRS, 31 Chemin Joseph Aiguier, 13009 Marseille, France

[2]Laboratoire de Biophysique, Université Louis Pasteur, Faculté de Pharmacie de Strasbourg, BL No.10, 67048 Strasbourg Cedex, France

[3]Department of Pharmacology, Vanderbilt University and Laboratory of Cellular and Molecular Physiology, Howard Hughes Medical Institute, Nashville, Tennessee 37232 USA

Calcium is one of several intracellular signal transducers utilized by eukaryotic cells as part of their integrated responses to external stimuli. For example, the occupancy of a cell surface receptor by a drug, hormone or a growth/differentiation factor often results, directly or indirectly, in a rise in the intracellular concentration of ionized calcium. This transient, localized rise in calcium sets off a discrete set of cellular responses. The ability of a divalent ion to serve as a selective and quantitative messenger in a complex biological system is based on calcium's ability to reversibly interact with and modulate the structure of a class of proteins referred to as calcium-modulated proteins (for a review see Van Eldik et al., 1982). One of these calcium modulated proteins that has been found in all eukaryotic cells examined is calmodulin.

Calmodulin is an integral subunit of several enzymes and is involved in the regulation of various subcellular structures such as cytoskeletal complexes and ion channels (for reviews and recent reports see Klee and Vanaman, 1982; Roberts et al., 1986; Schaefer et al., 1987a,b; Cohen and Klee, 1988). Because calmodulin is the calcium responsive element of multiple protein structures within the same cell, it provides a common motif to one set of pathways in the cell's response to transient rises in the concentration of intracellular ionized calcium. However, the calmodulin structural features required for regulation of enzyme activity can vary with each enzyme (Craig et al., 1987b, Putkey et al., 1988); and the calcium binding properties of calmodulin can, in turn, be altered by interaction with an enzyme (Olwin and Storm, 1985). Thus, an element of qualitative and quantitative selectivity can be introduced into the common motif as a result of protein-protein interactions. Knowledge of how the structures of calmodulin and calmodulin regulated enzymes are related to function are required for a full understanding of how eukaryotic cells are able to transduce calcium signals into qualitatively distinct biological responses and might provide insight into common features of the molecular mechanisms utilized in ion modulation of protein function. One general approach that has provided considerable insight into the structure, function and mechanism of calmodulin regulation is mutant analysis. There are four types of mutant analysis studies that have been done on calmodulin: 1. gene deletion studies (Davis et al., 1986, Takeda et al., 1987); 2. biochemical analyses of mutant organisms altered in their calcium mediated responses (Saimi et al., 1983; Burgess-Cassler et al., 1987; Schaefer et al., 1987a,b); 3. in vitro site-specific mutagenesis combined with in vitro functional analysis (Roberts et al., 1985a,b, 1986; Craig et al 1987b; Putkey et al., 1988; Maune et al, 1988); and 4. in vitro site-specific mutagenesis combined with in vivo functional analysis (Craig et al., 1987a).

Table 1. Examples of calmodulins generated by site-specific mutagenesis

I. Helix Mutations

A. Charge

SYNCAM-8	EEE82-84 —>KKK
SYNCAM-11	K75 —>P
SYNCAM-12	DEE118-120 —>KKK, M124 —>I
SYNCAM-12A	DEE118-120 —>KKK
SYNCAM-18	EEE82-84 —>KKK, DEE118-120 —>KKK, M124 —>I
SYNCAM-18A	EEE82-84 —>KKK, DEE118-120 —>KKK
SYNCAM-24	DEQ6-8 —>KKK
SYNCAM-26	E11 —>K
SYNCAM-28	E84 —>K
SYNCAM-29	E120 —>K
SYNCAM-30	S81 —>R
SYNCAM-35	K75 —>P, KGK insert between D80 and S81
SYNCAM-39	E82 —>K
SYNCAM-40	E84 —>K, E120 —>K
SYNCAM-43	EAE45-47 —>KKK
SYNCAM-44	E47 —>K
SYNCAM-45	E83 —>K
SYNCAM-46	EEE82,83 —>KK
SYNCAM-47	EEE82,84 —>KK
SYNCAM-48	EEE83,84 —>KK
SYNCAM-49	DEQ6-8 —>KKK, EAE45-47 —>KKK
SYNCAM-50	DEQ6-8 —>KKK, EEE82-84 —>KKK
SYNCAM-51	DEQ6-8 —>KKK, DEE118-120 —>KKK, M124 —>I
SYNCAM-51A	DEQ6-8 —>KKK, DEE118-120 —>KKK
SYNCAM-52	EAE45-47 —>KKK, EEE82-84 —>KKK
SYNCAM-53	EAE45-47 —>KKK, DEE118-120 —>KKK, M124 —>I
SYNCAM-53A	EAE45-47 —>KKK, DEE118-120 —>KKK
SYNCAM-54	DEQ6-8 —>KKK, EAE45-47 —>KKK, EEE82-84 —>KKK, DEE118-120 —>KKK
SYNCAM-55	E87 —>K
SYNCAM-56	E87 —>T
SYNCAM-57A	K75 —>A
SYNCAM-57B	K75 —>E
SYNCAM-57C	K75 —>G
SYNCAM-57D	K75 —>V
SYNCAM-62	E84 —>Q
SYNCAM-63A	E120 —>Q
SYNCAM-65	E84 —>Q, E120 —>Q

44

II. Calcium Binding Loop Mutations

A. Charge

SYNCAM-13 E67 —>A
SYNCAM-14 E140 —>A
SYNCAM-60 E31 —>A
SYNCAM-61 E104 —>A

B. Reporter Groups

SYNCAM-9 F99 —>W
SYNCAM-32 T26 —>W
SYNCAM-33 T62 —>W
SYNCAM-34 Q135 —>W

C. Other Calcium Binding Loop Mutations

SYNCAM-10 G96 —>Q
SYNCAM-17 S101 —>F
SYNCAM-36 S101 —>A
SYNCAM-37 S101 —>G
SYNCAM-38 S101 —>Y
SYNCAM-73A V136 —>T
SYNCAM-69A D118 —>K
SYNCAM-69B D118 —>Q
SYNCAM-69C E119 —>K
SYNCAM-69D E119 —>Q

B. Reporter Group

SYNCAM-31 S81 —>W
SYNCAM-67 E84 —>W
SYNCAM-68 E120 —>W
SYNCAM-7CA F141 —>W
SYNCAM-72A M145 —>W

C. Other Helix Mutations

SYNCAM-2 G33 —>V, K75 —>P, M76 —>L
SYNCAM-5 G33 —>V
SYNCAM-6 K75 —>P, M76 —>L
SYNCAM-7 A88 —>P
SYNCAM-15 KGK insert between D80 and S81
SYNCAM-64A Deletion E82,E83
SYNCAM-64B Deletion E84
SYNCAM-64C Deletion EEE82-84
SYNCAM-71A M145 —>S

III. Other Calmodulin Mutations

A. Charge

SYNCAM-3 K115 —>R
SYNCAM-4 K115 —>I
SYNCAM-16 K115 —>Y
SYNCAM-58A K115 —>D
SYNCAM-58B K115 —>H
SYNCAM-58C K115 —>N

B. Domain Shuffling

SYNCAM-19 SYNCAM-1 1-136, Chlamydomonas 137-159
SYNCAM-20 SYNCAM-1 1-101, Chlamydomonas 102-159
SYNCAM-21 SYNCAM-1 1-3, Chlamydomonas 4-159
SYNCAM-22 D78 —>E, S81 —>H, E83 —>D
SYNCAM-23 SYNCAM-1 1-136 (D78 —>E, S81 —>H, E83 —>D),
Chlamydomonas 137-159
SYNCAM-41 SYNCAM-1 1-77
SYNCAM-42 SYNCAM-1 78-148
SYNCAM-59A SYNCAM-1 1-84, Caltractin 85-152

Gene deletion studies have been reported only for yeast as of this date. The results show that calmodulin is essential for cell survival (Davis et al., 1986; Takeda et al., 1987). These studies also provide the tools for gene deletion-rescue experiments in which new and mutant calmodulin genes can be introduced into eukaryotic cells and the activity evaluated in vivo.

Biochemical analyses of mutant organisms in which the causative calcium response defect is a mutation of calmodulin have been reported only for Paramecium. The generation of mutant organisms requires that the mutation in calmodulin be strong enough such that the phenotypic defect is evident, but mild enough to allow cell survival. The results with Paramecium provide a formal proof of the model that mutations of calmodulin gene exons can be selective and non-lethal, insight into in vivo roles, and the first clear evidence that calmodulin might be involved in ion channel regulation.

In vitro site-specific mutagenesis and protein engineering combined with in vitro functional analyses has allowed a detailed dissection of the molecular mechanisms of calmodulin mediated calcium signal transduction, helped establish which possible mechanisms are thermodynamically and kinetically feasible, and, as a result, have provided a predictive aspect to biological studies (i.e., what could or might happen in vivo). Site-specific mutagenesis and protein engineering experiments provide the tools and the approach allows the efficient testing of hypotheses. Many detailed mechanistic studies can only be done in vitro; in some cases, experiments would not be possible or feasible without this approach (e.g., the design and production of isofunctional structural mutants with pre-selected placement of internal reporter groups). Overall, the site-specific mutagenesis/protein engineering approach has had the broadest impact and provided the most detailed mechanistic insight of all the mutant analysis approaches.

In vitro site-specific mutagenesis and protein engineering combined with in vivo functional analyses is a hybrid of the other three mutant analysis approaches, and is designed to take advantage of the strengths of each individual approach as well as being an attempt to integrate the knowledge derived from each approach. Examples include: 1. the microinjection of various genetically engineered calmodulins into mutant Paramecia and examining the ability of the purified proteins to temporarily restore the mutant phenotype back toward wild-type; and 2. the transfection and rescue of calmodulin-minus yeast with various calmodulin genes encoding proteins with mutations that selectively alter a set of calmodulin-mediated pathways.

OVERVIEW: SITE-SPECIFIC MUTAGENESIS AND PROTEIN ENGINEERING

The focus of this article is on how site-specific mutagenesis and protein engineering are providing insight into how calcium signals are transduced into biological responses. The emphasis is on how this approach allows the molecular dissection of the early steps in the cellular response mechanism: 1. calcium binding to calmodulin; 2. calcium dependent conformational changes and intramolecular signaling within the calmodulin molecule; and 3. activation of enzymes that have calmodulin as a calcium response element. Within the context of this article, we have used the terms site-specific mutagenesis and protein engineering based on the following working definitions: 1. site-specific mutagenesis, the alteration of the coding potential of DNA at pre-selected sites or regions; 2. protein engineering, the design and production of desired proteins by using recombinant DNA technology. While these definitions are somewhat restrictive, they will hopefully avoid confusion when these terms, which have been used in various contexts in the literature, are used in this article.

The initial site-specific mutagenesis and protein engineering studies of calmodulin and calmodulin related enzymes (Alvarado-Urbina et al., 1985; Asselin et al., 1988; Chabbert et al., 1989; Craig et al., 1987a,b,c; Haiech et al., 1988; Haiech and Watterson, 1988; Hurwitz et al., 1988; Kilhoffer et al., 1988; Kilhoffer et al., 1989; Lukas et al, 1987, 1988; Maune et al., 1988; Persechini et al., 1988; Putkey et al., 1986, 1988; Roberts et al., 1985a,b; Roberts et al., 1987; Watterson and Roberts, 1985; Watterson et al., 1985a,b; Weber et al., 1989; Wilson et al., 1988a,b) have been based on the previously accummulated knowledge about the primary (Figure 1) and tertiary (Figure 2) structures of naturally occurring calmodulins, the general calcium binding and enzyme activator properties of these calmodulins from various phylogenetic species, and the correlations of structure with function based on comparative, chemical modification, and limited proteolysis studies of calmodulin and certain calmodulin regulated enzymes. Regardless of the starting point, the subsequent mutagenesis studies have been of two general types: 1. a comparative, or horizontal, type of study in which an initial database of structure-function correlates are generated and the selectivity of mutations are addressed; or 2. a more detailed, or vertical, type of study in which the structural features of calmodulin or a specific calmodulin regulated enzyme are studied. In either type of study, a family of mutant structures are required. Some specific examples for calmodulin mutagenesis are listed in Table 1.

```
                                            *   *   *   *   *    *              39
A)    Ac-A-D-Q-L-T-E-E-Q-I-A-E-F-K-E-A-F-S-L-F-D-K-D-G-D-G-T-I-T-T-K-E-L-G-T-V-M-R-S-L-
B)                        |
C)      X-(A,Z,Z)     D   |                                    C
D)      X-(A,S₂,Z₂)       |                                    S
E)                        |
F)    Ac-A-A-N-T-E        |              A
G)           A    S-N   S |
H)           A    D       |
I)      Ac-(A,Q,E)        | K"           A
J)           S-S-N        |              A      N-N   S   S-S-S   A
K)      T-T-R-N   D       | R            R      Q     N   S-N     V
L)           A            |
```

```
          40                            *   *   *   *   *    *               75
A)    G-Q-N-P-T-E-A-E-L-Q-D-M-I-N-E-V-D-A-D-G-N-G-T-I-D-F-P-E-F-L-T-M-M-A-R-K-
B)                  |                              D
C)                  |                                          N-L
D)                  |                          N
E)                  |                                          S-L
F)                  |            S                             M-L
G)                  |                  Q       S              L
H)                  |                                          N-L
I)                  |                                          S-L
J)        L-S   S   | V-N   L-M       I   V        H-Q   E   S  A-L     S     Q
K)          S     A |                              T
L)                  |
```

```
      76                                *   *   *   *   *    *              112
A)    M-K-D-T-D-S-E-E-E-I-R-E-A-F-R-V-F-D-K-D-G-N-G-Y-I-S-A-A-E-L-R-H-V-M-T-N-L-
B)               |                          D   F
C)               | L-K                Q   F
D)        Q    T |           K                                             S
E)             T | L-I       K     R       L
F)        E    H D L         K             F
G)        Q  S   | K                       F                 I
H)               | L-K                     F
I)        E-Q    | L-I       K     R       L
J)      L  S-N  Q| L-L       K         N D L          K         L    S-I
K)          N    | V         K                   T-V-E         T    L     S
L)               |                         F
```

```
      113                               *   *   *   *   *    *              148
A)    G-E-K'L-T-D-E-E-V-D-E-M-I-R-E-A-D-I-D-G-D-G-Q-V-N-Y-E-E-F-V-Q-M-M-T-A-K-
B)             |                                          T       S
C)             |                 V           I            K-V     M
D)        K   N|                 L                D       K       I-V-R-
E)             |                       H-I                R       M
F)        K  S-E                 V                        R         S-G-
G)        K   |                  V           I            K       M-S
H)        K   |                  V                        V       M
I)             D|                            H-I          R       V-S
J)        K    A| D   L   V-S #       S   E-I   I-Q-Q  A-A-L-S-K  #
K)        R  S-Q| A-D         T           V-I         S-R-V-I-S-S
L)        K     |                         V-I            T       S
```

```
      149                 159
D)    N
F)            A-T-D-D-K-D-K-K-G-H-K
```

Fig. 1. Amino acid sequences of naturally occurring and synthetic (SYNCAM-1) calmodulins compared. Amino acid sequence of bovine brain calmodulin (Watterson et al., 1980; 1984) is shown in line A. Amino acid residues which differ from bovine brain calmodulin are shown for all other sequences shown: B) Scallop muscle (Toda et al., 1981); C) Spinach (Lukas et al., 1984); D) *Dictyostelium* (Marshak et al., 1984); E) *Tetrahymena* (Watterson et al., manuscript in preparation); F) *Chlamydomonas* (Lukas et al., 1985; Zimmer et al., 1988); G) *Trypanosoma* (Tschudi et al., 1985); H) Synthetic calmodulin (SYNCAM-1, originally VU-1 calmodulin) (Roberts et al., 1985a); I) *Paramecium* (Schaefer et al., 1987a,b); J) Yeast calmodulin - *Saccharomyces cerevisiae* (Davis et al., 1986); K) Yeast calmodulin - *Schizosaccharomyces pombe* (Takeda and Yamamoto, 1987); L) *Drosophila melanogaster* (Smith et al., 1987). The vertical dashed line indicates the conserved glutamic acid present in all four calmodulin domains. The asterisks show the calcium binding residues. The overlined residues correspond to alpha helical regions as determined from X-ray crystallographic results by Babu et al. (1988). Ac, acetyl group; X, unknown amino-terminal blocking group; K', trimethyllysine: K", dimethyllysine; #, gap in sequence inserted for optimal sequence alignment.

The calmodulin examples given in Table 1 reflect the two different avenues of research, or types of study, that can be pursued with site-specific mutagenesis and protein engineering, and indicate the interrelatedness of such studies. For example, as discussed in more detail below, the central helix cluster mutation produced in VU-8 (SYNCAM-8, Table 1) calmodulin was based on considerations of phylogenetic conservation of amino acid sequence, the possible role of electrostatics in the interaction of calmodulin with some enzymes, and the novel tertiary structural features of calmodulin and troponin C. This mutant calmodulin formed the basis of a horizontal type of study in which several calmodulin regulated enzymes were studied, was the starting point for a vertical type of study in which 28 mutant calmodulins (mostly alterations of acidic residues in alpha helices) were utilized in a study of myosin light chain kinase (MLCK) activation, and was a key mutant in studies of the coupling between calcium binding sites in calmodulin. Relatedly, the isofunctional mutant VU-9 (SYNCAM-9, Table 1)) calmodulin was a key mutant in the dissection of the calcium binding mechanism of calmodulin, but it is also proving to be an important tool in the study of calmodulin interaction with MLCK. The following sections outline how a generalized model of calcium signal transduction with a potential of response specificity is evolving from site-specific mutagenesis and protein engineering studies.

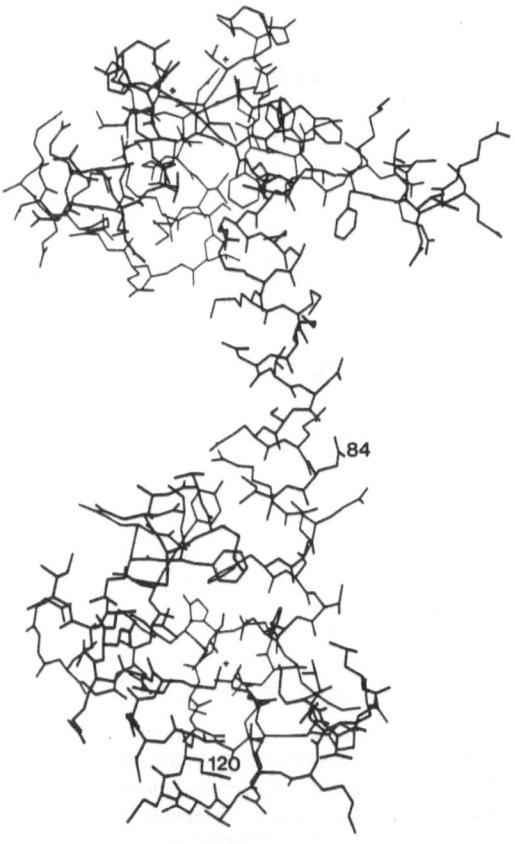

Fig. 2. Tertiary structure of the SYNCAM-1 calmodulin. A model of SYNCAM-1 calmodulin was constructed (Weber et al., 1989) based on the structure of rat testes calmodulin (Babu et al., 1988). Residues Ala-1, Asp-2, Gln-3, Leu-4 and Lys-148 not defined in the vertebrate structure have been added to extend the a-helices at the amino-terminus and carboxy terminus of the SYNCAM-1 model. Key residues Glu 84 and Glu 120 are labeled. The location of calcium ions is indicated with a +.

CALCIUM BINDING TO CALMODULIN

Mechanisms of calcium binding to calmodulin have been studied for more than 10 years by various techniques (e.g. equilibrium dialysis, flow dialysis, calcium electrode). Conformational changes associated with the calcium occupancy of the sites have been followed by UV spectrum, fluorescence spectrum, NMR, small angle neutron diffraction. Finally, kinetics of calcium dissociation have been recorded using tyrosine fluorescence as an intrinsic probe or quinII as a calcium sequestering agent and as an extrinsic probe. The main results can be summarized as follows: 1) The Scatchard representation of the calcium binding isotherm is almost linear and intercepts the abscissa at four bound calciums (n=4 sites, Kd= 1-4x10^{-6} M). 2) The calcium titration curve is in general biphasic, with a signal between 0 and 2 bound calciums and another one between 2 and 4 bound calciums. When using NMR, modifications of resonances assigned to residues from the COOH half of the molecule, occur between 0 and 2 bound calciums whereas modifications of resonances assigned to residues from the NH_2 half of the molecule appear in the second phase. 3) Finally, kinetics of calcium dissociation is biphasic with a rapid phase (k off = 650 s^{-1}) and a slow phase (k off = 9 s^{-1}).

Therefore, several models have been proposed to explain the mechanism of calcium binding to calmodulin (Milos et al., 1986; Forsen et al., 1986; Haiech et al., 1981; Wang et al., 1985). Briefly, these models can be placed into three basic groups: a) in the first one, calmodulin is considered as a protein with 4 calcium binding sites which are equivalent and independent. The binding isotherm is well explained but not the sites occupancy nor the kinetics. b) in the second one, calmodulin possesses two high affinity sites and two low affinity sites with a difference of about 100 between the two Kd values. These models take into account the NMR results and the kinetic results, but cannot explain the binding isotherm. c) in the third one, calmodulin presents 4 sites and the occupancy of one site modifies the properties of the other sites. Two variations of this model have been proposed. In the first model, calmodulin is composed of two pair of sites. The two sites of a given pair present a strong positive cooperativity, i.e., the occupancy of one site favors the occupancy of the other one. The mean Kd value for the COOH terminus pair of sites is 10 times lower than the mean Kd value for the NH_2 terminus pair of sites (the "pair-of-sites" model). The second interacting sites model is the sequential model. It assumes that calmodulin has four sites which are occupied in an ordered and sequential manner : a specific site of the COOH terminus is first occupied and that induces a modification of the second site of the COOH terminus ; then the occupancy of the two sites of the COOH terminus induces a modification of the NH_2 terminus such that one of the NH_2 sites can then bind calcium; this induces a modification of the last site which can then be occupied by calcium. Both interacting sites models can explain the binding isotherm and the calcium occupancies of the different sites but do not take into account the kinetic results.

What is proposed in Figure 3 is a refinement of the sequential model which allows the explanation of the kinetic results; it is a composite model that is consistent with all of the data. Using the set of values from Table 2, simulation of the model (Figure 4) shows that all previous facts are explained.

What are the main characteristics of this model ? 1) In the absence of calcium, the COOH half attenuates the binding of calcium to the NH_2 half. 2) This constraint between the two halves of the molecule is released through a conformational step (step 5) which is a limiting step in the kinetic process (the slow phase in the kinetics of the calcium dissociation reflects this step). 3) Finally, during the removal of calcium, the rapid phase is associated with the NH_2 half of the molecule whereas the slow phase is going to be associated with the COOH half of the molecule and any part of the molecule involved in this conformational step.

Notice that this conformational step allows a transmission of information between the COOH half of the molcule and the NH_2 half of the molecule (indeed, the NH_2 half of the molecule through this step "knows" that the COOH half is fully occupied by calcium ions).

How can this model be examined experimentally ? The model has been built to explain all the data gathered on calmodulin to date (Figure 4 and Table 2 illustrate this point). From the model, we may deduce some new facts and try to check for these. One of the main properties of the model is the coupling of the two halves of the molecule. Therefore, if this is correct, then it should be possible to generate mutants of calmodulin that have different kinetic properties but similar equilibrium binding properties. When SYNCAM-8 calmodulin (a mutant in which three glutamyl residues have been replaced by three lysines) was analyzed, we noticed that the binding properties of this mutant were similar to those of VU-1 (SYNCAM-1) calmodulin whereas the slow phase of the kinetics of calcium dissociation was 14 s^{-1} for SYNCAM-1 and 84 s^{-1} for SYNCAM-8 (unpublished data).

Table 2. Set of constants used for the simulation of Figure 4

$$k_1 = 6\times10^8 \; M^{-1} s^{-1} \qquad g_1 = 200 \; s^{-1}$$
$$k_2 = 1\times10^8 \; M^{-1} s^{-1} \qquad g_2 = 1000 \; s^{-1}$$
$$k_3 = 6\times10^8 \; M^{-1} s^{-1} \qquad g_3 = 2 \; s^{-1}$$
$$k_4 = 1\times10^6 \; M^{-1} s^{-1} \qquad g_4 = 1000 \; s^{-1}$$
$$k_5 = 110 \; s^{-1} \qquad g_5 = 11 \; s^{-1}$$
$$k_6 = 6\times10^8 \; M^{-1} s^{-1} \qquad g_6 = 1094 \; s^{-1}$$
$$k_7 = 2\times10^8 \; M^{-1} s^{-1} \qquad g_7 = 1150 \; s^{-1}$$
$$k_8 = 6\times10^8 \; M^{-1} s^{-1} \qquad g_8 = 5 \; s^{-1}$$
$$k_9 = 1\times10^6 \; M^{-1} s^{-1} \qquad g_9 = 1150 \; s^{-1}$$

Macroscopic constants : $K_1 = 3\times10^6$ M, $K_2 = 1\times10^6$ M, $K_3 = 0.5\times10^6$ M, $K_4 = 0.18\times10^6$ M
Scatchard Constant : 0.75×10^6 M
Kinetic of calcium dissociation : Fast Phase: $663 \; s^{-1}$, Slow Phase: $9 \; s^{-1}$

Trying to discriminate between the sequential model and the "pair-of-sites" model, we decided to insert reporter groups (tryptophan residues) at specific places in the protein. One of these (tryptophan at position 99 (SYNCAM-9)) allows us to see that the two sites of the COOH terminus do not present a positive cooperativity as suggested by the "pair-of-sites" model and are the first to be occupied (Kilhoffer et al., 1988; Kilhoffer et al., 1989; Chabbert et al., 1989). Moreover, using this mutant and two other ones (tryptophan in position 26 (SYNCAM-32) or 81 (SYNCAM-31)), we were able to show that the rapid kinetic phase is associated with the NH_2 terminus of the protein and the slow phase with the COOH terminus and the central helix. This information confirms the importance of the central helix in the conformational step as already suggested by the SYNCAM-8 calmodulin mutant.

Finally, using mutants in which glutamyl residues in the 12th position of the calcium loop have been changed for an alanine (SYNCAM-13 corresponds to the Glu 67 replaced by Ala and in SYNCAM-14 the Glu 140 replaced by Ala), we found that this change induces a loss of two calcium binding sites. It seems that the mutation perturbs one half of the molecule and not only a specific calcium binding site. This putative conformational modification of one half of the molecule seems to uncouple partially the two halves of the protein.

From our initial studies, it appears that : 1) The calcium binding mechanism to calmodulin is not simple and implies a strong coupling between the different sites. 2) The two halves of the molecule exert some constraints on each other. 3) The central helix plays an important role in allowing communication between the two halves of the molecule. 4) Disturbing one of the two halves allows an uncoupling of the COOH part from the NH2 part of calmodulin. 5) In absence of calcium, calmodulin exists as a family of conformations which interchanges fairly rapidly (ns range). Calcium selects for specific conformations and slows down the exchange rate between conformations (Kilhoffer et al., 1989; Chabbert et al., 1989).

In conclusion, the use of site directed mutagenesis and protein engineering techniques has allowed us to quickly test a model by creating a specific mutant which is designed to measure a given parameter of a model and thereby, to approach the fine details of a mechanism .

ENZYME ACTIVATION BY CALMODULIN

The initial family of site-specific mutations introduced into calmodulin, since the use of recombinant DNA approach to the study of calmodulin structure and function was first introduced in 1985 (Roberts et al., 1985a), have been mostly changes in the helices, especially the elongated central helix. As indicated in Table 1 and various reports (Alvarado-Urbina et al., 1985; Asselin et al., 1988; Chabbert et al., 1989; Craig et al., 1987a,b,c; Haiech et al., 1988; Haiech and Watterson, 1988; Hurwitz et al., 1988; Kilhoffer et al., 1988; Kilhoffer et al., 1989; Lukas et al, 1987, 1988; Maune et al., 1988; Putkey et al., 1986, 1988; Roberts et al., 1985a,b; Roberts et al., 1987; Watterson and Roberts, 1985; Watterson et al., 1985a,b; Weber et al., 1989; Wilson et al., 1988a,b) other mutations have also been introduced into the molecule. However, the more extensive insight has been derived from the helix mutations; therefore, the following discussion will be limited to this family of mutations.

The focus of early studies on the central helix was based on precedents in protein chemistry and the accumulated knowledge about calmodulin structure and function. First, studies of other proteins containing alpha-helices have shown that mutagenesis of solvent-exposed residues in an alpha-helix do not, in general, perturb the overall structure of the protein (Alber et al., 1987, Alber et al., 1988; Matsumura et al., 1988;). In cases where a slight perturbation of the helix was observed, the protein was able to compensate locally for this structural perturbation. Second, certain acidic amino acids are phylogenetically invariant and internally conserved (i.e., calmodulin is composed of a sequence motif that occurs four times in various forms within the primary structure and certain acidic amino acids are invariant among the repeated motifs, see Fig. 1).

Although naturally occurring calmodulins vary approximately 54% in their amino acid sequence, they do not have major differences in their ability to activate some enzymes, such as vertebrate myosin light chain kinase, suggesting that some conserved amino-acids may be functionally important. A third consideration in initiating mutagenesis studies in helices of calmodulin is the fact that some of the conserved acidic amino acids are in regions of uncompensated negative charge that are asymmetrically distributed on the surface of calmodulin (Weber et al., 1989). Fourth, positive charge clusters are found in calmodulin binding domains of some enzymes and appear to be important for a selective, high affinity interaction (Lukas et al., 1986; Blumenthal et al., 1985).

Computation of the possible effects of changing two of the acidic amino acid clusters in calmodulin that are in regions of uncompensated negative charge to lysines, or amino acid side chains with a net positive charge but similar helix forming potential, showed a disruption of the negative electrostatic potential surface of calmodulin (Weber et al., 1989). When these mutations (EEE 82-84 to KKK, or DEE 118-120 to KKK) were introduced into calmodulin, there was a selective effect on the ability to regulate certain calmodulin regulated enzymes, especially MLCK which has a required positive charge cluster in its calmodulin binding domain. As expected from precedents with other proteins and computational studies, there was little or no detectable effect on the gross physical properties of the calmodulins. When similar mutations were done in helices which occur at a similar position in the other internally homologous domains, but lack the relatively high density or uncompensated negative charge, there was no effect on activity. As summarized in Table 1, various combinations of these charge cluster reversal mutations were made. Only the combination of the 82-84 and 118-120 charge reversal mutations had a major effect on activity, and this effect was essentially the sum of the individual cluster mutation effects.

Individual charge reversal mutations at residue 84 and 120 (E to K) affect calmodulin activity in a similar manner to the two cluster charge reversal mutations, indicating that the phylogenetically invariant glutamic acid residues at 84 and 120 are functionally important for calmodulin binding and activation of some enzymes (e.g., MLCK), but not all calmodulin regulated enzymes (e.g., calmodulin protein kinase II). When glutamic acid 84 or glutamic acid 120 was changed to glutamine, there was no discernible effect on activity under the same conditions, although changing both E84 and E120 to glutamines did have a functional effect. This indicates that it is mostly the electrostatic properties of glutamic acid residues 84 and 120 that are important for MLCK activation.

Additional control experiments involved deleting either Glu 82, or Glu 83 and Glu 84, or the entire clusters of these three glutamyl residues. These changes had a minor or no effect on activity, at least none comparable to the Glu-84 to Lys-84 or Glu-120 to Lys-120 point mutations. Although it is difficult to interpret the results from deletion mutagenesis experiments of proteins when they are in the middle of the molecule, a simple inspection of the calmodulin structure (Fig. 1) reveals that each deletion could result in the presence of another acidic amino acid side at a similar position occupied by glutamic acid 84 in the native structure. A further control mutation, in which the sequence Gly-Lys-Gly was inserted into the central helix sequence, was constructed in order to further probe the ability of the central helix to be perturbed. The Gly-Lys-Gly sequence corresponds to the additional three amino-acids found in the troponin C central helix compared to that of calmodulin. While this mutation did have a discernible effect on activity with MLCK, the effect was slight compared to that seen with the glutamic acid 84 to lysine 84 point mutation. These results indicate that the central helix has enough flexibility, and/or the MLCK is able to adapt to these changes, to accomodate these mutations and raises questions about the importance of the relative orientation of the two lobes of the calmodulin structure.

Overall, the site-specific mutagenesis experiments have demonstrated a functional role for glutamic acid 84 and glutamic acid 120 in MLCK activation that appear to be based on their electrostatic properties and possibly their spatial relationship to a hydrophobic cleft in the carboxy-terminal half of the molecule.

It is intriguing to note that the calmodulin binding domain of MLCK has a basic amino acid cluster - hydrophobic amino acid cluster - basic amino acid motif that appears to form an amphiphilic helix (Lukas

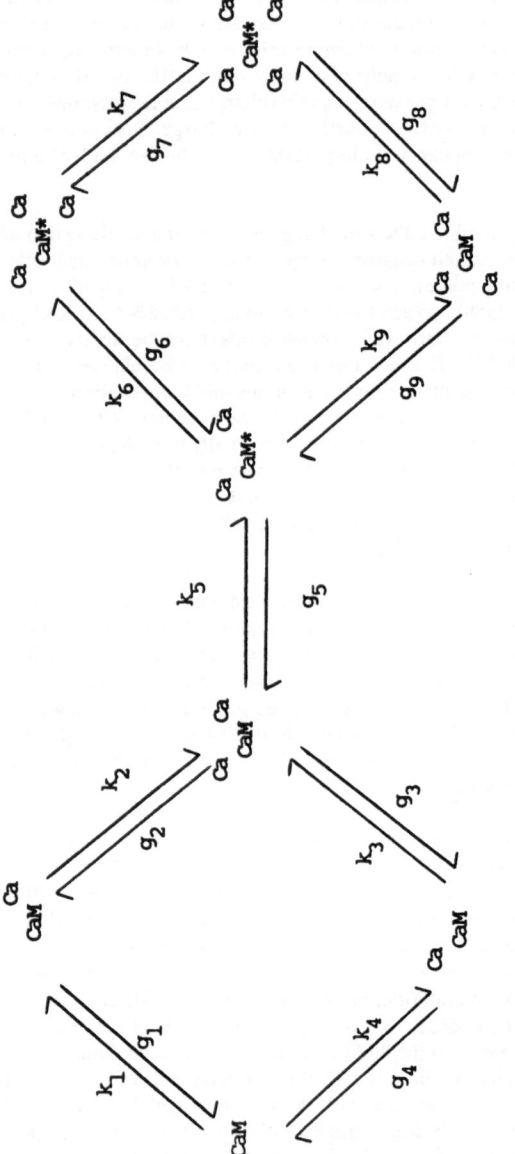

Fig. 3. Schematic representation of the calcium binding mechanism of calmodulin.

52

et al., 1986). Clearly, one logical next step would be to proceed to mutagenesis of these motifs in the cal-modulin binding domain of MLCK to see if the calmodulin mutation could be relieved, i.e. , a suppressor mutation. Thus, an initial exploratory cluster mutation of calmodulin was surveyed for its effects on several enzymes (horizontal study), which led, in turn, to a detailed investigation (vertical study) of the effects of a series of calmodulin mutations on one enzyme. With this detailed analysis of calmodulin, the information required for initiating site-specific mutagenesis analysis of the MLCK is now available, thus allowing further probing of the molecular mechanism.

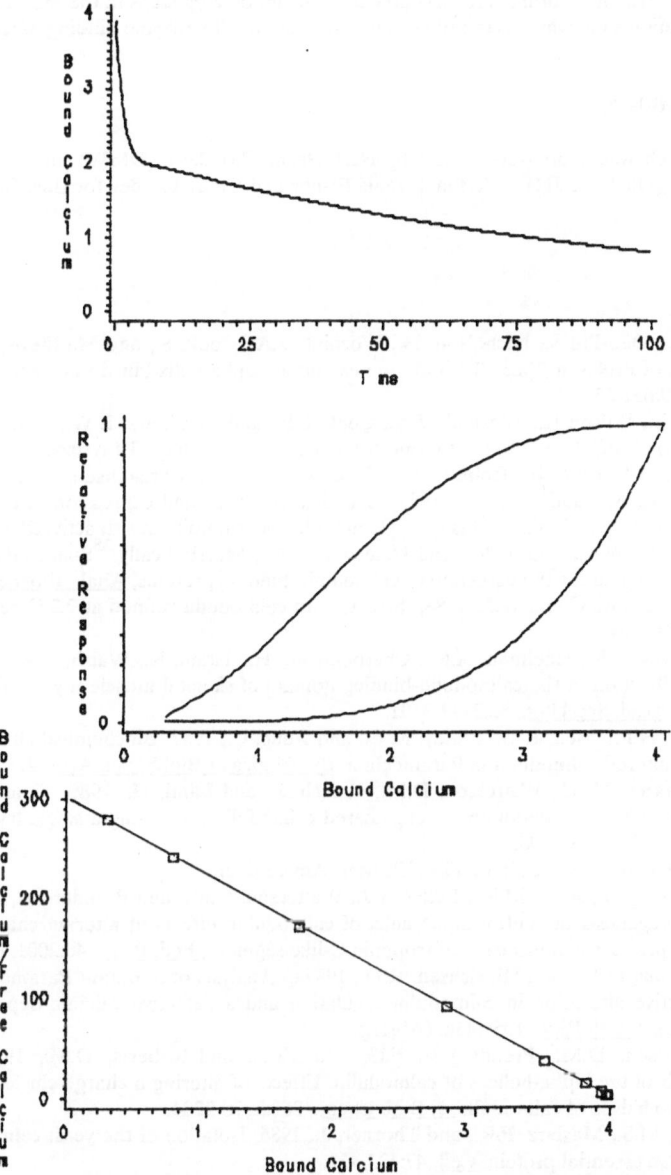

Fig. 4 . Simulation of the kinetic scheme from Figure 3 using the constants from Table II. The top figure represents the kinetic calcium dissociation and presents a biphasic response (fast phase 663 s^{-1} and slow phase 9 s^{-1}). The middle figure represents the titration of calmodulin by calcium. The signal rising between 0 and 2-3 calcium is associated with the occupancy of the sites from the COOH ter-minus and the other signal is associated with the occupancy of the NH$_2$ sites. The lower figure rep-resents a Scatchard representation of a direct calcium binding isotherm.

SUMMARY AND CONCLUSIONS

An example set of site-specific mutagenesis studies of calmodulin has been discussed in terms of strategy and how the results can provide insight into the functioning of calmodulin. A set of common examples for the study of calcium binding and enzyme activation were discussed.

Essentially, site-specific mutagenesis in these initial studies is a perturbation approach. From these perturbation studies, structural features can be correlated in future studies with function and mechanisms of action proposed. More importantly, the approach allows efficient testing of proposed mechanisms and further probing of the molecular aspects of the signal transduction pathways. Clearly, the key functional feature that must be addressed in future studies is how the calcium binding steps in the mechanism are coupled to the enzyme activation step, which is the final step of the calmodulin-enzyme binding mechanism.

ACKNOWLEDGEMENTS

This research was supported in part by NIH Grant GM30861 (DMW) and a joint NSF-CNRS Grant INT-8815276 (DMW & JH). We thank Janis Elsner and Cindy Reeder for their help in assembling this document.

REFERENCES

Alber, T., Bell, J.A., Dao-Pin S., Nicholson, H., Wozniak, J.A., Cook, S., and Matthews, B.W., 1988, Replacements of Pro86 in Phage T4 lysozyme extend an alpha-helix but do not alter protein stability, Science, 239:631-635.

Alber, T., Dao-Pin S., Wilson K., Wozniak, J.A., Cook, S.P., and Matthews, B.W., 1987, Contributions of hydrogen bonds of Thr 157 to the thermodynamic stability of phage T4 lysozyme, Nature, 330:41-46.

Alvarado-Urbina, G., Chiarello, R., Roberts, D., Vilain, G., Jurik, F., Christensen, L., Carmona, C., Fang, L., Watterson, M., and Crea, R., 1985, Chemical synthesis and expression of a calmodulin gene. Rapid automated synthesis via diisopropylaminophosphoramidite in situ activation, DNA, 4:94.

Asselin, J., Phaneuf, S., Watterson, D.M., and Haiech, J., 1989, Metabolically ^{35}S-labeled recombinant calmodulin as a ligand for the detection of calmodulin-binding proteins, Anal. Biochem., 178:141-147.

Babu, Y.S., Bugg, C.E., and Cook, W.J., 1988, Structure of calmodulin refined at 2.2 Ð resolution, J. Mol. Biol., 204:191-204.

Blumenthal, D.K., Takio, K., Edelman, A.M., Charbonneau, H., Titani, K., Walsh, K.A., and Krebs, E.G., 1985, Identification of the calmodulin-binding domain of skeletal muscle myosin light chain kinase, Proc. Natl. Acad. Sci. USA, 82:3187-3191.

Burgess-Cassler, A., Hinrichsen, R.D., Maley, M.E., and Kung, C., 1987, Biochemical characterization of a genetically altered calmodulin in Paramecium, Biochimica et Biophysica Acta, 913:321-328.

Chabbert, M., Kilhoffer, M.-C., Watterson, D.M., Haiech, J., and Lami, H., 1989, Time-resolved fluorescence study of VU-9 calmodulin, an engineered calmodulin possessing a single tryptophan residue, Biochemistry, 28:6093-6098.

Cohen, P., and Klee, C.B., 1988, "Calmodulin", Elsevier, Amsterdam.

Craig, T.A., Roberts, D.M., King, M.M., Lukas, T.J., Watterson, D.M., and Prendergast, F.G., 1987c, Site-specific mutagenesis of central alpha-helix of calmodulin: effects of altering charge clusters, substitution of prolines and insertion of troponin C-like segment, Fed. Proc., 46:2001. (Abstr.)

Craig, T.A., Watterson, D.M., and Hinrichsen, R.D., 1987a, Analysis of a mutant Paramecium with a non-lethal selective alteration in calmodulin regulation and a defective calcium-dependent potassium conductance, J. Cell Biol., 105:143a. (Abstr.)

Craig, T.A., Watterson, D.M., Prendergast, F.G., Haiech J., and Roberts, D.M., 1987b, Site-specific mutagenesis of the alpha-helices of calmodulin. Effects of altering a charge cluster in the helix that links the two halves of calmodulin, J. Biol. Chem., 262:3278-3284.

Davis, T.N., Urdea, M.S., Masiarz, F.R., and Thorner, J., 1986, Isolation of the yeast calmodulin gene: calmodulin is an essential protein, Cell, 47:423-431.

Forsen, S., Vogel, H.J., and Drakenberg, T., 1986, Biophysical studies of calmodulin, Calcium and Cell Function, VI:113-157.

Haiech, J., Klee, C.B., and Demaille, J.G., 1981, Effects of cations on the affinity of calmodulin for calcium: ordered binding of calcium ions allows the specific activation of calmodulin stimulated enzymes, Biochemistry, 20:3890-3897.

Haiech, J., Predeleanu, R., Watterson, D.M., Ladant, D., Bellalou, J., Ullmann, A., and Barzu, O., 1988, Affinity-based chromatography utilizing genetically engineered proteins, J. Biol. Chem., 263:4259-4262.

Haiech, J., and Watterson, D.M., 1988, Site-specific mutagenesis and protein engineering approach to the molecular mechanism of calcium signal transduction by calmodulin, In: "Calcium and Calcium Binding Proteins", Ch. Gerday, R. Gilles, and L. Bolis, eds., Springer-Verlag, Berlin Heidelberg., 191-200.

Hurwitz, M.Y., Putkey, J.A., Klee, C.B., and Means, A.R., 1988, Domain II of calmodulin is involved in activation of calcineurin, FEBS Lett., 238:82-86.

Kilhoffer, M.-C., Roberts, D.M., Adibi, A., Watterson, D.M., and Haiech, J., 1989, Fluorescence characterization of VU-9 calmodulin, an engineered calmodulin with one tryptophan in calcium binding domain III, Biochemistry, 28:6086-6092.

Kilhoffer, M.-C., Roberts, D.M., Adibi, A.O., Watterson, D.M., and Haiech, J., 1988, Investigation of the mechanism of calcium binding to calmodulin. Use of an isofunctional mutant with a tryptophan introduced by site-directed mutagenesis, J. Biol. Chem., 263:17023-17029.

Klee, C.B., and Vanaman, T.C., 1982, Calmodulin, Adv. Pro. Chem., 35:213-321.

Lukas, T.J., Burgess, W.H., Prendergast, F.G., Lau, W., and Watterson, D.M., 1986, Calmodulin binding domains: characterization of a phosphorylation and calmodulin binding site from myosin light chain kinase, Biochemistry, 25:1458-1464.

Lukas, T.J., Craig, T.A., Roberts, D.M., Watterson, D.M., Haiech, J., and Prendergast, F.G., 1987, An interdisciplinary approach to the molecular mechanisms of calmodulin action : Comparative biochemistry, site-specific mutagenesis, and protein engineering, In: "Calcium-Binding Proteins in Health and Disease", A.W. Norman, T.C. Vanaman, and A.R. Means, eds., Academic Press, Inc., 533-543.

Lukas, T.J., Haiech, J., Lau, W., Craig, T.A., Zimmer, W.E., Shattuck, R.L., Shoemaker, W.O., and Watterson, D.M., 1988, Calmodulin and calmodulin-regulated protein kinases as transducers of intracellular calcium signals, Cold Spring Harbor, Symposia on Quantitative Biol., 53:185-193.

Lukas, T.J., Iverson, D.B., Schleicher, M., and Watterson, D.M., 1984, Structural characterization of a higher plant calmodulin, Plant Physiol., 75:788-795.

Lukas, T.J., Wiggins, W.E., and Watterson, D.M., 1985, Amino acid sequence of a novel calmodulin from the unicellular alga Chlamydomonas, Plant Physiol., 78:477-483.

Marshak, D.R., Clarke, M., Roberts, D.M., and Watterson, D.M., 1984, Structural and functional properties of calmodulin from the eukaryotic microorganism Dictyostelium discoideum, Biochemistry 23:2891-2899.

Matsumura, M., Becktel, W.J., and Matthews, B.W., 1988, Hydrophobic stabilization in T4 lysozyme determined directly by multiple substitutions of Ile 3, Nature, 334:406-410.

Maune, J.F., Klee, C.B., and Beckingham, K., 1988, Calmodulin point mutations which affect Ca^{2+}-binding and calcineurin activation, J. Cell Biol., 107:287a. (Abstr.)

Milos, M., Schaer, J.-J., Comte, M., and Cox, J.A., 1986, Calcium-proton and calcium-magnesium antagonisms in calmodulin: microcalorimetric and potentiometric analyses, Biochemistry, 25:6279-6287.

Olwin, B.B., and Storm, D.R., 1985, Calcium binding to complexes of calmodulin and calmodulin binding proteins, Biochemistry, 24:8081-8086.

Persechini, A., Hardy, D.O., Blumenthal, D.K., Jarrett, H.W., and Kretsinger, R.H., 1988, The effects on enzyme activation of genetically-engineered amino-acid deletions in the calmodulin long helix, Biophys. J., 53:252a (Abstr.)

Putkey, J.A., Draetta, G.F., Slaughter, C.B., Klee, C.B., Cohen, P., Stull, J.T., and Means, A.R., 1986, Genetically engineered calmodulins differentially activate target enzymes, J. Biol. Chem., 261:9896-9903.

Putkey, J.A., Ono, T., Van Berkum, M.F.A and Means, A.R., 1988, Functional significance of the central helix in calmodulin, J. Biol. Chem., 263:11242-11249.

Roberts, D.M., R. Crea, M. Malecha, G. Alvarado-Urbina, R.H. Chiarello, and D.M. Watterson., 1985a, Chemical synthesis and expression of a calmodulin gene designed for site-specific mutagenesis, Biochemistry, 24:5090-5098.

Roberts, D.M., Crea, R., Malecha, M., and Watterson, D.M., 1985b, The chemical synthesis and expression of a gene coding for calmodulin, Fed. Proc., 44:1050. (Abstr.)

Roberts, D.M., Rowe, P.M., Siegel, F.L., Lukas, T.J., and Watterson, D.M., 1986, Trimethyllysine and protein function. Effect of methylation and mutagenesis of lysine 115 of calmodulin on NAD kinase activation, J. Biol. Chem., 261:1491-1494.

Roberts, D.M., Zimmer, W.E., and Watterson, D.M., 1987, The use of synthetic oligodeoxyribonucleotides in the examination of calmodulin gene and protein structure and function, Methods Enzym., 139:290-303.

Saimi, Y., Hinrichsen, R.D., Forte, M., and Kung, C., 1983, Mutant analysis shows that the Ca^{2+}-induced

K$^+$ current shuts off one type of excitation in Paramecium, Proc. Natl. Acad. Sci. USA, 80:5112-5116.

Schaefer, W.H., Hinrichsen, R.D., Burgess-Cassler, A., Kung, C., Blair, I.A., and Watterson, D.M., 1987a, A mutant Paramecium with a defective calcium-dependent potassium conductance has an altered calmodulin: A nonlethal selective alteration in calmodulin regulation, Proc. Natl. Acad. Sci. USA, 84:3931-3935.

Schaefer, W.H., Lukas, T.J., Blair, I.A. Schultz, J.E., and Watterson, D.M., 1987b, Amino acid sequence of a novel calmodulin from Paramecium tetraurelia that contains dimethyllysine in the first domain, J. Biol. Chem., 262:1025-1029.

Smith, V.L., Doyle, K.E., Maune, J.F., Munjaal, R.P., and Beckingham, K., 1987, Structure and sequence of the Drosophila melanogaster calmodulin gene, J. Mol. Biol., 196:471-485.

Takeda, T., and Yamamoto, M., 1987, Analysis and in vivo disruption of the gene coding for calmodulin in Schizosaccharomyces pombe, Proc. Natl. Acad. Sci. USA, 84:3580-3584,

Toda, H., Yazawa, M., Kondo, K., Honma, T., Narita, K., and Yagi, K., 1981, Amino acid sequence of calmodulin from scallop (Patinopecten) adductor muscle, J. Biochem., 90:1493-1505.

Tschudi, C., Young, A.S., Ruben, L., Patton, C.L., and Richards, F.F., 1985, Calmodulin genes in trypanosomes are tandemly repeated and produce multiple mRNAs with a common 5' leader sequence, Proc. Natl. Acad. Sci. USA, 82:3998-4002.

Van Eldik, L.J., Zendegui, J.G., Marshak, D.R., and Watterson, D.M., 1982, Calcium-binding proteins and the molecular basis of calcium action, International Review of Cytology, 77:1-61.

Wang, C.-L.A., 1985, A note on Ca^{2+} Binding to Calmodulin, Biochem. Biophys. Res. Comm., 130:426-430.

Watterson, D.M., Burgess, W.H., Lukas, T.J., Iverson, D., Marshak, D.R., Schleicher, M., Erickson, B.W., Fok, K.-F and Van Eldik, L.J., 1984, Towards a molecular and atomic anatomy of calmodulin and calmodulin-binding proteins, In: "Advances in Cyclic Nucleotide and Protein Phosphorylation Research", S.J. Strada, and W.J. Thompson, eds., Raven Press, New York. 205-226.

Watterson, D.M., Lukas, T.J., Roberts, D.M., and Crea, R., 1985a, Molecular analysis of calmodulin's enzyme activator and drug binding activities, J. Cell Biochem., Supplement 9B:139. (Abstr.)

Watterson, D.M., and Roberts, D.M., 1985b, Analysis of the contribution of trimethyllysine/lysine-115 to calmodulin activity by using site-specific mutagenesis, J. Cell Biol., 101:474a. (Abstr.)

Watterson, D.M., Roberts, D.M., and Lukas, T.J., 1985, Calmodulin molecular mechanisms: an interdisciplinary approach employing comparative biochemistry, site-directed mutagenesis and molecular biology, 13th Intl. Congress Biochemistry 272.(Abstr.)

Watterson, D.M., Sharief, F., and Vanaman, T.C., 1980, The complete amino acid sequence of the Ca^{2+}-dependent modulator protein (calmodulin) of bovine brain, J. Biol. Chem., 255:962-975.

Weber, P.C., Lukas, T.J., Craig, T.A., Wilson, E., King, M.M., Kwiatkowski, A.P., and Watterson, D.M., 1989, Computational and site-specific mutagenesis analyses of the asymmetric charge distribution on calmodulin, Proteins: Structure, Function and Genetics, in press.

Wilson, E., Craig, T.A., and Watterson, D.M., 1988a, Similarities between the ion channel associated calmodulin binding protein and the beta subunit of GTP-binding proteins, J. Cell Biol., 107:142a. (Abstr.)

Wilson, E., Hinrichsen, R.D., and Watterson, D.M., 1988b, Molecular mechanism of calmodulin regulation of calcium-dependent potassium channels, Proc. 4th Intl. Congress Cell Biol., p.5.5.7. (Abstr.)

Zimmer, W.E., Schloss, J.A., Silflow, C.D., Youngblom, J., and Watterson, D.M., 1988, Structural organization, DNA sequence, and expression of the calmodulin gene, J. Biol. Chem., 263:19370-19383.

PARVALBUMIN, MOLECULAR AND FUNCTIONAL ASPECTS

Claus W.Heizmann,[1] Jürg Röhrenbeck,[2] and Willem Kamphuis[3]

[1]Department of Pediatrics, Division of Clinical Chemistry, University of Zürich
Steinwiesstrasse 75, 8032-Zürich, Switzerland

[2]Max-Planck Institute for Brain Research, D-6000 Frankfurt 71, FRG

[3]Department of General Zoology, University of Amsterdam, Amsterdam, The Netherlands

INTRODUCTION

Parvalbumin was the first calcium-binding protein to be crystallized (Kretsinger 1980). From these data, the EF-hand structural arrangement, common to many calcium-binding proteins, was deduced. Parvalbumin contains 3 EF-hands but only the two COOH-terminal domains bind calcium. Parvalbumins were first isolated from the skeletal muscle of lower vertebrates (Hamoir, 1968; Pechère et al., 1971; Gerday, 1988) and later from skeletal muscles, testis, kidney and brain of various species including man (Heizmann, 1984; Heizmann and Berchtold, 1987).

Rat and human parvalbumin genes have been cloned (Berchtold et al.,1987) and compared to the genes of other EF-hand proteins (Means et al., 1988; Heizmann and Hunziker, 1989). While mammalian genomes contain a single parvalbumin gene, some lower vertebrates (e.g. fish and frog) must possess multiple copies since the muscles of these species contain several isoforms of parvalbumin with distinct amino acid sequences. These genes may be linked as found for the multiple genes in sea urchins that encode the ectoderm specific calcium-binding proteins (Hardin et al., 1985; Hardin et al., 1988). The rat parvalbumin gene was also utilized to assign its human counterpart to chromosome 22 by the use of heterologous cell hybrids (Berchtold et al., 1987).

Generally, introns divide functional domains. In the case of parvalbumin and several other members of the EF-hand protein family, introns are located within EF-hand domains. According to these criteria, the genes of Ca^{2+}-binding proteins may be grouped into different classes:

The first group includes parvalbumin, calmodulin, the sea urchin Spec I protein and myosin alkali light chain. Common to all these genes is that their coding sequence is interrupted by 5 introns with four of them localized at identical positions. This suggests that the common introns were acquired before the divergence of the different members of this group. The fifth intron, on the other hand, was inserted after the divergence.

The second group contains members of the S-100 protein family including calcyclin and the macrophage migration inhibitory factor related proteins (MRP-8 and MRP-14). Common to these genes are two introns, the first in the 5' nontranslated region and the second in the linker region separating the two EF-hands. In this case, the intron does separate functional domains.

A third group includes the 6 EF-hand domain protein calbindin D-28K. Coding sequences are interrupted by 10 introns with identical locations. Seven of these introns are located within an EF-hand coding structure. A more detailed description of these and other groups of genes have been reviewed recently (Means et al., 1988; Heizmann and Hunziker, 1989).

In addition to the TATA box and CAAT boxes, the parvalbumin promoter region contains other potentially functional regions, one with a striking homology to a promoter sequence found in the chicken myosin light chain 3F gene (Berchtold et al., 1987). Both parvalbumin and this myosin light chain are present in adult fast-twitch skeletal muscle fibres. So this promoter region could be important in the tissue specific regulation of parvalbumin. Further studies are necessary to establish the functional significance of the various sequences in the promoter regions of parvalbumin. In further studies, vectors will be constructed for transfection studies of suitable cells or for the generation of transgenic animals.

The distribution of parvalbumin and other calcium-binding proteins in various tissues, including the brain of birds, mammals and man, have been reviewed recently (Heizmann and Braun, 1989).

Here we would like to summarize two recent approaches describing: (A) the distribution of parvalbumin in the cat retina and (B) changes of parvalbumin immunoreactivity in the hippocampal kindling model of epilepsy.

On the basis of these results, a possible functional role of parvalbumin was proposed.

DISTRIBUTION OF PARVALBUMIN IMMUNOREACTIVITY IN THE CAT RETINA

Immunohistochemical localization of calcium-binding proteins labels several cell types in the vertebrate retina. Subclasses of photoreceptors, horizontal, bipolar, amacrine, and ganglion cells are labelled with antibodies against calbindin D-28K in many species such as frog, chick, mouse, rabbit (Schreiner et al., 1985), pigeon (Pasteels et al., 1987), rat (Rabié et al., 1985), and man (Verstappen et al., 1986). On the other hand, parvalbumin immunoreactivity is restricted to amacrine and horizontal cells in rat, monkey, and human (Endo et al., 1985). A more detailed classification of the parvalbumin- and calbindin-positive cell populations is possible using morphological criteria and double labelling on semithin sections.

RESULTS

Staining of the cat retina with antibodies against cat parvalbumin (Stichel et al., 1986) reveals both A- and B-type horizontal cells (Fig. 1). The A-type horizontal cell has a large cell body and thick primary dendrites. The smaller B-type has more delicate branches and is about three times more numerous than the A-type. A monoclonal antibody against a recombinant rat calbindin D-28K (Pinol et al., 1989) reveals staining of the A- but not the B-type horizontal cell (Fig. 2). Here the smaller somas of bipolar or amacrine cells appear out of focus.

Consecutive semithin sections of the cat retina can be stained alternately with antibodies against parvalbumin and GABA (Fig. 3 and 4). This procedure enables the staining pattern in individual cells to be compared. All horizontal cells show both parvalbumin- and GABA-like immunoreactives. In contrast, very few parvalbumin-positive amacrine cells are stained with GABA-antibodies. Fig. 5 shows a retinal whole mount labelled with both parvalbumin and anti-tyrosine hydroxylase (TH) antibodies. The peroxidase antiperoxidase method was used to detect TH, and parvalbumin was localized with a FITC-linked secondary antibody. The TH-positive cells form a dense dendritic network in the inner plexiform layer with characteristic "rings", which contact the glycinergic AII amacrine cell (Pourcho, 1982, Pourcho and Goebel, 1985, Voigt and Wässle, 1987). Many parvalbumin-positive amacrine cells are located in these rings, suggesting that the AII cells form a subpopulation of the labelled amacrine cells. Ganglion cells and the optic fibre layer are stained by parvalbumin but do not have a GABA-like immunoreactivity.

Localization and function of parvalbumin and calbindin D-28K in the retina

The intense staining of cat horizontal cells with antibodies against parvalbumin and calbindin indicates a high level of these proteins. Interestingly, parvalbumin and calbindin are co-localized in the A- but not in the B-type horizontal cell. Horizontal cells show GABA-like immunoreactivity (as do most neurons of the rat cerebral cortex (Celio 1986)), but this is not the case for the other parvalbumin-positive cells of the retina. At least half of the labelled amacrine cells are glycinergic. Ganglion cells are presumed to use a dipeptide (Anderson et al., 1987) as neurotransmitter. The heterogeneity of parvalbumin stained cell populations means that the functional role of parvalbumin remains unclear.

Calcium-binding proteins as neuronal markers

Parvalbumin and calbindin are reliable markers for subpopulations of retinal neurons. This enables the B-type horizontal cell to be stained quantitatively (Röhrenbeck et al., 1987). Until recently this cell type

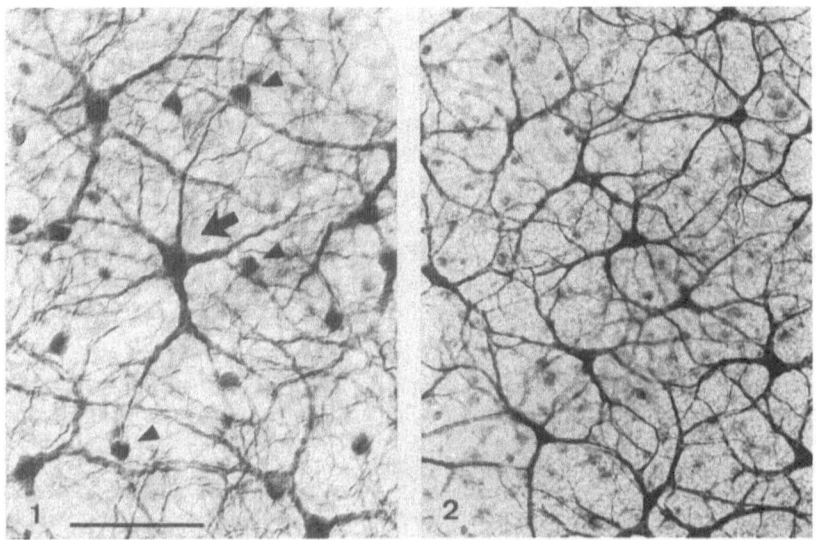

Fig. 1 and 2. Whole mounted retina. The focus is at the outer plexiform layer. The retina is stained with antibodies against parvalbumin (Fig. 1) and calbindin (Fig. 2). Parvalbumin-staining is detected in A-type (arrow) and B-type (arrow heads) horizontal cells but only the A-type shows calbindin immunoreactivity Bar = 50 μm in Fig. 1 and 100 μm in Fig. 2.

could only be identified by its relatively large nucleus (Wässle et al., 1978). In the monkey retina, the two horizontal cell populations are also (to a lesser extent) differentially stained by parvalbumin and calbindin (Röhrenbeck et al., 1989).

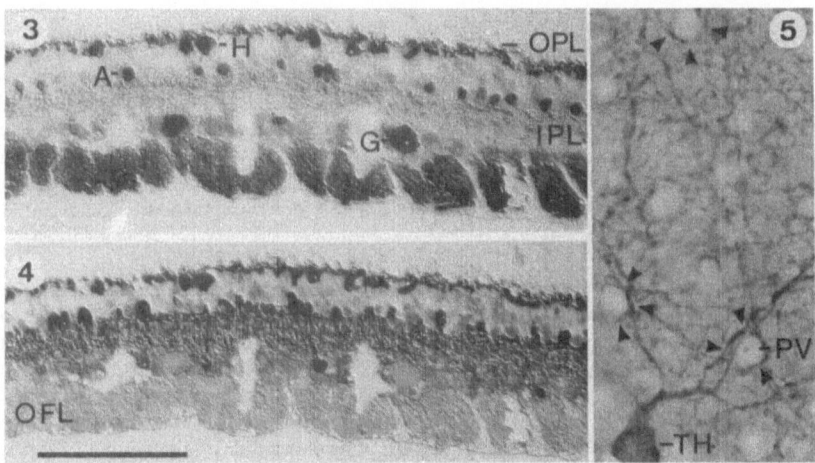

Fig. 3 and 4 . Adjacent semithin section stained for parvalbumin- (Fig. 3) and GABA-immunoreactivity (Fig. 4). Fig. 3, horizontal cells (H), some amacrine cells (A), ganglion cells (G) and the optic fibre layer (OFL) are labelled. Fig. 4, horizontal cells also show GABA-LI. However, the labelled amacrine cell population is different from the one in Fig. 3. OPL, outer plexiform layer, IPL, inner plexiform layer. Bar = 100 μm.

Fig. 5. Parvalbumin/tyrosine hydroxylase (TH) double labelling. Whole mount preparation focused at the IPL/amacrine cell border. Several parvalbumin-positive amacrine cells, which appear as bright spots, are located in the rings (arrows) of the TH-positive dendrites. One TH-cell is labelled. Bar = 50 μm.

Double labelling with TH and parvalbumin reveals that the glycinergic AII amacrine cells form a large part of the parvalbumin-positive amacrine cell population. The AII, which is interposed between rod and cone pathways, is probably the most numerous amacrine cell type in the cat retina.

These two examples illustrate that the immunohistochemical localization of calcium-binding proteins is a powerful tool which can reveal distinct cell populations within the vertebrate retina.

CHANGES OF PARVALBUMIN IMMUNOREACTIVITY IN THE HIPPOCAMPAL KINDLING MODEL OF EPILEPSY

Parvalbumin and GABAergic inhibition

Parvalbumin is present in a distinct neuron population in the central nervous system. It has been demonstrated that the spatial distributions of parvalbumin- and GABA-immunopositive cell bodies show a close correspondence (Celio, 1986; Kosaka et al., 1987a; 1987b; Stichel et al., 1988). Regarding a possible functional role for parvalbumin, alterations in the parvalbumin content of certain brain areas may be related to an alteration of the local neuronal activity (Braun et al., 1985a; 1985b; Heizmann and Celio, 1987). This was suggested by a correlation between parvalbumin labelling and high levels of cytochrome oxidase, an indicator of metabolic rate (Braun et al., 1985a; 1985b). Perhaps changes in parvalbumin content might modify the local neuronal activity by a regulation of the firing activity of inhibitory GABAergic interneurons.

A proposed mechanism for such a regulation is the following : depolarization of GABAergic interneurons results in an influx of calcium ions which causes an increase of the intracellular calcium concentration. This increase activates the calcium dependent potassium current, leading to a hyperpolarization of the interneuron, thus contributing to the termination of action potential generation. It was postulated that enhancement of the calcium buffering capacity by an increased parvalbumin content would lead to a limited rise in the intracellular calcium resulting in a prolonged firing activity and hence an increased inhibitory control on the local network (Kawaguchi et al., 1987).

The kindling model of epilepsy

A profound alteration in the stability of a neuronal circuit is found in epileptic brain tissue as evidenced by the occurrence of paroxysmal synchronous firing. In general, a disturbed balance between excitatory and inhibitory synaptic transmission underlies epileptic activity. This shift in balance may result either from an increase in activity of excitatory neurons or from a decrease of inhibitory control (Meldrum, 1975). Indeed, a pharmacologically reduced GABAergic transmission, for instance by blocking the GABA receptors with specific antagonists, leads readily to seizure activity. From the animal models currently available to study epileptogenesis, the kindling model of epilepsy has received considerable attention by many investigators because of its unique properties (Goddard et al., 1969). For reviews see McNamara et al. (1985), and Peterson and Albertson (1982).

Kindling is the term to indicate the development of behavioral and electroencephalographic (EEG) epileptiform manifestations in response to repeated application of short electrical tetanic stimulations. To produce kindling, stimulations are given via permanently implanted electrodes, once or twice daily to a certain brain region over an extended period of time. The behavioral expression of kindling epileptogenesis proceeds through characteristic stages on which seizure intensity classification (stage 0 - 5) is based. Seizures evolve in the course of kindling from an almost indiscernible change in behavior to characteristic generalized tonic-clonic seizures (stage 5), lasting several minutes. EEG recordings show a continuous increase in the length of the epileptiform seizure pattern recorded following the tetanus, the afterdischarge. The increase in seizure activity reaches a plateau when the tetanus provokes a generalized tonic-clonic seizure. From that point on, each tetanus will result in a similar type of seizure and the animal is said to be kindled. However, in the kindled state no spontaneous seizures occur. Kindled animals left unstimulated for as long as one year will respond to the first or second electrical stimulation with a class 5 seizure. The kindled state reflects therefore a long-lasting and probably permanent alteration of neuronal stability (Goddard et al., l969).

Many neurotransmitter and putative neurotransmitter systems have been implicated in the mechanisms leading to kindling. Hitherto, no conclusive evidence has been presented that one of the known neurotransmitters, such as catecholamines, acetylcholine, or opiates, is involved in kindling (Peterson and Albertson, l982). During the past few years there has been an increasing number of studies on the possible role of excitatory (e.g. glutamate and aspartate) and inhibitory (e.g. GABA) amino acid neurotransmitters in the kindling model.

The hypothesis that a decreased GABAergic inhibition underlies the process of kindling development and the kindled state was the basis for a series of immunocytochemical investigations carried out in our laboratory. The number of GABA-immunoreactive somata in the CA1 area of the dorsal rat hippocampus was quantified at different points during kindling development and in the kindled state. First, an increase (10%) was found after the 6th afterdischarge and a significant increase of 38%, in comparison to controls, after the 14th afterdischarge. This result suggests an enhanced activity of GABAergic neurons in the initial phase of kindling development. Later, at session 34 when animals were fully kindled, the cell density of GABA-immunoreactive somata had decreased in comparison to session 14. When daily stimulation was stopped after 10 generalized seizures, the cell density decreased and after 28 days a significantly decreased cell density compared to unstimulated controls was observed (Kamphuis et al., 1986; 1987; 1989b).

Parvalbumin co-exists with GABA in a subpopulation of CA1 interneurons with their somata located in the stratum oriens (SO) and stratum pyramidale (SP) with terminals on the pyramidal neuron somata and their initial axon segments (Kosaka et al., 1987b; Katsumaru et al., 1988). Most notably, fast spiking interneurons in the SP of the CA1 region were demonstrated to contain parvalbumin (Kawaguchi et al., 1987). In other studies similar fast-spiking interneurons were shown to be selectively localized in the SO and SP and to be morphologically similar to parvalbumin-immunoreactive interneurons (Seress and Ribak, 1985; Kawaguchi and Hama, 1987). Since the neuronal activity in the hippocampus and fascia dentata is strongly controlled by GABAergic interneurons either by feedforward or feedback inhibitory pathways (Roberts, 1986), we hypothesized that the levels of parvalbumin might serve as a marker for metabolic and/or electrical activity of this subpopulation of GABAergic interneurons. Moreover, the presence of parvalbumin in a GABAergic neuron may act as a calcium buffer and protect these interneurons against an excessive calcium influx, possibly underlying the observed reduction in the number of GABAergic somata in the kindled state. We investigated whether a change in parvalbumin immunoreactivity took place in kindled rats, first, in the CA1 region of the dorsal hippocampus and second, whether the presence of parvalbumin in GABAergic neurons was influencing the long-term reduction in cell density of GABAergic somata.

Experimental procedures and results

Stimulation and recording electrodes were placed in the CA1 area of the left dorsal hippocampus of male Wistar rats. After two weeks of postoperative recovery, kindling stimulations (1-2 s, 50 Hz, 150-300 uA) were given at a rate of two sessions a day to the left hippocampus. EEG recordings, started immediately after the stimulus, revealed an afterdischarge in response to the kindling stimulation. A weight and age matched group of control animals was also implanted and treated identically to the kindled animals except for the tetanic stimulations.

Two series of experiments were carried out. The aim of the first series was to study changes in the parvalbumin immunoreactivity during (13th session, stage-3 group) and immediately after the completion of the kindling process (33rd session, fully kindled stage-5 group). Control animals were fixed simultaneously with the kindled rats.

Animals were anaesthetized and perfused with a fixative. Vibratome sections (50μm) were cut and incubated overnight in parvalbumin antiserum and the immunoreaction visualized by the PAP technique and quantified as described previously (Kamphuis et al., 1987).

In control animals labelled perikarya showing marked variation in the optical density of the DAB label were observed in the hippocampus and dentate gyrus. In hippocampal regions CA1 and CA3 heterogenously shaped parvalbumin-immunoreactive perikarya were found almost exclusively in SO and SP, and rarely in stratum radiatum (SR) (Fig. 6). Long varicose dendritic processes, without apparent spines, extended from perikarya located in CA1 SP into SR. Perikarya of pyramidal neurons were surrounded by punctate immunoreactivity indicating parvalbumin-immunoreactive axon terminals. In dentate gyrus immunoreactive cell bodies were located in the hilus, within the granular cell layer and in the subgranular zone immediately below the stratum granulosum, facing the hilus. Immunoreactive processes originating from these cell bodies extended to the molecular layer and into the hilus. Perikarya of granular neurons were also surrounded by immunoreactive punctae.

The pattern of parvalbumin immunoreactivity itself was not changed in any of the kindled groups when compared to the controls. However, faintly stained perikarya as observed in control groups were al-

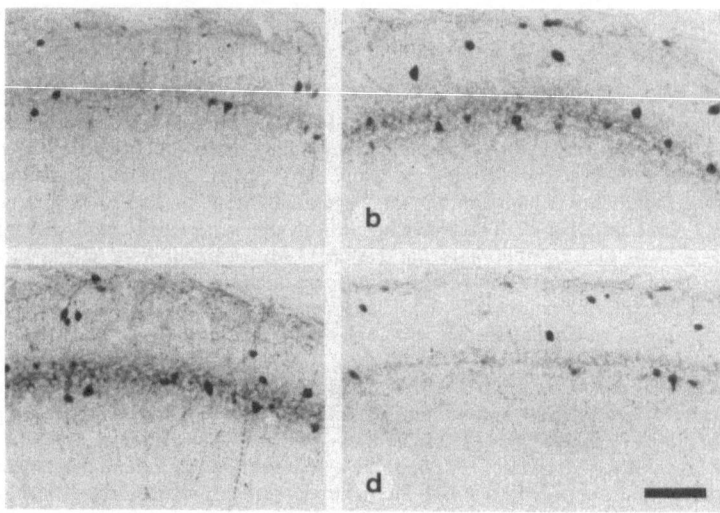

Fig. 6. Photomicrographs illustrating parvalbumin-immunoreactive somata in stratum oriens (SO) and stratum pyramidale (SP) in the CA1 region of the rat dorsal hippocampus. Parvalbumin-immunoreactive processes can be found in stratum oriens and in stratum radiatum (SR). Axon terminal-like immunoreactivity is observed surrounding the unstained somata of pyramidal neurons. a: control implanted animal, b: stage-3, c: stage-5, d: long-term. All photomicrographs are of the ipsilateral, implanted side. Note the increased parvalbumin immunoreactivity in stage-3 and stage-5. Scale bar: 100 μm.

most absent in stage-3 and stage-5 groups (Fig. 6). Perikarya in hippocampus and in fascia dentata in these kindled groups were more uniform and more intensely labelled when compared to controls. A similar observation was made for the labelling of dendrites and immunoreactive punctae surrounding the pyramidal and granular somata.

Mean cell density was significantly increased in the stage-3 group and in the stage-5 group when compared to the control group. The relative increase was 18% and 21%, respectively. In addition, a column of at least 1 mm^2 of cortex overlying the hippocampus was quantified in each animal but no significant differences could be found in this region.

Animals of the second series, long-term group, were kindled in an identical way. This group and a control group were fixed 31 days after the last of at least 7 generalized tonic-clonic class-5 seizures. The parvalbumin-immunocytochemical study in this group was carried out in combination with an investigation on changes in GABA-immunocytochemistry (Kamphuis et al., 1989b).

The long-term group showed neither an increase in parvalbumin immunoreactivity (Fig. 6) nor a significant difference in parvalbumin-immunoreactive cell density when compared to nonkindled controls (Table 1).

For co-localization in the long-term group, serial sections were immunostained alternately for GABA and parvalbumin. A precise overlap in position of a parvalbumin- and GABA-immunoreactive cell body in two consecutive sections was scored as co-localization in a cell bisected by the Vibratome knife (Fig. 7).

Co-localization of GABA in parvalbumin-immunoreactive somata in the controls was compared with that of kindled animals of the long-term group. We quantified the cell density of parvalbumin-, and of GABA-immunoreactive cell bodies, and the co-localization of parvalbumin and GABA in somata in the SO and SP region of CA1. The results are given in Table 2. The mean cell density of parvalbumin-immunoreactive cell bodies in SO and SP in the control group did not differ significantly from the long-term kindled group. In contrast, the mean cell density of GABA-IR cell bodies in the kindled group was significantly ($P < 0.004$) decreased when compared to the control group, 39.1 vs 62.2, a decrease of 37%. The

Table 1 . Number of parvalbumin-immunoreactive somata in hippocampal CA1 region (SO and SP and SR) per mm^2

```
FIRST SERIES

                control      33.7 ± 1.4   (n=10)
                stage-3      39.8 ± 1.4*  (n=14)
                stage-5      40.7 ± 1.7*  (n=13)


SECOND SERIES

                control      34.6 ± 2.0   (n=7)
                long-term    33.6 ± 2.6   (n=7)
```

```
Values are the mean cell density ± S.E.M. from n
animals. Statistical comparison of the kindled
groups with its matched control groups was made
using the Mann-Whitney U-test. * P < 0.02.
```

mean cell density of the parvalbumin-immunoreactive somata that also showed GABA-IR was the same for controls and kindled rats (10.1 vs 9.5).

From the cell densities obtained we calculated for each animal the cell density of somata which were only parvalbumin-immunoreactive or only GABA-immunoreactive. The results are given in Table 2. It showed that the major change occurred in the population of exclusively GABA-immunoreactive somata. This cell density was significantly reduced (50%) in the kindled group (P < 0.002.).

CONCLUSIONS AND DISCUSSION

Our observations show that the cell density of parvalbumin-immunoreactive cell bodies quantified in the hippocampal CA1 region is significantly increased in partially kindled animals after a class-3 seizure (Table 1). The cell density remains enhanced as long as further afterdischarges are elicited to obtain a complete kindling development. The intensity of the DAB label showed that kindling induced an overall increase in parvalbumin staining in CA1 as well as in the fascia dentata. Related to our observations is an increased parvalbumin content found in a genetically epileptic mice strain by Baimbridge and Miller (1988).

We interpret the increase in parvalbumin immunoreactivity and parvalbumin-immunoreactive cell density as the result of a kindling induced increased parvalbumin; cell bodies and terminals that were not identified in controls because they remained below the detection threshold level for parvalbumin, could be visualized in kindled rats. The previously described enhanced GABA-immunoreactivity in the early stages of the kindling process could originate from a similar mechanism. The observed increase in parvalbumin-immunoreactivity during kindling may therefore correlate with an increase in the activity of this subpopulation of inhibitory neurons. By an increased buffering of the Ca^{2+} influx through voltage dependent channels, these cells may reduce the Ca^{2+} dependent K^+ after- hyperpolarization (Kawaguchi et al., 1987). According to this view, the elevated levels of parvalbumin may be interpreted as being involved in a compensatory mechanism to increase stability of the neuronal network. However, this mechanism seems unable to block other, so far unknown, kindling stimulation induced alterations that lead to a decreased neuronal stability.

The long-lasting decrease in neuronal stability characteristic of the kindling state is not accompanied by an alteration in parvalbumin level in CA1 following the observation that the parvalbumin immunoreactivity and parvalbumin-immunoreactive cell density values in CA1 were back to control levels in the animals fixed long after the last elicited generalized convulsion.

We conclude that, in relation to the kindling model of epilepsy in the rat hippocampus (CA1 region), alterations in parvalbumin level are neither an indicator for the decreased stability of the local network nor responsible for the decreased stability. Enhanced parvalbumin levels are limited to the period when seizures are elicited by the tetanic stimulations and are possibly a consequence of seizure activity (Kamphuis et al., 1989a).

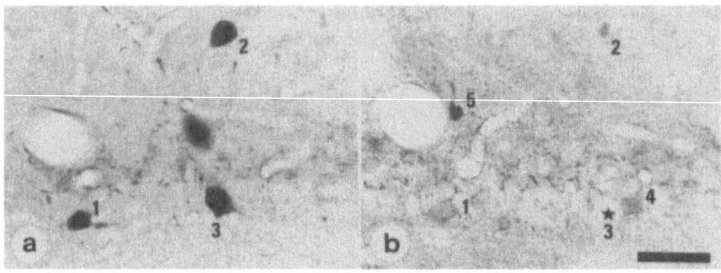

Fig. 7. Photomicrographs from the same CA1 region of the hippocampus in two consecutive sections in which the localization of parvalbumin (a) and GABA (b) was visualized. Comparison of both surfaces shows that cell bodies 1 and 2 are parvalbumin- and GABA-immunoreactive, cell body 3 is only parvalbumin-immunoreactive (* indicates position of cell body 3 in the GABA-immunoreactive section) and somata 4 and 5 are only GABA-immunoreactive. Scale bar : 50 μm.

Our co-localization study showed that there was no reduction in the number of parvalbumin containing GABA cells long periods after kindling, while the reduction in the non-parvalbumin containing GABA somata in SO and SP was 50%. The cellular mechanism that leads to a persistent reduction of GABA immunoreactivity after the establishment of a kindled focus has not yet been clarified; but it seems from our results that the presence of parvalbumin in GABAergic interneurons renders them lesslikely to lose their GABA immunoreactivity during kindling. In view of the calcium binding properties of parvalbumin and the assumed neurotoxicity of high intracellular calcium concentrations (Meldrum, 1986), it may be speculated that the increase in parvalbumin content fulfils a protective role. By creating a larger intracellular calcium buffer capacity, parvalbumin-containing GABAergic interneurons are better able to deal with a potentially neurotoxic overload of intracellular Ca^{2+} resulting from an enhanced influx during seizures (Kamphuis et al., 1989a).

ACKNOWLEDGMENTS

We like to thank Dr. A. Rowlerson and M. Killen for critical reading and E. Kuemin for typing of this manuscript. This research was supported by the Swiss National Science Foundtion (grant 3.139.0.88), EMDO- and Sandoz-Stiftungen, the Foundation for Medical Research (MEDIGON, grant 550-054) of the

Table 2 . Co-localization of GABA in parvalbumin-positive somata in hippocampal CA1 region (SO and SP) at ipsilateral side in controls and in rats of the long-term kindled group.

PV	GABA	PV+GABA (n/mm^2)	s-PV	s-GABA
CONTROLS				
78.9 $\pm$ 6.3	62.2 $\pm$ 4.6	20.2 $\pm$ 3.3	58.7 $\pm$ 3.3	42.0 $\pm$ 2.4
LONG-TERM KINDLED				
78.1 $\pm$ 2.1	39.1 $\pm$ 1.5[**]	18.1 $\pm$ 2.1	60.0 $\pm$ 1.9	20.9 $\pm$ 2.1[***]

PV: cell density of parvalbumin positive somata. GABA: cell density of GABA positive somata. PV+GABA: cell density of somata in which co- existence of PV and GABA was shown. s-PV and s-GABA: cell density of somata, respectively, containing singly PV or singly GABA. Means $\pm$ S.E.M. for the whole group are presented. [**] P < 0.004, [***] P < 0.002 (Mann-Whitney U-test).

Organization for Pure Scientific Research (ZWO) and by the National Committee for Epilepsy Research (CLEO-TNO, grant A-67).

REFERENCES

Anderson, K. J., Borja, M. A., Cotman, C. W., Moffett, J. R., Namboodiri, M.A.A., Neale, J.H., 1987, N-Acetylaspartylglutamate identified in the rat retinal ganglion cells and their projections in the brain, Brain Res., 411:172.

Baimbridge, K. G., and Miller, J. J., 1988, Calcium binding proteins and experimental models of epilepsy, in: "Synaptic plasticity in the hippocampus," H. L. Haas and G. BuzsÈki, eds., Springer-Verlag, Berlin, Heidelberg.

Berchtold M. W., Epstein, P., Beaudet, A. L., Payne, E. M., Heizmann, C. W., and Means, A. R., 1987, The structural organization of the rat parvalbumin gene, J. Biol. Chem., 262:8696

Braun, K., Scheich, H., Schachner, M., and Heizmann, C. W., 1985a, Distribution of parvalbumin, cytochrome oxidase activity and ^{14}C-2-deoxyglucose uptake in the brain of the zebra finch. I. Anatomy and vocal motor systems, Cell and Tissue Res., 240:101

Braun, K., Scheich, H., Schachner M., and Heizmann, C. W.,1985b, Distribution of parvalbumin, cytochrome oxidase activity and ^{14}C-2-deoxyglucose uptake in the brain of the zebra finch. II. Visual system, Cell and Tissue Res., 240:117.

Celio, M. R., 1986, Parvalbumin in most gamma-aminobutyric acid containing neurons of the rat cerebral cortex, Science, 231:995.

Endo, T., Kobayashi, M., Kobayashi, S., and Onaya, T., 1986, Immunocytochemical and biochemical localization of parvalbumin in the retina, Cell Tissue Res., 243:213.

Gerday, Ch., 1988, Soluble calcium-binding proteins in vertebrates and invertebrate muscles, in:" Calcium and calcium binding proteins. Molecular and functional aspects", Ch. Gerday, L. Bollis and R. Gilles, eds., pp 23, Springer-Verlag, Heidelberg.

Goddard, G. V., McIntyre, P.C., and Leech, D. K., 1969, A permanent change in brain function resulting from daily electrical stimulation, Expl. Neurol., 25:295.

Hamoir, G., 1968, The comparative biochemistry of fish sarcoplasmic proteins, Acta Zool. Pathol., Antwerpen, 46:69.

Hardin, S. H., Carpenter, C. D., Hardin, P. E., Bruskin A. M., and Klein, W. H., 1985, Structure of the Spec 1 gene encoding a major calcium-binding protein in the embryonic ectoderm of the sea urchin Strongylocentrotus purpuratus, J. Mol. Biol., 186:243.

Hardin, P. E., Angerer, L. M., Hardin, S. H., Angerer, R. C., and Klein, W. H., 1988, Spec 2 genes of Strongylocentrotus purpuratus. Structure and differential expression in embryonic aboral ectoderm cells, J. Mol. Biol., 202:417.

Heizmann, C. W., and Celio, M. R., 1987, Immunolocalization of parvalbumin, Methods in Enzymology, 139: 552.

Heizmann, C. W., 1984, Parvalbumin, an intracellular calcium-binding protein; distribution, properties and possible roles in mammalian cells, Experientia, Basel, 40:910.

Heizmann, C. W., and Berchtold, M. R., 1987, Expression of parvalbumin and other calcium-binding proteins in normal and tumor cells: a topical review, Cell Calcium, 8:1.

Heizmann, C. W., and Hunziker, W., 1989, Intracellular calcium-binding molecules in: "Intracellular calcium regulation," F. Bronner, ed., Alan R. Liss Inc., in press.

Heizmann, C. W., and Braun, K., 1989, Calcium binding proteins. Molecular and functional aspects, in: "The role of calcium in biological systems," L. J. Anghileri, ed., CRC-Press, Boca Rato, Florida, Vol V in press.

Kamphuis, W., Wadman, W. J., Buijs, R. M., and Lopes da Silva, F. H., 1986, Decrease in number of hippocampal gamma-aminobutyric acid (GABA) immunoreactive cells in the rat kindling model of epilepsy, Exp. Brain Res., 64:491.

Kamphuis, W., Wadman, W. J., Buijs, R. M., and Lopes da Silva, F. H., 1987, The development of changes in hippocampal gamma-aminobutyric acid (GABA) in the rat kindling model of epilepsy: A light microscopic study with GABA-antibodies, Neuroscience, 23:433.

Kamphuis, W., Wadmann, W. J., Huisman, E., Heizmann, C. W., and Lopes da Silva, F. H., 1989a, Kindling induced changes in parvalbumin immunoreactivity in rat hippocampus and its relation to long-term decrease in GABA-immunoreactivity, Brain Res., 479:23.

Kamphuis, W., Huisman, E., Wadman, W. J., and Lopes da Silva, F. H., 1989b, Decrease in GABA immunoreactivity and alteration of GABA metabolism after kindling in the rat hippocampus, Exp. Brain Res., 74:375.

Katsumaru, H., Kosaka, T., Heizmann, C. W., and Hama, K., 1988, Immunocytochemical study of

GABAergic neurons containing the calcium-binding protein parvalbumin in the rat hippocampus, Exp. Brain Res., 72:347.

Kawaguchi, Y., and Hama, K., 1987, Two subtypes of non-pyramidal cells in rat hippocampal formation identified by intracellular recording and HRP injection, Brain Res., 411:190.

Kawaguchi, Y., Katsumaru, H., Kosaka, T., Heizmann, C. W., and Hama, K., 1987, Fast spiking cells in rat hippocampus (CA1 region) contain the calcium-binding protein parvalbumin, Brain Res., 416: 369.

Kosaka, T., Kosaka, K., Heizmann, C. W., Nagatsu, I., Wu, J.-W., Yanaihara, N., and Hama, K., 1987a, An aspect of the organization of the GABAergic system in the rat main olfactory bulb: laminar distribution of immunocytochemistry defined subpopulations of GABAergic neurons, Brain Res., 411:373.

Kosaka, T., Katsumaru, H., Hama, K., Wu, J-Y., and Heizmann, C. W., 1987b, GABAergic neurons containing the Ca^{2+}-binding protein parvalbumin in the rat hippocampus and the dentate gyrus, Brain Res., 419:119.

Kretsinger, R. H., 1980, Structure and evolution of calcium-modulated proteins, CRC Critical Reviews Biochem., 8:119.

Means, A. R., Putkey, J. A., and Epstein, P., 1988, Organisation and evolution of genes for calmodulin and other calcium-binding protein, in: "Calmodulin," P. Cohen and C. B. Klee, eds., pp 17, Elsevier, N.Y.

Meldrum, B. S., 1975, Epilepsy and gamma-aminobutyric acid-mediated inhibition, Int. Rev. Neurobiol., 17:1.

Meldrum, B. S., 1986, Cell damage in epilepsy and the role of calcium in cytotoxicity, in: Advances in Neurology, vol 44, A. V. Delgado-Escueta, J. J. Ward Jr., D. M. Woodbury, and R. J. Porter, eds., pp 849, Raven Press, N.Y.

McNamara, J. O., Bonhaus D. W., Shin, C., Crain, B. J., Gellman, R. L., and Giacchino, J. L., 1985, The kindling model of epilepsy: a critical review, CRC Critical Reviews Clin. Neurobiol., 1:341.

Pasteels, B., Parmentier, M., Lawson, E. M., Verstappen, A., and Pochet, R., 1987, Calcium binding protein immunoreactivity in the pigeon retina, Inv. Ophthal. Vis. Sci., 28:658.

Pechère, J. F., Demaille, J., and Capony, J. F., 1971, Muscular parvalbumins: preparative and analytical methods of general applicability, Biochem. Biophys. Acta, 236:391.

Peterson, S. L., and Albertson, T. E., 1982, Neurotransmitter and neuromodulator function in the kindled seizure and state, Progr. Neurobiology, 19:237.

Pourcho, R. G., 1982, Dopaminergic amacrine cells in the cat retina, Brain Res., 252:101.

Pourcho, R. G., and Goebel, D. J., 1985, A combined Golgi and autoradiographic study of (3H)glycine accumulating amacrine cells in the cat retina, J. Comp. Neurol., 233:473.

Rabié, A., Thomasset, M., Parkes, C.O., and Clavel, M.C., 1985, Immunocytochemical detection of 28000-MW calcium-binding protein in horizontal cells of the rat retina, Cell Tissue Res., 240:493.

Roberts, E., 1986, GABA: The road to neurotransmitter status, in: Benzodiazepine/GABA Receptors and Chloride Channels: Structural and Functional Properties, R. W. Olsen and J. G. Venter, eds., pp 1, Alan R. Liss, N.Y.

Röhrenbeck, J., Wässle, H., and Heizmann, C. W., 1987, Immunocytochemical labelling of horizontal cells in mammalian retina using antibodies against calcium-binding proteins, Neurosci. Letters, 77:255.

Röhrenbeck, J., Wässle, H., and Boycott, B. B., 1989, Horizontal cells in the monkey retina, Europ. J. Neurosci., in press.

Schreiner, D. A., Jande, S. S., and Lawson, D. E. M., 1985, Target cells of vitamin D in the retina, Acta anat., 121:153.

Seress, L, and Ribak, C. E., 1985, A combined Golgi-electron microscopic study of non-pyramidal neurons in the CA1 area of the hippocampus, J. Neurocyt., 14:717.

Stichel, C. C., Kaegi, U., and Heizmann, C. W., 1986, Parvalbumin in the cat brain: isolation, characterization, and localization, J. Neurochem., 47:46.

Stichel, C. C., Singer, W., and Heizmann, C. W., 1988, Light and electron microscopic immunocytochemical localization of parvalbumin in the dorsal lateral geniculate nucleus of the cat: evidence for coexistence with GABA, J. Comp. Neurol., 268:29.

Verstappen, A., Parmentier, M., Chirnoaga, M., Lawson, D. E. M., Pasteels, J. L., and Pochet, R., 1986, Vitamin D-dependent calcium binding protein immunoreactivity in human retina, Ophthalmic Res., 18:209.

Voigt, T., and Wässle, H., 1987, Dopaminergic innervation of AII amacrine cells in mammalian retina, J. Neurosci. 7:4115.

Wässle, H., Peichl, L., and Boycott, B. B., 1978, Topography of horizontal cells in the retina of the domestic cat, Proc. R. Soc. Lond. B., 203:269.

UNIQUE CALCIUM BINDING PROTEINS IN INVERTEBRATES

Jos A. Cox

Department of Biochemistry, University of Geneva, 30, Quai Ernest Ansermet, 1211 Geneva 4, Switzerland

The research on sarcoplasmic calcium-binding proteins (SCP) started in 1973 in Geneva and has its roots in the, at-that-time, intense research on parvalbumins. Parvalbumins form a rather homogeneous sub-family of the EF hand calcium-binding proteins with a distribution in nearly all vertebrates. In search of parvalbumins in invertebrate muscle we discovered another type of soluble sarcoplasmic calcium-binding protein in crustacea, annelids, mollusks and protochordates. Recently it became evident that a marked homology exists between SCP's and other members of the EF-hand family, namely with aequorin and luciferin-binding protein and with a Streptomyces erythraeus calcium-binding protein, the first sequenced EF-hand protein of a prokaryote. All these proteins can be classified in one EF-hand subfamily, i.e. the SARC*AEQ double family.

The research on SCP's led in 1986 to the discovery of a new sarcoplasmic calcium-binding protein, abundant in extracts of Amphioxus muscle. Since it is present in a tight complex with an endogeneous 36 kDa protein and its interactive properties with hydrophobic peptides resemble those of proteins directly involved in the intracellular Ca^{2+}-signal, we called it calcium-vector protein (CaVP). It is a unique protein, although part of its structure resembles that of calmodulin and troponin C.

I. SARCOPLASMIC CALCIUM-BINDING PROTEINS

A. General Characteristics and distribution

The similarity of invertebrate SCP's to the parvalbumins in vertebrates points to their functional analogy: 1° They can be very easily extracted with physiological ionic strength buffers. 2° They are abundant in the sarcoplasm (0.5 mM per kg wet weight). 3° They show pronounced polymorphism. Polymorphism is due to a limited number of amino acid substitutions: in the case of amphioxus in a short 16-residue-long segment in the first Ca^{2+}-binding domain (Takagi et al., 1986; Takagi and Cox, submitted), in the other phyla, point mutations are spread over the whole length of the protein sequence (Takagi and Konishi, 1984; Takagi et al., 1984). In crustacea, polymorphism is increased due to homo- and heterodimer formation (Wnuk and Jauregui-Adell, 1983). 4° They do not interact with hydrophobic matrices such as phenyl-Sepharose or fluphenazine-Sepharose. Nor do they interact with amphiphilic peptides such as melittin, which often serves as model to monitor the interactions of calmodulin with ITS target proteins. These data suggest that SCP's are incapable of protein-protein interaction.

The distribution pattern of proteins belonging to the SARC*AEQ double family and parvalbumins is schematically represented in Fig. 1. Four general observations deserve comment: 1° Parvalbumin and SCP's have not yet been found to coexist in comparable amounts in the same animal. 2° SCP's are present in various muscle types (cross- and obliquely-striated, smooth) of invertebrate phyla. In some phyla and classes no SCP could be detected. 3° Some of the Coelenterates possess SCP-related proteins, i.e. the bioluminescent aequorin and luciferin-binding protein. 4° The only prokaryotic Ca^{2+}-binding protein of the EF-hand family with a known sequence, i.e. S. erythraeus CaBP, displays strong sequence homology with annelid SCP. It is thus not unlikely that the SARC*AEQ double family may have emerged

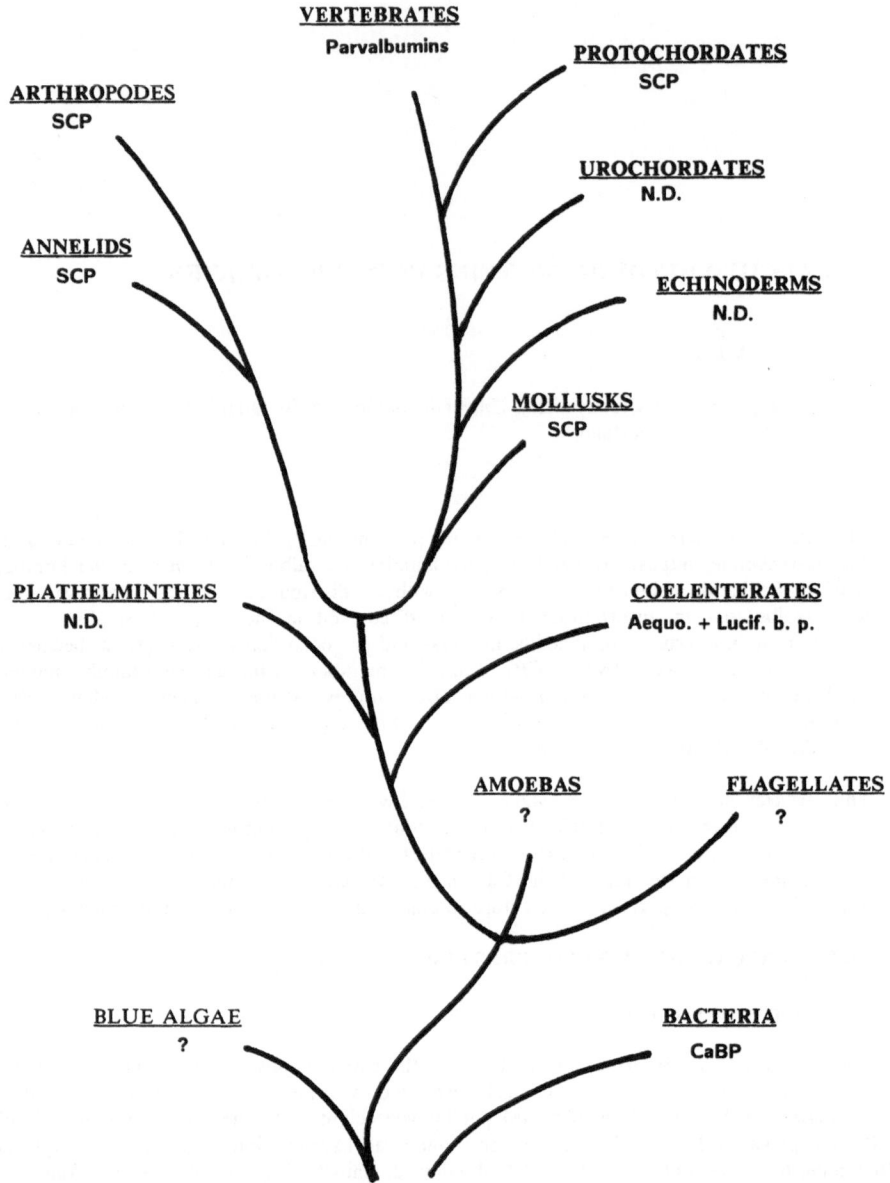

Fig. 1 . Distribution of sarcoplasmic calcium-binding proteins. CaBP is the calcium-binding protein of *Streptomyces*; Aequo. + Lucif. b.p. stands for aequorin and luciferin binding protein; N.D. means that no sarcoplasmic calcium-binding protein has been detected; ? means that the phylum has not been investigated.

at least 1400 million years ago and is the oldest of the EF-hand subfamilies. The interphylum distribution shown in Fig. 1 does not reflect the variability of the presence of SCP's within each phylum. In the Arthropoda phylum no SCP's have been detected in insects (Cox et al., 1981) although anti-Amphioxus SCP II stains very specifically about 15 neurons in the optic lobe in D. melanogaster (Buchner et al., 1988). Molluscan gastropods and cephalopods do not contain significant amounts of SCP's, but bivalves are rich in SCP's. In fact, their distribution seems as unpredictible as that of parvalbumins (Heizmann and Berchtold, 1987) and may depend on the individual and specific need for ion buffering of each cell, organ and animal species.

B. Structure

The complete amino acid sequence of 11 SCP's from 4 different phyla has recently been reported (Takagi and Konishi, 1984a; Takagi and Konishi, 1984b; Takagi et al., 1984; Kobayashi et al., 1984; Takagi et al., 1986; Collins et al., 1988; Jauregui-Adell et al., 1989). Within the Ca^{2+}-binding domains conservation is dictated by the constraints of the EF-hand motif, though some are still functional while others have degenerated (Table I). It appeared that for the classification in the SCP subfamily a useful criterion, a "fingerprint", is the length of the linkers between the EF-domains. In all SCP's domains I and II are separated by a 18 to 20 residue-long linker, II and III by a 11 to 12 , III and IV by 1 to 2 residues. This irregular spacing of the Ca^{2+}-binding domains has not been observed in the other subfamilies of the EF-hand family, such as calmodulins and troponin C's. However, the "SCP fingerprint" has been found in the amino acid sequences of two bioluminescent Ca^{2+}-binding proteins from coelenterates and of S. erythraeus Ca^{2+}-binding protein (Cox and Bairoch, 1988). It was therefore concluded that the bioluminescent proteins, the prokaryotic protein and the SCP's belong to the same subfamily, named the SARC*AEQ double family according to the classification of Kretsinger et al. (1988).

Nereis SCP crystallizes at pH 7.6 in the space group $P2_1$ with two monomers per asymmetric unit (Babu et al., 1987). Amphioxus SCP II crystals are long orthorombic prisms with one monomer per asymmetric unit (Cook, personal communication).

C. Calcium and magnesium binding properties of SCP's

Ion binding properties of the SCP's of the different phyla are distinct, as can already be inferred from the differences in the location and number of functional Ca^{2+}-binding sites (Table I). In general Ca^{2+}-binding properties are complex and involve positive cooperativity, sites of different affinity and pronounced Mg^{2+} antagonism. Crustacean SCP's possess 1 Ca^{2+}-specific and 2 Ca^{2+}-Mg^{2+} sites per monomer (Wnuk et al., 1979); protochordate SCP's contain 2 Ca^{2+}-specific and one Ca^{2+}-Mg^{2+} site (Takagi et al., 1986); annelid SCP's contain 3 Ca^{2+}-Mg^{2+} sites (Cox and Stein, 1981) and mollusc SCP's 1 Ca^{2+}-specific and 1 Ca^{2+}-Mg^{2+} site (Collins et al., 1983). Ca^{2+}-binding in the absence of Mg^{2+} displays little or no positive cooperativity, whereas the presence of millimolar concentrations of Mg^{2+} leads to a strong positive cooperativity. At physiological ionic strength and in the presence of 1 mM Mg^{2+} the apparent Ca^{2+} dissociation constants for the different SCP's are similar (30 to 100 nM) and comparable to those of parvalbumins (ca 50 nM) (Wnuk et al., 1982).

Pronounced conformational changes usually accompany the binding of divalent cations to SCP's, most conformational changes being sequential, i.e. they closely follow the appearance of one SCP.Ca_n species, with n being different for different types of conformational probes: in crayfish SCP, the major structural changes occur when the second (a-helix content), third (tryptophan fluorescence) or fourth (thiol reactivity) Ca^{2+} binds to the dimer (Wnuk et al., 1981).

D. Physiological implications

In the discussion on the biological role of SCP's some of the most salient characteristics have to be reiterated : 1° They usually are abundant in fast moving animals (Wnuk et al., 1982; Gerday, 1988) and more concentrated in fast contracting than in slow contracting muscles (Cox et al., 1976). 2° Their concentrations are sufficiently high, i.e. 0.2 mmol/l, to mop up nearly all the Ca^{2+} that flows in during a single activation cycle. 3° All SCP's posses Ca^{2+}-Mg^{2+} sites and exchange at these sites may lead to important fluctuations of the intracellular Mg^{2+} concentration.

It should be noted that all these characteristics also apply to parvalbumins and that SCP's can therefore be considered as the functional counterparts of parvalbumins in invertebrates. Two different hypotheses have been put forward for the exact role of these proteins in the contraction-relaxation process in muscle:

1. Soluble relaxing factors : According to this idea parvalbumins and SCP's act as shuttles transporting Ca^{2+} from the myofibrils to the sarcoplasmic reticulum during the relaxation phase (Pechère et al., 1975). Direct arguments were provided by Pechère et al. (1977) and by Somlyo et al. (1981). For the case of parvalbumins and SCP's that contain solely Ca^{2+}-Mg^{2+} sites the hypothesis however does not withstand objections of kinetic order (Cox et al., 1979; Robertson et al., 1981). Indeed, Ca^{2+} and Mg^{2+} bind and dissociate slowly from these type of sites as compared to the kinetics of the Ca^{2+} tran-

Table I. Functional E-F hand domains in the SARC*AEQ double family

	Domains			
	I	II	III	IV
Aequorin	+	−	+	+
Luciferin-binding protein	+	−	+	+
Streptomyces CaBP	+	−	+	+
Mollusk SCP				
Scallop	+	−	+	−
Annelid SCP's				
Nereis	+	−	+	+
Perinereis	+	−	+	+
Crustacean SCP's				
Shrimp $\alpha 1$	+	+	+	−
$\alpha 2$	+	+	+	−
β	+	+	+	−
Crayfish α	+	+	+	−
Protochordate SCP's				
Amphioxus I	+	+	+	−
II	+	+	+	−
III	+	+	+	−
IV	+	+	+	−

+ : functional domains   − : abortive domains

sients in contracting muscle. However, Ca^{2+}-specific sites respond fast enough to sense the rapid changes of sarcoplasmic free Ca^{2+} concentrations and can act as a relaxing factor.

2. *Protectors against high Ca^{2+} levels during prolonged contractions + regulators of intracellular Mg^{2+}* : Only repeated stimulations of muscle (smooth tetanus) can modify the occupancy of the Ca^{2+}-Mg^{2+} sites, in other words these mixed sites act as a slow responding Ca^{2+} sink. In this way SCP's can a protect against lethal rises in cytosolic Ca^{2+} and probably also accelerate energy provision. The latter point is better appreciated when one realizes that: 1° the free sarcoplasmic Mg^{2+} concentration is 0.5 mM and that of mixed sites in fast skeletal muscle amount to up to 2 mM. Thus potentially important amounts of Mg^{2+} can be liberated from these sites during muscle tetanization. 2° Mg^{2+} is a cofactor for different glycolytic enzymes and a co-substrate in most reactions involving phosphorus-containing metabolites

II. AMPHIOXUS CALCIUM VECTOR PROTEIN

A. General characteristics, Ca^{2+}-binding and distribution

Extraction of amphioxus muscle with isotonic buffer liberates, in addition to a mixture of SCP's, impressive amounts of another calcium-binding protein (ca 2 mg per g of wet tissue), named CaVP (Cox, 1986). CaVP has a molecular weight of 18.300 kDa and is highly asymmetric. The protein binds 2 Ca^{2+} ions to so-called Ca^{2+}-specific sites in a noncooperative way with an intrinsic dissociation constant of 0.11 μM (Cox, 1986). Ca^{2+}-binding increases its a-helical content and favors the interaction with melittin and with phenyl Sepharose (Cox, 1986). Qualitatively these interactions are similar to those of most well known intracellular Ca^{2+} vectors. The ability of CaVP to expose hydrophobic surfaces and to become involved in Ca^{2+}-dependent protein-protein contacts allows us to classify it as a calcium vector protein, even though its role is actually not known.

Affinity-purified polyclonal antibodies against CaVP showed no cross-reactivity with calmodulin or troponin C. Although in Western blotting experiments no cross-reactivity is observed with any component in extracts of different organs from rat and fish (Cox, 1986), in immunolocalization experiments (Muntener, data not published) positive staining with anti-CaVP is observed in rat skeletal muscle. In

Amphioxus anti-CaVP stains intensively the body wall muscles and the neuronal chord. A weakly, but specific and very particular staining pattern has been observed in other organs such as the liver and the gills of the pharynx, indicating that CaVP is not exclusively a muscular protein.

B. Structure

The amino acid sequence of CaVP (Kobayashi et al. 1987) revealed the 4 repeats of an EF-hand domain. Domains III and IV are canonical according to well-established rules and show high sequence identity with the corresponding domains of calmodulin and of chicken skeletal muscle troponin C. Domain III contains 2 e-trimethyllysin residues in the a-helices flanking the Ca^{2+}-binding loop. The amino-acid sequences of domains I and II deviate from the EF-hand canonical rules and no longer bind Ca^{2+}. In its native form CaVP contains a disulfide bridge linking the first a-helix of domain I (Cys_{16}) to the second of domain II (Cys_{78}). The segment between the second and third Ca^{2+}-binding loop shows 47% sequence identity with the corresponding segment in calmodulin and does not contain a-helix breaking residues. The presence of a long central a-helix in CaVP is thus likely.

The high sequence homology of CaVP with calmodulin and troponin C allows us to deduce its tridimensional structure from template modelization calculations starting from the calmodulin and/or troponin C structures (Schaad and Cox, work in progress). Preliminary conclusions are that the structure of the C-terminal half is very similar to that of calmodulin and troponin C and that the formation of the disulfide bridge is energetically favorable provided domains I and II are in the Ca^{2+}-configuration, even though they do not bind Ca^{2+}. This comforts the hypothesis that the disulfide bridge stabilizes the Ca^{2+}-like structure in the N-terminal half and that reduction leads to a rearrangment similar to the structure encountered in the metal-free N-terminal half of TNC.

C. Function

To evaluate the exact function of CaVP in Amphioxus muscle, we must rely on its most characteristic properties: 1° Its abundance (2 mg/g wet tissue) is comparable to that of troponin C in rabbit skeletal muscle. 2° It does not substitute for troponin C or for calmodulin in the adequate functional assays (Cox, 1986). Phylogenetically it neither belongs to the calmodulin nor to the troponin C subfamilies (Kretsinger et al., 1988). 3° It forms Ca^{2+}-regulated complexes with amphiphilic peptides. 4° It is isolated as a 1 to 1 complex with a 36 kDa protein (which we named T36 for target of 36 kD) and urea or heat is required for the dissociation of that complex.

We thus assume that CaVP is not merely a buffer and regulator of intracellular Ca^{2+}, but acts as a Ca^{2+} activator of T36. In this respect it is instructive to examine in some detail the properties of T36. The amino acid sequence of 193 residues at the C-terminal end of T36 has been elucidated by Dr Takagi, Sendai. With the Dayhoff MDM-78 matrix analysis no significant homology could be found with any protein of known primary structure. Hence, both CaVP and its natural target are unique proteins and may be at the basis of a new limb of Ca^{2+}-regulation in Amphioxus. Whether this system also occurs in higher vertebrates remains an open question.

III. CONCLUSIONS

Besides classical calmodulin and troponin C, a panoply of new Ca^{2+}-binding proteins has been found in invertebrate muscle and other tissues. Despite detailed knowledge accumulated about their occurrence, structure and interactive properties, no precise role in cell functioning could be assigned. Some are reminiscent of parvalbumins and thought to regulate the Ca^{2+} and Mg^{2+} fluxes. Physiological experiments, such as tension measurements of muscle fibers under twitch or tetanos stimulation while changing the SCP concentrations by artificial or genetic means, will be neccessary for the understanding of the function of these natural Ca^{2+}-buffers. In some invertebrates Nature used these proteins to develop a inter-species communication system based on light emission. It is clear that other new Ca^{2+}-binding proteins, such as the 18 kDa protein in Amphioxus, are more directly involved in the processing of the Ca^{2+} message. The question now is which response system is activated.

REFERENCES

Babu, Y.S., Cox, J.A., and Cook, W.J., 1987, Crystallization and preliminary X-ray investigation of sarcoplasmic calcium-binding protein from Nereis diversicolor. J. Biol. Chem. 262: 11884-11885.

Buchner, E., Bader, R., Buchner, S., Cox, J.A., Emson, P.C., Flory, E., Heizmann, C.W., Hemm, S., Hofbauer, A., and Oertel, W.H., 1988, Cell-specific immuno-probes for the brain of normal and mutant Drosophila melanogaster. I Wildtype visual system. Cell Tissue Res. 253: 357-370.

Collins, J., Cox, J.A., and Theibert, J.L., 1988, Amino acid sequence of a sarcoplasmic calcium-binding protein from the sandworm Nereis diversicolor. J. Biol. Chem. 263: 15378-15385.

Collins, J., Johnson, J.D., and Szent-Gyorgyi, A.G., 1983, Purification and characterization of a scallop sarcoplasmic calcium-binding protein. Biochemistry 22: 341-345.

Cox, J.A., and Bairoch, A., 1988, Sequence homologies in prokaryote and invertebrate calcium-binding proteins. Nature 331:491-492.

Cox, J.A., and Stein, E.A., 1981, Characterization of a new sarcoplasmic calcium-binding protein with magnesium-induced cooperativity in the binding of calcium. Biochemistry 20: 5430-5436.

Cox, J.A., Kretsinger, R.H., and Stein, E.A., 1980, Sarcoplasmic calcium-binding proteins in insect muscle. Isolation and properties of locust calmodulin. Biochim. Biophys. Acta 670: 441-444.

Cox, J.A., Wnuk, W., and Stein, E.A., 1976, Isolation and properties of a sarcoplasmic calcium-binding protein from crayfish. Biochemistry 15: 2613-2618.

Cox, J.A., 1986, Isolation and characterization of a new M_r 18000 protein with calcium vector properties in amphioxus muscle and identification of its endogenous target protein. J. Biol. Chem. 261: 13173-13178.

Gerday, Ch., 1988, in: "Calcium and Calcium Binding Proteins", Ch. Gerday, L. Bolis, and R. Gilles, eds., pp. 23-39, Springer-Verlag, Berlin Heidelberg.

Heizmann, C.W., and Berchtold, M.W., 1987, Expression of parvalbumin and other Ca^{2+}-binding proteins in normal and tumor cells: a topical review. Cell Calcium 8: 1-41.

Jauregui-Adell, J., Wnuk, W., and Cox, J.A., 1989, Complete amino acid sequence of the sarcoplasmic calcium-binding protein (SCP-I) from crayfish (Astacus leptodactylus). FEBS Lett. 243: 209-212.

Kobayashi, T., Takagi, T., Konishi, K., and Cox, J.A., 1987, The primary structure of a new M_r 18,000 calcium vector protein from Amphioxus. J. Biol. Chem. 262: 2613-2623.

Kretsinger, R., Moncrief, N.D., Goodman, M., and Czelusniak, J., 1988, Homology of calcium-modulated proteins: Their evolutionary and functional relationships. in: "The Calcium Channel: Structure, Function and Implications", M. Morad, W. Nayler, S. Kazda, and M. Schramm, eds., pp. 16-38, Springer-Verlag, Berlin Heidelberg.

Kretsinger, R.H., Rudnick, S.E., Smeden, D.A., and V.B. Schatz, 1980, Calmodulin, S-100 and crayfish sarcoplasmic calcium binding protein crystals suitable for X-ray diffraction studies. J. Biol. Chem. 255: 8154-8156.

Pechère, J.-F., Derancourt, J., and Haiech, J., 1977, The participation of parvalbumins in the activation-relaxation cycle of vertebrate fast skeletal muscle. FEBS Lett. 75: 111-114.

Robertson, S.P., Johnson, J.D., and Potter, J.D., 1981, The time course of Ca^{2+}-exchange with calmodulin, troponin, parvalbumin and myosin in response to transient increases in Ca^{2+}. Biophys. J. 34: 559-569.

Somlyo, A.V., Gonzales-Serattos, H., Schuman, H., McCleilan, G., and Somlyo, A.P., 1981, Calcium release and ionic changes in the sarcoplasmic reticulum. J. Cell Biol. 90: 577-594.

Swan, D.G., Hale, R.S., Dhillon, D., and Leadlay, P.F., 1987, A bacterial calcium-binding protein homologous to calmodulin. Nature 329: 84-85.

Takagi, T., and Konishi, K., 1984a, Amino acid sequence of a chain of sarcoplasmic calcium binding protein obtained from shrimp tail muscle. J. Biochem. (Tokyo) 95: 1603-1615.

Takagi, T., and Konishi, K., 1984b, Amino acid sequence of the b chain of sarcoplasmic calcium bindning protein (SCP) obtained from shrimp tail muscle. J. Biochem. (Tokyo) 96: 59-67.

Takagi, T., Konishi, K., and Cox, J.A., 1986, The amino acid sequence of two sarcoplasmic calcium-binding proteins from the protochordate Amphioxus. Biochemistry 25: 3585-3592.

Takagi, T., Kobayashi, T., and Konishi, K., 1984, Amino acid sequence of sarcoplasmic calcium-binding protein from scallop (Patinopecten yessoensis) adductor striated muscle. Biochim. Biophys. Acta 787: 252-257.

Wnuk, W., and Jauregui-Adell, J., 1983, Polymorphism in high-affinity calcium-binding proteins from crustacean sarcoplasm. Eur. J. Biochem. 131: 177-182.

Wnuk, W., Cox, J.A., and Stein, E.A., 1982, Parvalbumin and other soluble sarcoplasmic Ca-binding proteins. in: "Calcium and Cell Function", W.Y. Cheung, ed., Vol II, pp. 243-278, Academic Press, New York.

Wnuk, W., Cox, J.A., and Stein, E.A., 1981, Structural changes induced by calcium and magnesium in a high affinity protein from crayfish sarcoplasm. J. Biol. Chem. 256: 11538-11544.

Wnuk, W., Cox, J.A., Kohler, L.G., and Stein, E.A., 1979, Calcium and magnesium binding properties of a high-affinity calcium-binding protein from crayfish sarcoplasm. J. Biol. Chem. 254: 5284-5289.

STRUCTURE OF THE CALCIUM RELEASE CHANNEL OF SKELETAL MUSCLE SARCOPLASMIC RETICULUM AND ITS REGULATION BY CALCIUM

F. Anthony Lai and Gerhard Meissner

Departments of Biochemistry and Physiology, University of North Carolina, School of Medicine, Chapel Hill, NC 27599, USA

Skeletal muscle contraction is initiated by an action potential-induced depolarization of the muscle plasma membrane. This electrical excitation originates at the neuromuscular synapse and spreads rapidly into the transverse tubule (T-) system of membrane infoldings, which extend inward from the surface membrane to surround each myofibril. In an incompletely understood process termed excitation-contraction coupling (Somlyo, 1985), T-system depolarization somehow triggers the rapid release of a large Ca^{2+} pool stored within the intracellular membrane system, sarcoplasmic reticulum (SR), thus providing an elevated free Ca^{2+} concentration that results in muscle contraction (Ebashi, 1976; Endo, 1977). The signal which induces opening of the SR Ca^{2+} release channels is believed to be transmitted at specialized junctions where the T-system and SR membranes become closely apposed to form a narrow 12 nm gap. Large protein structures projecting from the SR membrane span this junctional gap to provide apparent continuity between the T-system and SR, and have previously been defined morphologically and termed either feet (Franzini-Armstrong, 1970), bridges (Somlyo, 1979), pillars (Eisenberg and Eisenberg, 1982) or spanning proteins (Caswell and Brunschwig, 1984). Biochemical and morphological analysis of SR fragmented by homogenization of muscle tissue has shown that both Ca^{2+} release activity and the feet structures are enriched in "heavy" SR vesicles, a subcellular microsomal fraction derived from the junctional SR (Nagasaki and Kasai, 1983; Meissner, 1984; Ferguson et al., 1984; Saito et al., 1984; Ikemoto et al., 1985; Meissner et al., 1986). Study of the Ca^{2+} release channel activity using isolated heavy SR vesicles has been approached by applying two complementary techniques; macroscopic $^{45}Ca^{2+}$ flux from passively loaded vesicles, and microscopic Ca^{2+} currents through single channels incorporated into planar lipid bilayers. Vesicle $^{45}Ca^{2+}$ flux studies have shown that the SR Ca^{2+} release channel can be activated by micromolar Ca^{2+} or millimolar adenine nucleotides to give enhanced release rates, but could be optimally activated only by the combined presence of Ca^{2+} and adenine nucleotides to give maximal release rates with first-order rate constants of $30\text{-}100s^{-1}$ (Meissner et al., 1986). The presence of millimolar Mg^{2+}, millimolar Ca^{2+} or micromolar ruthenium red inhibited channel activity. The Hill coefficient for adenine nucleotide activation of Ca^{2+} release was 1.6-1.9, whereas that for Ca^{2+} activation was 0.8 -2.1 depending on Mg^{2+} and nucleotide concentration, and for Mg^{2+} inhibition was 1.1 - 1.6 depending on Ca^{2+} concentration. Observations of the behavior of single Ca^{2+} channels upon fusion of heavy SR vesicles into a planar lipid bilayer have corroborated the vesicle flux studies, and further indicated that the channel comprised a high conductance pathway with unitary conductance of 100 pS (in 50 mM Ca^{2+} trans), and was highly selective for cations over anions (Smith et al., 1985; 1986). The channel's fraction of open time (P_o) was close to zero in nanomolar Ca^{2+} cis, and could be increased to 0.02-0.2 by addition of Ca^{2+} cis to 1-10 micromolar. The presence of micromolar Ca^{2+} and millimolar adenine nucleotide resulted in a Po close to 1.0. Additional vesicle flux and single channel recordings have further shown that caffeine and related methyl-xanthines can also activate the Ca^{2+} release channel by acting at the Ca^{2+} activation site (Rousseau et al., 1988), while calmodulin, a ubiquitous Ca^{2+} binding protein, partially inhibited channel activity (Meissner, 1986a; Smith et al., 1989).

Structural definition of the SR Ca^{2+} release channel has been greatly facilitated by the identification of ryanodine, a neutral plant alkaloid insecticide obtained from the South American shrub Ryania

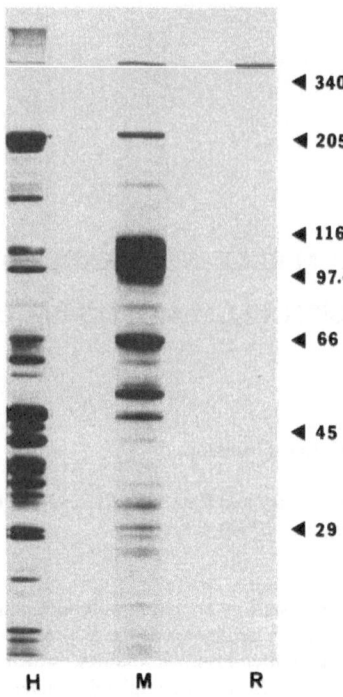

Fig. 1 . Silver stained gel of purified skeletal ryanodine receptor. SDS-polyacrylamide gradient (5-12%) gel analysis of whole muscle homogenate (H), heavy SR membranes (M) and purified 30S ryanodine receptor complex (R) from rabbit skeletal muscle (Lai et al., 1988). Sizes of molecular weight standards are shown (x 10^{-3}).

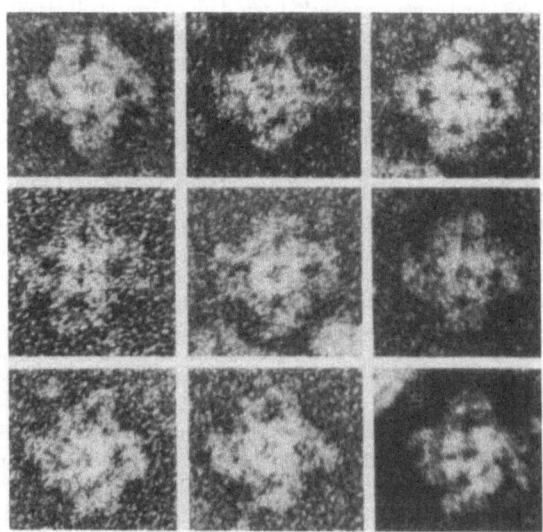

Fig. 2 . Negative-stain electron microscopy of the 30S ryanodine receptor complex. A selected panel of particles with the characteristic clover-leaf shape of the feet structures (Ferguson et al., 1984), stained by uranyl acetate. Dimensions of the quatrefoil are 34 nm from tip of one leaf to the tip of the opposite one, with each leaf 14 nm wide. Central electron dense region is 14 nm diameter with central hole 1-2 nm diameter. (Taken with permission from Lai et al., 1988).

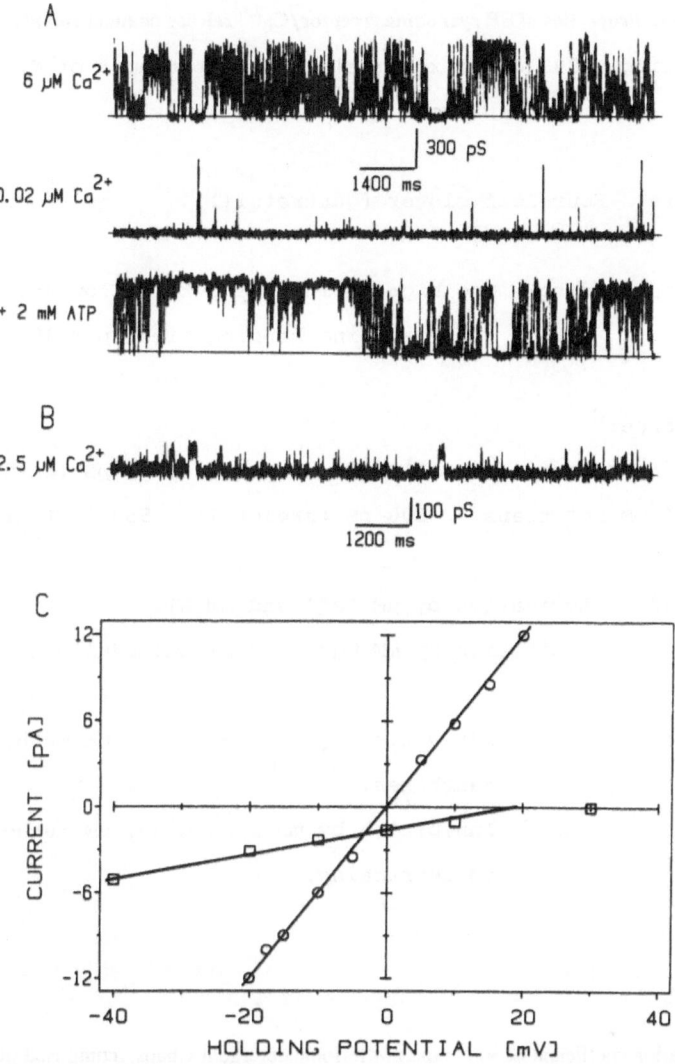

A

6 μM Ca²⁺

300 pS

1400 ms

0.02 μM Ca²⁺

+ 2 mM ATP

B

2.5 μM Ca²⁺

100 pS

1200 ms

C

CURRENT [pA]

HOLDING POTENTIAL [mV]

Fig. 3 . Reconstitution of the 30S ryanodine receptor complex into planar lipid bilayers. a, Single-channel Na^+ currents, shown as upward deflections, were recorded in symmetric 0.5 M NaCl, 10 mM NaPIPES, pH 7 with 6 μM free Ca^{2+} cis (100 μM EGTA, 100 μM $CaCl_2$) (top trace), or with 0.02 μM free Ca^{2+} cis (2.1 mM EGTA, 100 μM $CaCl_2$) (second trace), or with 0.02 μM free Ca^{2+} plus 2 mM ATP in the cis chamber (bottom trace). Holding potential (HP) = -15 mV. b, Single-channel current recorded after perfusion with 50 mM $Ca(OH)_2$/250 mM HEPES, pH 7.4, 10% glycerol trans, and 125 mM Tris/250 mM HEPES, pH 7.4, 10% glycerol, plus 2.5 μM free Ca^{2+} (100 μM EGTA, 100 μM $CaCl_2$) cis. HP = 0 mV. c, Current-voltage relationship for recordings a (top trace) and b. Values of unit conductance: $_{Na}$, 595 pS with 0.5 M Na^+ (o); and $_{Ca}$, 91 pS with 50 mM Ca^{2+} ( ) as the conducting ion. (Taken with permission from Lai et al., 1988).

speciosa. Ryanodine was found, using vesicle flux and single channel studies, to specifically affect the Ca^{2+} release channel by activating release at low (nanomolar) concentrations but inhibiting the channel at high (> micromolar) concentrations (Meissner, 1986b; Rousseau et al., 1987). High-affinity binding of [3H]ryanodine was enriched in heavy SR preparations, with a nanomolar K_D and B_{max} up to 20 pmol [³H]ryanodine bound/mg protein. Subsequent approaches to isolate the ryanodine receptor using immunoaffinity chromatography (Imagawa et al., 1987), column chromatography (Inui et al., 1987) and density gradient centrifugation (Lai et al., 1987; 1988) have revealed that the receptor exists as a complex with

Table 1. Properties of SR ryanodine receptor/Ca^{2+} release channel complex

1. Composition: 30S complex comprising homotetramer of M_r ~400,000 polypeptides.

2. Structure: Four-leaf clover (quatrefoil)

3. Ryanodine binding: One high-affinity and three low-affinity [³H]ryanodine binding sites per 30S complex

4. Conductance:

50 mM Ca^{2+} trans	100 pS (skeletal)	75 pS (cardiac)
500 mM Na^+ trans	600 pS (skeletal)	550 pS (cardiac)

5. Regulation: Activation by μM Ca^{2+} and mM ATP

   Inhibition by mM Mg^{2+} and μM calmodulin

6. Exogenous ligands: Activation by nM ryanodine, mM methyl xanthines.

   Inhibition by mM ryanodine, μM ruthenium red, mM tetracaine.

apparent sedimentation coefficient of 30S in the zwitterionic detergent Chaps, composed of polypeptides of apparent relative molecular mass 400,000 (Fig. 1). Further studies have shown that the complex most likely comprises a homotetramer of negatively charged and cooperatively coupled subunits (Lai et al., 1989a,b). Negative-stain electron microscopy of the isolated ryanodine receptor (Fig. 2) has shown a striking morphological resemblance to the four-leaf clover appearance of the feet structures obtained from images of rotary-shadowed heavy SR vesicles (Ferguson et al, 1984).

Reconstitution of the ryanodine receptor complex, isolated in the absence of ryanodine, into planar lipid bilayers, has revealed intrinsic Ca^{2+} and Na^+ conductances with unitary conductance of 90 and 600 pS respectively, and pharmacological profile characteristic of the Ca^{2+} release channel in native heavy SR vesicles (Fig. 3). Fig. 3a (top trace) shows a single channel recording of the reconstituted purified ryanodine receptor with Na^+ as the permeant ion (600 pS). Reduction of free Ca^{2+} to nanomolar concentrations reduced the fraction of channel open time to 0.03 (Fig. 3a, middle trace) and addition of millimolar adenine nucleotide (ATP) increased open time to 0.47 (Fig. 3a, lower trace). Perfusion of the Na+ buffer with Ca^{2+}-containing buffer revealed a Ca^{2+} current of 90 pS (Fig. 3b). The likelihood that these channel properties were the result of incorporation of a minor contaminant present in the ryanodine receptor preparation was dispelled by the observation that the typical effect of ryanodine on the native Ca^{2+} release channel could also be obtained using the purified reconstituted receptor (Lai et al., 1988; Liu et al., 1989; Lai and Meissner, 1989). Thus, the 30S ryanodine receptor complex comprises the SR Ca^{2+} release channel and is identical to the feet structures that span the T-SR junctional gap. The structural and functional properties of the SR ryanodine receptor - Ca^{2+} release channel complex so far identified, are summarized in Table 1.

ACKNOWLEDGEMENTS

Supported by USPHS grants and a fellowship from the Muscular Dystrophy Association to F.A. Lai.

REFERENCES

Caswell, A.H., and Brunschwig, J.P., 1984, J. Cell Biol., 99, 929-939.

Ebashi, S., 1976, Annu. Rev. Physiol., 38, 293-313.

Eisenberg, B.R., and Eisenberg, R.S., 1982, J. Gen. Physiol., 79, 1-19.

Endo, M., 1977, Physiol. Rev., 57, 71-108.

Ferguson, D.G., Schwartz, H.W., and Franzini-Armstrong, C., 1984, J. Cell Biol., 99, 1735-1742.

Ikemoto, N., Antoniu, B., and Meszaros, L.Y., 1985, J. Biol. Chem., 260, 14096-14100.

Imagawa, T., Smith, J.S., Coronado, R., and Campbell, K.P., 1987, J. Biol. Chem., 262, 16635-16643.

Inui, M., Saito, A., and Fleischer, S., 1987, J. Biol. Chem., 262, 15637-15642.

Lai, F.A., Erickson, H., Block, B.A., and Meissner, G., 1987, Biochem. Biophys. Res. Commun., 143, 704-709.

Lai, F.A., Erickson, H.P., Rousseau, E., Liu, Q.Y., and Meissner, G., 1988, Nature, 331, 315-319.

Lai, F.A., and Meissner, G., 1989, J. Bioenerg. Biomembs., 21, 227-246.

Lai, F.A., Smith, H.A., and Meissner, G., 1989a, Biophys. J., 55, 207a.

Lai, F.A., Missa, M., Xu, L., Smith, H.A., and Meissner, G., 1989b, J. Biol. Chem., in press.

Liu, Q.Y., Lai, F.A., Rousseau, E., Jones, R.V., and Meissner, G., 1989, Biophys. J., 55, 415-424.

Meissner, G., 1984, J. Biol. Chem., 259, 2365-2374.

Meissner, G., 1986a, Biochemistry, 25, 244-251.

Meissner, G., 1986b, J. Biol. Chem., 260, 6300-6306.

Meissner, G., Darling, E., and Eveleth, J., 1986, Biochemistry 25, 236-244.

Nagasaki, K. and Kasai, M., 1983, J. Biochem. (Tokyo), 94, 1101-1109.

Rousseau, E., LaDine, J.K., Liu, Q.Y., and Meissner, G., 1988, Arch. Biochem. Biophys., 267, 75-86.

Rousseau, E., Smith, J.S., and Meissner, G., 1987, Am. J. Physiol., 253, C364-C368.

Saito, A., Seiler, S., Chu, A., and Fleischer, S., 1984, J. Cell Biol., 99, 875-885.

Smith, J.S., Coronado, R., and Meissner, G., 1985, Nature, 316, 446-449.

Smith, J.S., Coronado, R., and Meissner, G., 1986, J. Gen. Physiol., 88, 573-588.

Smith, J.S., Rousseau, E., and Meissner, G., 1989, Circ. Res., 64, 352-359.

Somlyo, A.P., 1985, Nature, 316, 298-299.

Somlyo, A.V., 1979, J. Cell Biol., 80, 743-750.

STRUCTURAL AND FUNCTIONAL INVESTIGATION OF p68 - A PROTEIN

OF THE LIPOCORTIN/CALPACTIN FAMILY

Stephen E. Moss and Michael J. Crumpton

Cell Surface Biochemistry Laboratory, Imperial Cancer Research Fund, Lincoln's Inn Fields, London WC2A 3PX, UK

INTRODUCTION

The recent discovery of a family of structurally related, calcium and phospholipid binding proteins has stimulated considerable interest in many laboratories, and invited wide-ranging hypotheses regarding their physiological functions (Crompton et al., 1988b, Klee 1988). The first proposals for the functions of two of these proteins were based on their apparent inducibility by glucocorticoids (as regulators of phospholipase A2 activity in inflammatory responses), and on their calcium-dependent association with actin (in bundling of actin filaments). The two proteins were thus named lipocortin and calpactin respectively, and these names have been adopted generically to describe the family. Six unique proteins have now been cloned and sequenced, and can be firmly placed in the lipocortin/calpactin family (Pepinsky et al., 1988). The elucidation of the structures of the individual members of the family, has come from groups with diverse interests, and this is reflected in the diversity of the proteins nomenclature (see Table 1). The five which fall into the 30 - 40kD molecular weight range are protein II, lipocortin I (p35), calpactin I (p36), lipocortin III and IBC (the inhibitor of blood coagulation). The sixth and largest member of the family, which will provide the focus of this manuscript, is p68 (67k-calelectrin).

Studies in this laboratory suggested that the lymphocyte plasma-membrane fraction included a group of related polypeptides (Mr 32,000, 36,000 and 68,000), in that these polypeptides exhibited a reproducible calcium-dependent association with the detergent-insoluble residue of B-lymphoblastoid plasma membrane (Davies et al., 1984). The largest of these proteins, namely p68, appeared as a closely-spaced polypeptide doublet when analysed by polyacrylamide gel electrophoresis in the presence of sodium dodecyl sulphate (SDS-PAGE), followed by Coomassie blue staining.

To investigate the structure of p68, and thereby gain possible insights into both it's function and relationships with other proteins, both human and murine p68 were cloned and sequenced (Crompton et al., 1988a; Moss et al., 1988).

CLONING AND SEQUENCING OF p68

The cloning of human p68 was achieved by the isolation of cDNA clones from a lambda-phage expression library of the leukaemia cell-line J6, using an antiserum to denatured purified p68. Murine p68 cDNA clones were subsequently isolated from plasmid cDNA libraries by cross-species hybridisation at low stringency with human p68 probes. The two nucleotide sequences were remarkable in their degree of homology, being 89% homologous in the coding region and 50% in the 3' and 5' non-coding regions. Both cDNA clones were approximately 2,500 nucleotides in length, comprising 5'-untranslated regions of 100 and 30 nucleotides in human and mouse respectively, an open reading frame of 2,019 nucleotides encoding a poly-peptide of 673 amino-acids, and a 3' untranslated region of approximately 400 nucleotides, terminating in each case in a string of adenosine residues. Close inspection of the cDNA sequences revealed several in

Table 1. The Lipocortin/Calpactin family

M Wt by SDS-PAGE	Synonyms
68,000	p68, 67k-calelectrin, lipocortin VI, protein III, chromobindin-20
38,000	lipocortin I, calpactin II, p35, chromobindin-9
36,000	calpactin I, lipocortin II, p36, protein I, chromobindin-8
34,000	inhibitor of blood coagulation (IBC), placental anticoagulant protein (PAP) vascular anticoagulant (VAC), anchorin CII, p34, lipocortin V, endonexin II, chromobindin-5
33,000	lipocortin III, p33
32,000	32k-calelectrin, protein II, endonexin, lipocortin IV, chromobindin-4
GENERIC NAMES	lipocortins, calpactins, endonexins, calelectrins, chromobindins, annexins etc.

teresting features. At the far 5' end, the untranslated regions (which in the human sequence extended further 5' than in the mouse) were notable in respect of their high GC content, and in particular the frequent incidence of the dinucleotide CpG. It is believed that cytosine residues which occur in this dinucleotide, are particularly susceptible to conversion to thymidine residues by methylation and deamination, and that this has led through evolution to a depletion of CpG dinucleotides throughout mammalian genomes (Bird 1986). However, islands rich in CpG dinucleotides have been located at the 5' ends of many genes which are frequently widely expressed, and which are, thus, often referred to as housekeeping genes. Although CpG islands are usually restricted to the far 5' intron, they occasionally extend into the first exon and may consequently appear during sequence analysis of cDNA clones. The presence of 16 CpG dinucleotides within the first 100 nucleotides of the human p68 cDNA sequence therefore identifies p68 as a probable housekeeping gene. Interestingly, none of the other members of the lipocortin/calpactin family exhibit this characteristic in their cDNA sequences, although full analysis of genomic clones for these proteins will be necessary to determine whether p68 is unique in this respect.

Inspection of the human and murine cDNA sequences determined in this laboratory, revealed no recognisable polyadenylation signal (AATAAA) or any of the known functional variants, immediately prior to the far 3' end of either non-coding region. This is in contrast to the findings of Sudhof et al., (1988) who reported an AATAAA polyadenylation signal in the sequence of human p68 from a retinal cDNA library. Since the size of p68 messenger RNA as judged by Northern blotting is somewhat larger than that predicted by cDNA cloning and sequencing in these studies, the most likely explanation for the lack of a polyadenylation signal, is that the string of adenosine residues identified in both species is actually a conserved midstream sequence of sufficient length to allow priming during construction of the libraries, suggesting that the genuine 3' end may lie further downstream. A second explanation is that polyadenylation in p68 is signalled by a hitherto unrecognised sequence. This would appear unlikely since there is no conserved hexa- or penta-nucleotide sequence identical in both species in the appropriate region, meaning that not one, but two novel polyadenylation signals would be required to substantiate this proposal. A third explanation is that cloning artefacts arising during the construction of the human and murine cDNA libraries employed in these studies, resulted in the deletion of nucleotides found in the polyadenylation signals.

The most interesting feature identified in the p68 open reading frame was the presence of an apparent alternate splice sequence, observed initially during the sequencing of non-identical human p68 cDNA clones. A sequence of 18 nucleotides encoding the hexapeptide Val-Ala-Ala-Glu-Ile-Leu (VAAEIL) was present in some but not all clones. Although the possibility exists that this sequence may have arisen artefactually during the construction of the library, three compelling points argue strongly against this. Firstly, the extra sequence occurs within recognised splice donor and acceptor sites and does not interrupt the reading frame. Secondly, and probably beyond coincidence, an identical 18 nucleotide sequence was found to be either absent or present (at the same position in the sequence) in non-identical murine cDNA clones. Thirdly, the identification of two p68 cDNAs differing by 18 nucleotides, implies two mRNAs (probably arising from a single primary transcript), and therefore two polypeptides with a molecular weight difference of approximately 800 Daltons, which would provide a convenient and plausible explanation for the presence of p68 as a closely spaced polypeptide doublet on SDS-PAGE. Furthermore, by using a fortuitously placed Bgl II restriction site within the splice sequence, it was possible to rapidly and easily assess the relative abundance of each form in a series of non-identical murine cDNA clones. Clones containing the splice sequence out-numbered those without by 5:1, which if reflected in relative amounts of protein would suggest a preponderance of the slightly larger of the two p68 polypeptides. This is supported by numerous studies on various human and murine cell types, in which the upper band of the p68 doublet is consistently heavier than the lower band when examined by either Western blotting or immunoprecipitation. The essentially hydrophobic character of the splice sequence, suggests that the larger of the two p68 species may possess an additional or alternative membrane-binding capability. In future studies it will clearly be important to investigate the possible functional significance of alternate splicing of p68, and whether this is associated with any degree of tissue specificity.

THE p68 AMINO-ACID SEQUENCE

The human and murine p68 cDNA sequences both encode proteins of 673 or 667 amino-acids (depending on the presence or absence of the splice sequence) with a predicted molecular weight of approximately 75,000 Daltons. This is somewhat larger than the size of the protein when estimated by SDS-PAGE (approximately 70kDa), but such discrepancies are not unusual and can generally be explained by protein folding. It is possible that post-translational modifications such as limited proteolytic cleavage could give rise to a smaller mature protein (in this respect, it is known that the N-terminal sequence of lipocortin I is proteolytically labile), although the integrity of the p68 N-terminus was demonstrated immunologically using antisera raised against a synthetic peptide corresponding to this region. The most remarkable feature of the derived human and murine amino-acid sequences was the high degree of sequence identity, at 95% overall with only 35 mismatches, of which 12 were of a conservative nature such that the physicochemical characteristics of the amino-acid side chains were retained. Apart from interspecies homology, p68 also exhibited a highly organised internal structural homology (Fig 1). A conserved sequence of approximately 60 residues was found to be repeated eight times, each repeat being separated by variable N-terminal sequences of 7 - 43 amino-acids. The longest of the variable regions lies between repeats 4 and 5, and essentially divides the protein into two halves which share 45% sequence identity. The observation that each half of p68 bears a close structural resemblance to the five other members of the lipocortin/calpactin family, and the presence within each repeated domain of a highly conserved 17 amino-acid sequence common to all these proteins (Geisow et al., 1986), confirms p68 as a member of this family. Alignment of the individual repeats with each other revealed that homology is highest between repeats 1 and 5, 2 and 6 etc, reinforcing the concept of p68 as a protein of two halves. This suggests that p68 may have arisen by the duplication of a gene encoding a four-repeat protein, which in turn may have arisen by two consecutive duplications of a DNA sequence encoding a protein resembling a single repeat (Crompton et al., 1988b). Although there is scant evidence for the existence of proteins comprising one or two repeats, there is one potential candidate for a single repeat protein, namely uteroglobin. Despite the generally poor homology between uteroglobin and members of the lipocortin/calpactin family, uteroglobin contains a nonapeptide sequence closely related to a sequence in lipocortin I, which may be responsible for the ability of both proteins to inhibit phospholipase A2 (Miele et al., 1988).

Inspection of the p68 amino-acid sequence revealed several interesting features. The alternate splice sequence (VAAEIL), was found to occur close to the start of the seventh repeat (see Fig 1). Alternate splicing has not been reported for any other members of the lipocortin/calpactin family, suggesting that it may have arisen following the hypothetical gene duplication event which created the p68 gene. It is unclear whether alternate splicing of p68 is functionally important, or if it occurs as a simple stochastic event. An antiserum raised against a synthetic peptide corresponding to this region, which recognises only the larger of the two p68 bands on Western blotting and immunoprecipitation, will clearly be a valuable tool for pursuing studies on the possible significance of p68 splicing.

Also contained within both the human and murine p68 sequences are two conserved potential N-glycosylation sites in the third and fourth repeats, although there is no evidence to suggest that p68 is normally glycosylated. Like all the members of the lipocortin/calpactin family, p68 lacks a characteristic

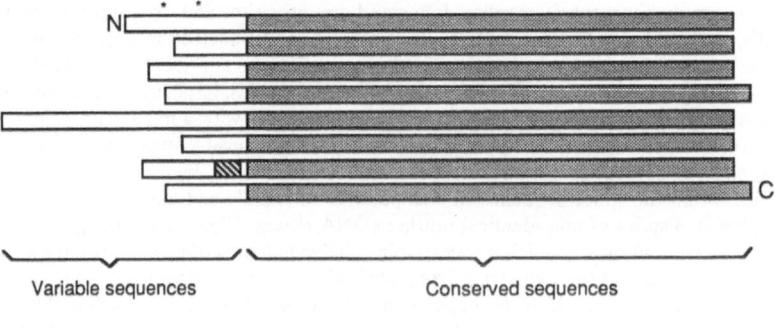

Variable sequences   Conserved sequences

▨ Alternate splice sequence (VAAEIL)
* Possible N-terminal tyrosine and serine phosphorylation sites

Fig. 1 . Structure of p68 showing internally repeated sequences.

N-terminal signal sequence, which is generally considered to be an absolute structural requirement for the navigation of nascent proteins through the Golgi apparatus, the site of N-glycosylation. The apparent non-glycosylation of p68 is consistent with it's almost exclusively intracellular localisation to the inner, or cytoplasmic face of the plasma-membrane. With the single exception of peritoneal fluid, neither p68 nor any of the related proteins have been described in an extracellular compartment. Indeed, the observation that the proteins of the lipocortin/ calpactin family are non-glycosylated, argues against their proposed extracellular roles either as inhibitors of phospholipase A2 in inflammatory responses, or as anticoagulants.

Studies aimed at understanding the physiological roles of the lipocortin/calpactin family proteins, have been largely based on in vitro experimentation, and are therefore subject to the caveats imposed by that approach. Thus, the proposed inhibition of phospholipase A2, the anticoagulant activity and the bundling of actin which are all readily demonstrated with many of these proteins in vitro, have yet to be shown as genuine physiological phenomena. One notable exception has been the observation that lipocortin I (p35) and calpactin I (p36) are both targets for phosphorylation in intact cells, by the epidermal growth factor receptor and pp60[v-src] tyrosine kinase activities respectively. The further discovery that phosphorylation of these proteins altered their calcium requirements for phospholipid binding, added weight to the potential functional significance of this event. Interestingly, site mapping of the phosphorylated tyrosine residues in p35 and p36, revealed them to be in the N-terminal regions of both proteins. Furthermore, serine residues located in the N-termini, were also shown to be targets for phosphorylation in vivo, in this case by protein kinase C. Therefore, the presence of tyrosine and serine residues in the N-terminus of p68 at positions 10 and 13 respectively (see Fig 1), clearly raised the possibility that p68 may also be a substrate for phosphorylation in whole cells.

To address this question, a series of experiments were performed in which p68 was immunoprecipitated from several different human and murine cell types grown in the presence of ^{32}P, and stimulated with a variety of mitogens and growth factors. Under standard labelling conditions, and irrespective of the state of cell growth and stimulation, it was not possible to detect p68 phosphorylation. There were two obvious explanations for this. Firstly, and despite the homology with p35 and p36, p68 may not necessarily be a target for phosphorylation in vivo. Secondly, if phosphorylation of p68 does occur in vivo, labile phosphorylated residues may become dephosphorylated immediately upon cell lysis, or alternatively, the phosphorylation may be turned over very rapidly. In investigating the second possibility, it was found that with judicious use of particular phosphatase inhibitors during cell culture, phosphorylated p68 could be readily immunoprecipitated from Swiss 3T3 cell lysates, independently of exogenously added growth factors, mitogens or other stimuli. Phosphoaminoacid analysis revealed that under these conditions, phosphorylation occurred at least partly on serine. Significantly however, addition of epidermal growth factor resulted in an increase in the overall level of p68 phosphorylation, providing the cells were grown in the presence of phosphatase inhibitors. At present, it is unclear whether any significance can be attached to the phosphorylation of p68 in vivo, but by analogy with p35 and p36, one possibility is that phosphorylation may change the calcium requirement for phospholipid binding. In physiological terms, this suggests that the Ca^{2+}-dependent association of p68 with cellular membranes may be regulated at least in part by phosphorylation.

In conclusion, we have demonstrated that the p68 Ca^{2+} and phospholipid binding protein, is a member of the lipocortin/calpactin family. Like the other members of the family, p68 has a highly ordered internal structure, comprising an eight-fold repeat of the conserved sequence which occurs four times in each of the other family members. The protein was found to be especially highly conserved between human and mouse, suggesting that functional constraints may have restricted the 'sequence drift' which normally occurs with evolution. Furthermore, an apparent splice sequence encoding six amino-acids which was identified in both human and murine p68 cDNAs, provides a possible explanation for the presence of p68 as a closely-spaced polypeptide doublet on SDS-PAGE. Finally, p68 was found to be a substrate for protein kinases in intact cells, and further studies will be aimed at determining whether the phosphorylation of p68 is functionally significant.

REFERENCES

Bird, A.P., 1988, CpG-rich islands and the function of DNA methylation, Nature, 321:209.

Crompton, M.R., Owens, R.J., Totty, N.F., Moss, S.E., Waterfield, M.D., and Crumpton, M.J., 1988a, Primary structure of the human membrane-associated Ca^{2+}-binding protein p68: a novel member of a protein family, EMBO J., 7:21.

Crompton, M.R., Moss, S.E., and Crumpton, M.J., 1988b, Diversity in the lipocortin/calpactin family, Cell, 55:1.

Davies, A.A., Wigglesworth, N.M., Allan, D., Owens, R.J., and Crumpton, M.J., 1984, Nonidet P-40 extraction of lymphocyte plasma-membranes. Characterisation of the insoluble residue, Biochem. J., 219:301.

Geisow, M.J., Fritsche, U., Hexham, J.M., Dash, B., and Johnson, T., 1986, A consensus amino acid sequence repeat in Torpedo and mammalian Ca^{2+}-dependent membrane-binding proteins, Nature, 320:636.

Klee, C.B., 1988, Ca^{2+}-dependent phospholipid- (and membrane-) binding proteins, Biochemistry, 27:6645.

Miele, L., Cordella-Miele, E., Facchiano, A., and Mukherjee, A.B., 1988, Novel anti-inflammatory peptides from the region of highest similarity between uteroglobin and lipocortin I, Nature, 335:726.

Moss, S.E., Crompton, M.R., and Crumpton, M.J. 1988, Molecular cloning of murine p68, a Ca^{2+}-binding protein of the lipocortin family, Eur. J. Biochem., 177:21.

Pepinsky, R.B., Tizard, R., Mattaliano, R.J., Sinclair, L.K., Miller, G.T., Browning, J.L., Chow, E.P., Burne, C., Huang, K-S., Pratt, D., Wachter, L., Hession, C., Frey, A.Z., and Wallner, B.P., 1988, Five distinct calcium and phospholipid binding proteins share homology with lipocortin I, J. Biol. Chem., 263:10799.

Sudhof, T.C., Slaughter, C.A., Leznicki, I., Barjon, P., and Reynolds, G.A., 1988, Human 67kDa-calelectrin contains a duplication of four repeats found in 35-kDa lipocortins, Proc. Natl. Acad. Sci. USA, 85:664.

STRUCTURE-FUNCTION RELATIONS IN TROPONIN C. CHEMICAL

MODIFICATION STUDIES

Zenon Grabarek, Yasuko Mabuchi and John Gergely

Department of Muscle Research, Boston Biomedical Research Institute; Department of Biological Chemistry and Molecular Pharmacology, Harvard Medical School; Neurology Service, Massachusetts General Hospital, Boston MA, USA

Troponin C (TnC) is the calcium binding component of thin filaments in skeletal and cardiac muscles. Calcium induced changes in the conformation of TnC transmitted to other components of the thin filament result in triggering of muscle contraction (for review see Leavis and Gergely 1984; Zot and Potter 1987). The 3-D crystal structure of TnC isolated from turkey and chicken skeletal muscle has been solved and refined recently at better than 0.2 nm resolution (Satyshur et al 1988; Herzberg and James 1988). The molecule appears to have two globular domains each containing a pair of Ca^{2+}-binding sites. Each site consists of a Ca^{2+}-binding loop flanked by two -helices, which are labeled A through H starting with the N-terminus. Sites I and II in the N-terminal domain are the "triggering ", low affinity, Ca^{2+}-specific sites, and sites III and IV in the C-terminal domain are the high affinity Ca^{2+}-Mg^{2+} sites believed to be always occupied by a divalent metal in the living muscle. The two domains are linked by a 29-residue -helix (D+E or central helix) a considerable portion of which is exposed to the solvent. Since under the conditions of crystallization only sites III and IV are filled with Ca^{2+} the crystal structure represents, at best, the "inhibited" form of TnC. Herzberg et al. (1986) speculated that the conformational changes associated with Ca^{2+}-binding to sites I and II involve a change in the relative disposition of helices in the N-terminal domain leading to a structure similar to that found at the Ca^{2+}-filled sites III and IV in the C-terminal domain of TnC and in other Ca^{2+}-binding proteins of known 3-D structure. Such transition would result in separation of helices B and C from the central helix. The Ca^{2+} induced increase in accessibility of Cys-84 in cardiac TnC (Ingraham and Hodges, 1988, Fuchs and Grabarek 1989) seems to be consistent with such a model .

The unusual feature of TnC, with the long central helix and the resulting elongated shape of the molecule, requires consideration. Sundaralingam et al. (1985) suggested that possible salt bridges along the central helix between negatively charged side chains of Glu and positively charged Lys and Arg side chains may stabilize the central helix. Recent work utilizing fluorescence energy transfer (Wang et al. 1987; Cheung et al. 1989) and low angle X-ray scattering (Heidorn and Trewhella 1988) suggest that the two domains are closer to each other than expected from the crystal structure and cast serious doubt on the identity of the solution structure with the crystal structure. It is not clear what effects the shape of TnC and the stability of the central helix would have on the interactions with other troponin components TnI and TnT.

Thus, in spite of a considerable amount of information about TnC the molecular mechanisms involved in the generation and transmission of the Ca^{2+}-induced regulatory signal within TnC and to the other components of the thin filament are not fully understood. Since various pieces of evidence suggest that charged side chains contribute significantly to the protein-protein interactions in troponin we have begun studies of the effects of chemical modification of selected charged side chains of TnC.

ACETYLATION OF TNC AT LYSINE RESIDUES

There are nine lysine residues in rabbit skeletal TnC, three of which, at positions 84, 88 and 91, are located in the middle of the central helix in the part presumably exposed to the solvent. Thus by

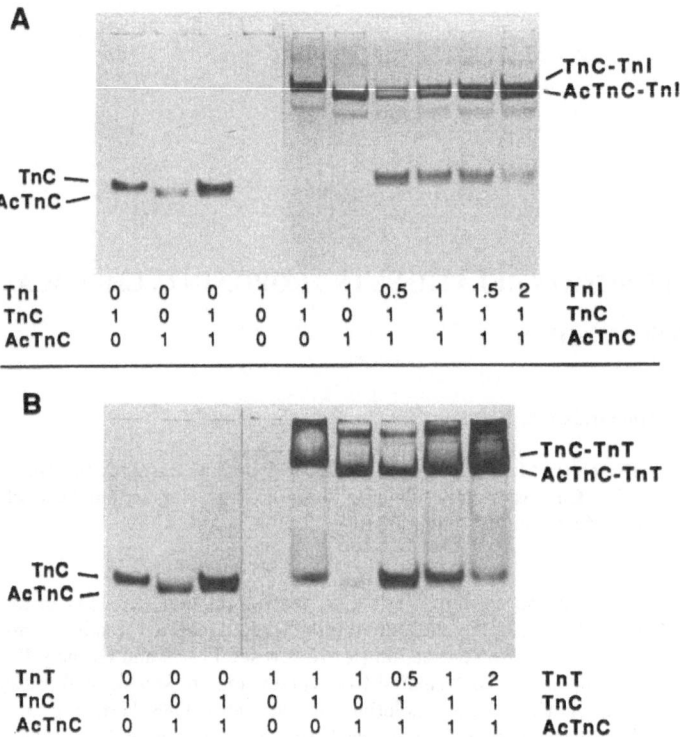

Fig. 1 . Acetylation does not change TnC's affinity for TnI but increases its affinity for TnT.
TnC fully acetylated with acetic anhydride according to Fraenkel-Conrat (1957), was mixed with unmodified TnC in a 1:1 molar ratio and various amounts of TnI (A) or TnT (B) were added as indicated. Samples were run on 10 % polyacrylamide gels in tris-glycine buffer, pH 8.6, containing 1mM $CaCl_2$. Note that under these conditions TnI and TnT do not enter the gel unless they are bound to TnC or AcTnC.

modification of these residues it should be possible to alter the charge-charge interactions in the central helix. Treatment of TnC with acetic anhydride results in acetylation of all lysine residues in TnC as judged from its inability to react with fluorescamine. Far UV circular dichroism spectra indicate that acetylation does not affect the secondary structure of TnC in the absence of Ca^{2+}, but in the presence of Ca^{2+} there is a slightly higher content of α-helical structure; the difference corresponds to approximately two turns of an α-helix. Inspection of the amino acid sequence indicates that there should be repulsion between the side chains of Lys-84 and Lys-88 if this segment has an α-helical conformation. Acetylation would decrease such repulsion by neutralization of the charges and thus increase stability in the region. While this interpretation still requires experimental verification it is interesting to note that the peptide bonds at Lys 84 and lys 88 become more susceptible to proteolysis with trypsin in the presence of Ca^{2+} which is consistent with local unfolding of this segment of the molecule (Drabikowski et al. 1977, Grabarek et al. 1981).

We have further studied the effect of acetylation on the ability of TnC to interact with the other troponin subunits TnI and TnT. Complex formation between subunits is indicated in Fig. 1 by formation of a band of intermediate mobility on polyacrylamide gel under nondissociating conditions (no SDS, no urea). A substantial increase in the net negative charge of TnC upon acetylation is reflected by the increased mobility on the gel. Similarly, the complexes of acetylated TnC (AcTnC) with TnI and TnT have higher mobility than the corresponding complexes of unmodified TnC. Fig 1 shows a titration of a 1:1 mixture of AcTnC and TnC with TnI or TnT. As judged from the essentially identical intensities of the bands corresponding to the AcTnC-TnI and TnC-TnI complexes acetylation does not affect the ability of TnC to interact with TnI. On the other hand the binding of AcTnC to TnT appears to be stronger than that of native TnC. Finally, it appears that the ability of AcTnC to activate the ATPase activity of the reconstituted acto-S1 complex is substantially higher than that of unmodified TnC.

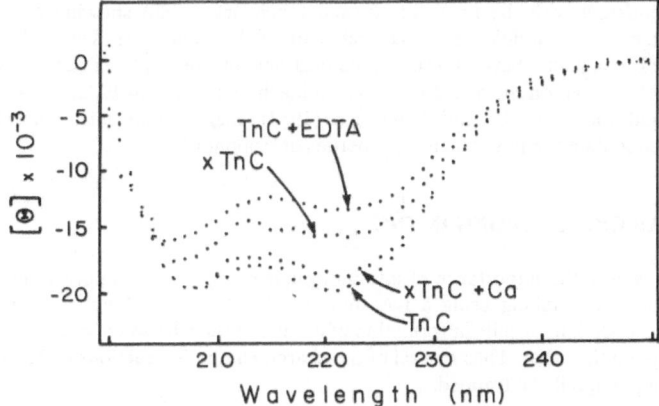

Fig. 2. Intramolecular crosslinking of TnC reduces the Ca^{2+}-sensitivity of its secondary structure.
For intramolecular crosslinking a sample of TnC was incubated with 5 mM EDC for 60 min, 25°C in a solution containing 0.1 M KCl, 50 mM 2(N-morpholino)ethanesulfonic acid (MES) pH 6.0 and 5 mM NHS. After addition of 20 mM 2-mercaptoethanol the sample was further incubated for 2 h, 25°C. Circular dichroism spectra of TnC or intramolecularly crosslinked TnC (xTnC) in 0.1 M KCl, 5 mM Hepes pH 7.5 were recorded in the presence of 0.2 mM EDTA or 0.5 mM $CaCl_2$.

It is remarkable that neutralization of 9 out of 16 positively charged groups had very little effect on the properties of TnC and, in fact, it enhanced its structure and the biological activity in vitro. In this respect

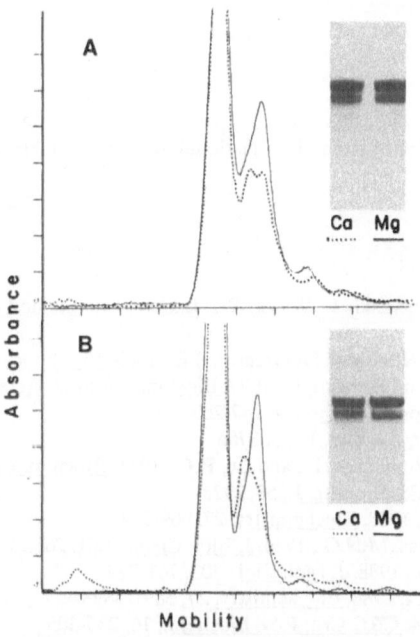

Fig. 3. Ca^{2+} or Mg^{2+} specific conformers of TnC are trapped upon intramolecular crosslinking.
Samples of TnC were activated with 5 mM EDC for 5 min 25°C in a solution containing 0.1 M KCl, 50 mM MES and 5 mM NHS. The activation step was terminated by addition of 20 mM 2-mercaptoethanol and the mixture further incubated for 2 h at 25°C. (A) - Divalent metals (0.2 mM $CaCl_2$ or 5 mM $MgCl_2$) were present only during the crosslinking step. (B) - Divalent metals were added before activation. Electrophoresis was carried out on 15% polyacrylamide gels (tris-gly buffer pH 8.6, 5 M urea, 1 mM EDTA); photographs of Coomassie blue - stained gels and their densitometric scans are shown.

these results are consistent with the result of Xu and Hitchcock (1988) showing that the removal of a tripeptide Lys-Gly-Lys in the middle of the central helix slightly enhances TnC's ability to activate the reconstituted acto-S1 complex. Several other modifications of the TnC sequence using recombinant technology had no effect at all on the activity of TnC (Reinach and Carlsson 1988; Dobrowolski et al. 1989). Both our results and those of Xu and Hitchcock (1988) suggest that intramolecular charge-charge interactions in TnC modulate the physiological properties of troponin C.

INTRAMOLECULAR CROSSLINKING IN TNC

To further explore the importance of intramolecular charge-charge interactions we have subjected TnC to intramolecular crosslinking using a two-step zero-length crosslinking procedure (Grabarek and Gergely 1988). The crosslinking results in formation of covalent bonds between side chains of Asp or Glu on the one hand and Lys on the other. Thus several of the charge-charge interactions in TnC, including those in the central helix, can potentially be trapped.

Treatment of TnC with 1-ethyl-3-[3-(dimethylamino) propyl]-carbodiimide (EDC) in the presence of N-hydroxysuccinimide (NHS) results in formation of intramolecular crosslinks in TnC. Crosslinking decreases the Ca^{2+}-sensitivity of the secondary structure of TnC by increasing the -helix content in the absence of Ca^{2+} and decreasing it in the presence of Ca^{2+} (Fig. 2). Polyacrylamide gel electrophoresis performed in the presence of urea and EDTA shows formation of several crosslinked species having higher mobility on polyacrylamide gel as compared with unmodified TnC. The use of the two-step procedure enabled us to study the effects of conformation on crosslinking by adding Ca^{2+} or Mg^{2+} after the activation step (Fig. 3). Different crosslinks are formed depending on whether Ca^{2+} or Mg^{2+} was present during the crosslinking step; Mg^{2+} favours the formation of the faster moving species. The difference between Ca^{2+} and Mg^{2+} with respect to their influence on crosslink formation is probably due to the differential effect on various potential crosslinking sites. We expect that the analysis of the crosslinked products will yield information about the difference in conformation between the inhibited $(+Mg^{2+})$ and the activated $(+Ca^{2+})$ forms of TnC.

ACKNOWLEDGEMENTS

This work was supported by grants from the National Institutes of Health (R-37-HL-05949) and the Muscular Dystrophy Association.

REFERENCES

Cheung, H.C., Wang, C.K., Gryczynski, I., Wiczk, W., Laczko, G. Steiner, R., and Lakowicz, J.R., 1989, Biophysical J., 55, 122a

Dobrowolski, Z., Xu, G.-Q., and Hitchcock DeGregori, S.E., 1989, Biophysical J., 55, 274a

Drabikowski, W., Grabarek, Z., and Barylko, B., 1977, Biochim. Biophys. Acta, 490, 216-224

Fraenkel-Conrat, H., 1957, Methods Enzymol., 4, 247-269

Fuchs, F., and Grabarek, Z., 1989, Biophys. J., 55, 275a

Grabarek, Z., Drabikowski, W. Vinokurov, L., and Lu, R.C., 1981, Biochim. Biophys. Acta, 671, 227-233

Grabarek, Z., and Gergely, J., 1988, Biophys. J., 53, 392a

Heidorn, D.B., and Trewhella, J., 1988, Biochemistry, 27, 909-914

Herzberg, O., Moult, J., and James, M.N.G., 1986, J. Biol. Chem., 261, 2638-2634

Herzberg, O., and James, M.N.G., 1988, J. Mol. Biol., 203, 761-779

Ingraham, R.H., and Hodges, R.S., 1988, Biochemistry, 27, 5891-5898

Leavis, P.C., and Gergely, J., 1984, CRC Crit. Rev. Biochem., 16, 235-305

Reinach, F.C., and Carlsson, R., 1988, J. Biol. Chem., 263, 2371-2376

Satyshur, K.A., Rao, S.T., Pyzalska, D, Drendel, W., Greaser, M., and Sundaralingam, M., 1988, J. Biol. Chem., 263, 1628-1647

Sundaralingam, M., Drendel, W., and Greaser, M., 1985, Proc. Natl. Acad. Sci. USA 82, 7944-7947

Wang, C.-L. A., Grabarek, Z., Tao, T., and Gergely, J., 1987, in: Calcium Binding Proteins in Health and Disease pp 440-442. A.W. Norman, T.C. Vanaman, A.R. Means, eds., Academic Press Inc., San Diego

Xu, G.-Q., and Hitchcock DeGregori, S.E., 1988, J. Biol. Chem., 263, 13962-13969

Zot, A.S., and Potter, J.D., 1987, Ann Rev. Bophys. Biophys. Chem., 16, 536-559

Ca^{2+}-DEPENDENT MOBILITY SHIFT OF PARVALBUMIN IN ONE-

AND TWO-DIMENSIONAL GEL-ELECTROPHORESIS

H.- J. Gregersen,[1] C. W. Heizmann,[2] U. Kaegi,[2] and M. R. Celio[1]

[1]Institut für Anatomie, Christian Albrechts Universität zu Kiel, Olshausenstraße 40, D-2300 Kiel, West-Germany

[2]Universitäts-Kinderklinik, Abtlg für Klin. Chemie, Sternwiesstraße 75 CH-8032 Zürich, Switzerland

ABSTRACT

Under Ca^{2+}-loaded conditions parvalbumin migrates in one- and two-dimensional gel-systems as a double-band or -spot whereas in Ca^{2+}-free condition it appears as one band or spot.

Parvalbumin (PV), a member of the family of calcium-binding proteins [1], was first described in 1934 [2] and occurs in fast-contracting muscles and in subpopulations of neurons in vertebrates and humans [3,4,5]. The physical characteristics of molecular weight (Mv 12 KD), isoelectric point (pI 4.9) and Ca^{2+}-binding properties are established (PV binds 2 Ca^{2+} per molecule) [5,6]. Physiological roles discussed for PV range from trigger- to buffer-protein of intracellular Ca^{2+}-ions [7]. In this paper we report an as yet not described mobility shift of PV in gel-electrophoresis after manipulation of the Ca^{2+} concentration, which may have implications for its physiological function.

MATERIALS AND METHODS

Rat parvalbumin was purified from the musculus extensor digitorum longus by high-performance liquid chromatography as described [8]. One-dimensional gel-electrophoresis was performed according to Laemmli [9] and two-dimensional gel-electrophoresis was performed following the procedure of O'Farrel [10] with minor modifications. Proteins were visualized with Coomassie-blue. To obtain Ca^{2+}-saturation, 0.1 mM CaCl$_2$ was added to proteins, all buffers and solutions used for electrophoresis. For Ca^{2+}-free condition 1 mM EGTA (ethylene-glycol-bis-(ß-aminoethylether)-N,N'-tetra-acetic-acid) were added. After electrophoresis the proteins were blotted onto nitrocellulose and incubated with the polyclonal anti-PV-serum No. 187i-MOI directed against rat muscle parvalbumin and produced in rabbit. As second antibody we applied horseradish peroxidase-conjugated goat anti-rabbit IgG (Bio-Yeda, Rehovot). The peroxidase was developed with 4-chloro-1-naphthol and H$_2$O$_2$.

RESULTS

Under Ca^{2+}-loaded conditions parvalbumin can be detected as a double-band (Fig. 1b) in a one-dimensional gel and in the corresponding immunoblot (Fig. 2b). The apparent molecular weight is 12 KD for the major band and 14 KD for the minor band. In two-dimensional gels under the same conditions PV appears as a double-spot (Fig. 3b). In contrast, under Ca^{2+}-free conditions PV migrates as one band (Fig. 1a and 2a) or one spot (Fig. 3a) with an apparent molecular weight of approximately 14.5 KD. Under both conditions no shift of pI can be observed. Calmodulin, another CaBP, shows similar mobility-shift [11] (Fig. 1), but no splitting of the protein-band is visible under both conditions. Calbindin D-28k, the vitamin D-dependent CaBP, does not alter its mobility in the presence or absence of Ca^{2+} (Fig. 1). So three structurally homologous proteins behave differently under these conditions.

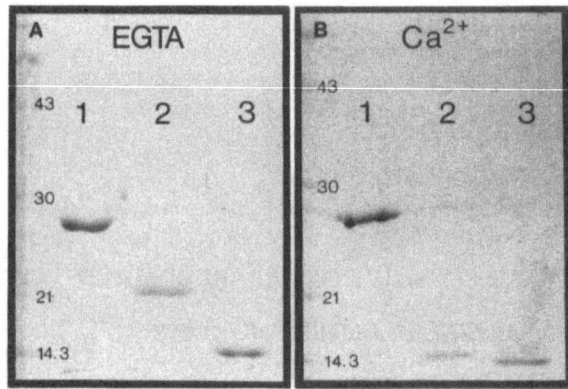

Fig. 1 . One-dimensional gels with identical probes under Ca^{2+}-free (panel A) and Ca^{2+}-loaded (panel B) conditions: Lane 1 - 1.43 µg recombinant rat brain Calbindin D-28k, Lane 2 - 1.34 µg rat Calmodulin, Lane 3 - 1.5 µg rat muscle Parvalbumin; numbers at the left side of both panels gives the mobility of molecular weight markers in Kd.

DISCUSSION

PV appears as a doublet of Mr 12 kD and 14 kD under Ca^{2+}-loaded conditions in contrast to a single band or spot of 14.5 kD under Ca^{2+}-free conditions. This electrophoretic analysis implicates that PV shows Ca^{2+}-dependent conformational changes under the described conditions. Similar observations are reported for Calmodulin whose two bands on SDS-gels thought to result from different shapes of the molecule in Ca^{2+}-loaded / or Ca^{2+}-free form [11].

The calcium-binding sites of intestinal CaBP D-9K, another vitamin D-dependent CaBP, showing some similarities with PV, are motile parts of the molecule [12] also leading to Ca^{2+}-occupation dependent conformational changes. Therefore, binding of two Ca^{2+}-ions to PV probably leads to a more tightly compact form with higher electrophoretic mobility as suggested by the major band at lower MW under Ca^{2+}-loaded conditions.

Each mole of PV is capable of binding 2 moles of Ca^{2+} with high affinity {$K_{Diss} = 10^{-6}$ M}. Consequently, 0.1 mM Ca^{2+} should have been sufficient to result in the occupation of the two calcium-binding sites on the molecule. However, during the electrophoretic run Ca^{2+} precipitates in the running buffer [11]

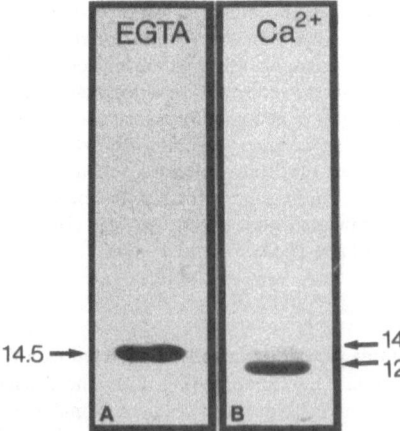

Fig. 2 . Immunoblots after one-dimensional gel-electrophoresis of 1.5 µg rat muscle parvalbumin corresponding to Fig 1 in Ca^{2+}-free (panel A) and Ca^{2+}-loaded (panel B) condition incubated with the polyclonal anti-PV-serum No. 187i-MOI. Notice the doublet in panel B consisting of a major band at MW 12 Kd and a minor at MW 14 Kd and the single band in panel A at approximately 14.5 Kd.

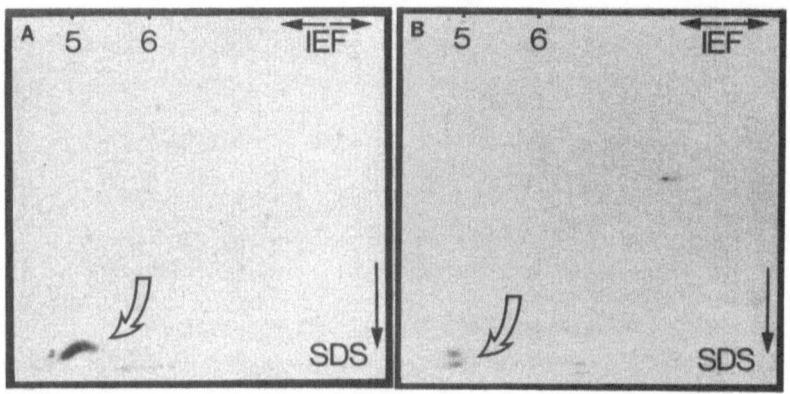

Fig. 3 . Immunoblots after two-dimensional gel-electrophoresis of 0.2 μg rat muscle parvalbumin in Ca^{2+}-free (panel A) and Ca^{2+}-loaded (panel B) conditions incubated with the polyclonal anti-PV serum No. 187i-MOI. Note the the double-spot under Ca^{2+}-loaded condition (right) in contrast to the single spot in Ca^{2+}-free condition (left). (a) EGTA; (b) Ca^{2+}

leading to an actual [Ca^{2+}] lower than 0.1 mM. As a result, not all PV-molecules present in the sample may have bound two Ca^{2+}. Thus, the lighter higher molecular mass band/spot (Fig. 2b and 3b) probably represents PV binding none or only one Ca^{2+}-ion. Under Ca^{2+}-free condition PV reveals a discrete higher apparent molecular weight accounted for a less compact and therefore less mobile molecule. Therefore three different Ca^{2+}-saturation stages of the PV molecule exist under these conditions: 1. Ca^{2+}-free, 2. one Ca^{2+}-bound and 3. two Ca^{2+}-bound. This may have implications for the physiological role of PV, if it can be demonstrated on native gels in the absence of SDS. Parvalbumin is assumed to be a buffer protein [13] which helps to decrease the fluctuations of intracellular calcium. Our results point to a possible role of PV as a "trigger" protein [14] which mediates Ca^{2+} effects on target proteins.

REFERENCES

1. C.W. Heizmann, and M.W. Berchtold, Expression of parvalbumin and other Ca^{2+}-binding proteins in normal and tumor cells: A topical review. Cell Calcium 8: 1-41 (1987).
2. H.J. Deuticke, Über die Sedimentationskonstante von Muskelproteinen. Hoppe-Seylers Zeitschrift für Physiol. Chem. 224: 216-228 (1934).
3. M.R. Celio, and C.W. Heizmann, Calcium-binding protein parvalbumin as a neuronal marker. Nature 293: 300-302 (1981).
4. M.R. Celio, and C.W. Heizmann, Calcium-binding protein parvalbumin is associated with fast contracting muscle fibres. Nature 297: 504-506 (1982).
5. M.W. Berchtold, M.R. Celio, and C.W. Heizmann, Parvalbumin in human brain. J. Neurochem. 45: 235-239 (1985).
6. H.E. Blum, P. Lehky, L. Kohler, E.A. Stein, and E.H. Fischer, Comparative properties of vertebrate parvalbumins. J. Biol.Chem. 252, 2834-2838 (1977).
7. C.W. Heizmann, Parvalbumin, an intracellular calcium-binding protein; distribution, properties and possible roles in mammalian cells. Experientia 40: 910-921 (1984).
8. M.W. Berchtold, C.W. Heizmann, and K.J. Wilson, Ca^{2+}-Binding Proteins : A Comparative Study of Their behavior during high-performance liquid chromatography using gradient elution on reverse-phase supports. Anal. Biochem. 129: 120-131 (1983).
9. U.K. Laemmli, Cleavage of structural proteins during the assembly of the head of bacteriophage T4. Nature 227: 680-685 (1970).
10. P.H. O'Farrel, High resolution two-dimensional electrophoresis of proteins. J. Biol. Chem. 250:(10), 4007-4021 (1975).
11. W.E. Heydorn, G.J. Creed, and D.M. Jacobowitz, Observations and implications on the migration of calmodulin in a two-dimensional gel system. Electrophoresis 8: 251-252 (1987).
12. D.M.E. Szebenyi, and K. Moffat, The refined structure of vitamin D-dependent Calcium-binding protein from bovine intestine. J. Biol. Chem. 261:(19) 8761-8777 (1986).
13. R.J. Miller, Calcium signalling in neurons. TINS 11: (10) 415-419 (1988).
14. M.R. Celio, Parvalbumin in most gamma-aminobutyric acid-containing neurons of the rat cerebral cortex. Science 231: 995-997 (1986).

ROLE OF CALCIUM IN SECRETION AND SYNTHESIS IN BOVINE

ADRENAL CHROMAFFIN CELLS

Marie-France Bader, Jean-Pierre Simon, Jean-Marie Sontag, Keith Langley and Dominique Aunis

Groupe de Neurobiologie Structurale et Fonctionelle, Unité INSERM U-44, Centre de Neurochimie du CNRS, 5 rue Blaise Pascal, 67084 Strasbourg Cedex, France

INTRODUCTION

Secretion of many neurotransmitters, hormones and enzymes has long been known to require extracellular calcium (Douglas 1968; Katz 1969) and this fact was important in formulating the idea that a rise in intracellular calcium is the trigger for secretion. Additional data corroborating the role of calcium in secretion were obtained from measurements of calcium influx associated with secretion and in triggering secretion by introducing calcium directly into the cytosol. To date, the role of calcium as the triggering and controlling agent in the secretory process has been firmly established, but the question of what this role entails still awaits elucidation.

The purpose of the present chapter is to show two candidates amongst others not yet identified that might be involved in calcium signalling: actin, which seems to control the movement of secretory granules through the cytosol in a calcium-dependent manner, and protein kinase C, a calciprotein which modulates the synthesis and release of secretory materials.

1) Calcium control of secretion in adrenal medullary chromaffin cells

a) Calcium controls the activity of voltage-dependent calcium channels

Chromaffin cells from the adrenal medulla store their secretory products in membrane-bound organelles, the chromaffin granules. Stimulation with acetylcholine or direct depolarization with high concentrations of potassium (59mM K^+) provoke a cascade of events leading to secretion of chromaffin granule contents. Rapid measurements of $^{45}Ca^{2+}$ fluxes permit monitoring of calcium transients induced by membrane depolarization. Using this technique, we have observed that during depolarization with 59mM K^+, $^{45}Ca^{2+}$ enters the chromaffin cell. Calcium accumulation is very fast, precedes catecholamine release and terminates earlier than the secretory response (Artalejo et al. 1986). The fact that calcium entry is inhibited by inorganic (La^{3+}, Co^{2+}, Mg^{2+}) and organic (nifedipine) calcium channel antagonists and is enhanced by the calcium channel activator, Bay-K-8644, indicates that calcium gains access to the chromaffin cell cytosol mainly through specific voltage-dependent calcium channels.

Calcium uptake evoked by 59mM K^+ was found to be linear during the first 5 sec of stimulation and then continued to rise during the next 60sec but at a slower rate. We observed that the rate of catecholamine release closely parallels the rate of calcium uptake suggesting that the activity of the voltage-sensitive calcium channels tightly modulates the kinetics of the secretory process during chromaffin cell depolarization. The inactivation of voltage-dependent calcium channels seems to be a calcium-dependent process, since cytosolic calcium chelation by Quin-2 prevented the inactivation of calcium uptake while inactivation was accelerated in cells preloaded with calcium using a ionophore (Artalejo et al. 1987). Thus, intracellular calcium levels control the influx of calcium during the secretory process by modulating the activity of the voltage-dependent calcium channels.

b) Calcium : a signal for secretion in chromaffin cells

Data obtained from $^{45}Ca^{2+}$ uptake experiments indicated a cytoplasmic free calcium concentration of 170 μM following depolarization of chromaffin cells with 59mM K^+ (Artalejo et al. 1987). This calculated calcium concentration is several orders of magnitude greater than the micromolar level of intracellular calcium measured with the fluorescent probe Quin-2 after stimulation of chromaffin cells with acetylcholine or high K^+ (Kao and Schneider 1986). This difference suggests that chromaffin cells possess a very powerful calcium-sequestering system which is able to reduce calcium entering into the cell almost instantaneously.

The actual ionic requirement for secretion has been determined recently by using cells with the plasma membrane made permeable to buffers placed in the extracellular medium (Gomperts and Fernandez 1984). One method to render cells leaky is to incubate them with -toxin from Staphylococcus aureus (Bader et al. 1986a; Grant et al. 1987). This technique has some advantages since the lesions in the plasma membrane appear to be very stable (Grant et al. 1987) with an effective diameter of 2 nm, which i) prevents large molecules like cytoplasmic proteins leaking out, and ii) preserves the cytoplasmic organization (Sarafian et al. 1987). In toxin-permeabilized chromaffin cells, the introduction of 20 μM free calcium directly into the cytoplasm triggers a secretory response similar to that obtained from intact cells in response to acetylcholine (Bader et al. 1986a). It is thus clear that the increase in intracellular calcium is the event triggering secretion.

Permeabilized cells have also been successfully used to determine the energy requirement of the secretory process. We observed in cells depleted of their cytosolic ATP by permeabilization in ATP-free media that calcium is unable to induce catecholamine release in the absence of exogenous Mg-ATP. This provides strong evidence of a crucial role played by Mg-ATP in the calcium-dependent release. Since non-hydrolyzable analogues of ATP cannot be substituted for Mg-ATP, it seems reasonable to assume that phosphorylation-dephosphorylation in chromaffin cells are essential events during the secretory process.

2) Relationship between intracellular calcium and cytoskeletal organization

a) Cytoskeletal organization in the subplasmalemmal space

Many secretory cell types possess a prominent cytoskeleton localized at their periphery, a region acknowledged to be strategic for secretion. Morphological analysis has revealed that the subplasmalemmal space of chromaffin cells is particularly rich in cytoskeletal elements connecting the plasma membrane and secretory granules. Using rhodamine-conjugated phalloidin, a probe specific for polymerized actin (Wulf et al. 1979), we found that actin filaments are exclusively localized at the cell periphery forming a continuous ring (Sontag et al. 1988). Interestingly, several actin-regulating proteins have also been exclusively localized at the periphery of chromaffin cells : fodrin (Langley et al. 1986) and caldesmon (Burgoyne et al. 1986), two actin-gelating proteins which crosslink actin filaments into networks, and gelsolin (Bader et al. 1986b; Sontag et al. 1988), an actin-filament length regulator which severs actin filaments. The presence of these calcium-regulated actin-binding proteins suggests that the organization of the three-dimensional actin network, and thereby the viscosity in the subcortical cytoplasm region, is likely to be correlated with changes in intracellular calcium concentration.

b) The role of actin filaments in calcium-mediated catecholamine release

The finding of a dense network of actin filaments at the cell periphery has resulted in the suggestion that actin disassembly may be required to allow the secretory granule access to exocytotic sites. Recent work showing that changes in the assembly state of actin accompany secretion in several cell types favours this idea (Bernstein et al. 1987; Pfeiffer et al. 1985; Cheek and Burgoyne 1986). We have examined actin organization in toxin-permeabilized chromaffin cells using rhodamine-conjugated phalloidin to decorate actin filaments; exocytotic sites were visualized with antibodies to dopamine-ß-hydroxylase (a-DBH), a protein marker of the inner side of the granule membrane and accessible only when the granule interior communicates with extracellular space. In resting cells, no staining of cells with a-DBH could be observed, while rhodamine-conjugated phalloidin was visible as a typical intense peripheral fluorescent ring indicating high concentration of actin filaments at the cell cortex. In contrast, stimulated chromaffin cells decorated with a-DBH displayed a patchy pattern of fluorescence on their surface, reflecting the accessibility of DBH

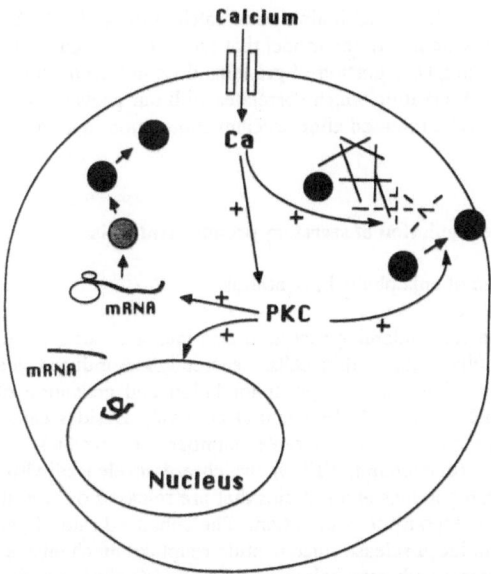

Fig. 1 . Model representing transient calcium as the second messenger in coupling stimulation, secretion and synthesis of secretory proteins in chromaffin cells. Cholinergic stimulation of chromaffin cells sets up a chain of events leading to the opening of voltage-dependent calcium channels. One possible role of the transient calcium is to control the organization of the subplasmalemmal actin network. Calcium inhibits the actin cross-linking protein fodrin and activates the severing activity of gelsolin, thereby inducing solation of the cytoplasm under the plasma membrane and the movement of secretory granules to undergo exocytosis. Calcium is the intracellular signal for both the exocytotic release and the synthesis of chromogranin A, the major secretory protein in chromaffin granules. Calcium controls the activation of protein kinase C, (PKC) which potentiates the secretory response and increases the synthesis of chromogranin A at a locus (transcription, translation) that remains to be determined. These observations emphasize the crucial role of calcium in stimulation-secretionsynthesis coupling.

during exocytosis. These stimulated cells, however, were virtually unstained with rhodamine-conjugated phalloidin. Thus, cortical actin filaments are disassembled following a rise in intracellular calcium in chromaffin cells.

In order to further evaluate the role of peripheral actin filaments in the secretory process, we examined the effect of actin-destabilizing agents on secretion from permeabilized cells (Sontag et al. 1988). Introduction into chromaffin cells of cytochalasin B or DNAse 1 was followed by a leftward shift in the calcium activation curve for secretion which occurred at the later stages of the secretory response. Thus, agents which shorten actin filaments and induce solation of the cytoplasm, potentiate calcium-evoked secretion of a certain population of secretory granules. The observation that a first population of granules can be mobilized by calcium in a cytoskeleton-independent way suggests that different populations of secretory granules exist in chromaffin cells, each population being controlled differently by the cytoskeleton.

c) Role of calcium-regulated actin-binding proteins in the secretory process

We propose that in resting chromaffin cells, fodrin cross-links actin filaments and forms a physical barrier preventing secretory granules from closely approaching the plasma membrane. On stimulation, the increase in intracellular calcium concentration might inactivate fodrin and concomitantly stimulate the severing activity of gelsolin, thereby inducing solation of the cytoplasm and the movement of granules to undergo exocytosis (Fig. 1).

In order to further evaluate this hypothesis, we took advantage of the large holes in the plasma membrane of detergent-permeabilized cells to introduce affinity-purified anti-α-fodrin antibodies and to test their effect on secretion. We found that intracellular binding of anti-α-fodrin immunoglobulins or of their Fab fragments resulted in the partial inhibition of calcium-evoked catecholamine release (Perrin et al. 1987). This observation is consistent with the model that the calcium-regulated protein fodrin is functionally involved in exocytosis. Interestingly, a portion of granules does not seem to be affected by the inhibition of fodrin with its antibody, an observation which correlates with our previous suggestion that a subpopulation of secretory granules is able to be released after calcium stimulation in a cytoskeleton-independent manner.

3) The role of calcium in the regulation of secretory product synthesis

a) Calcium and the regulation of enkephalin biosynthesis

Events that promote catecholamine secretion also increase catecholamine biosynthesis in adrenergic tissues. In adrenal medullary chromaffin cells, secretagogue-induced acceleration of catecholamine biosynthesis has been attributed to specific phosphorylation and activation of the rate limiting enzyme, tyrosine hydroxylase (Haycock et al. 1982; Meligeni et al. 1982). Besides catecholamines, chromaffin cells store in their secretory granules a considerable number of enzymes (i.e. DBH), proteins (i.e. chromogranins) and peptides (enkephalin, VIP,...) which are coreleased with catecholamines. They must have a mechanism to replenish peptides and proteins that are released on stimulation in order to maintain a continuously secreted level in response to stimulation. The enhanced rate of peptide biosynthesis is likely to be coupled to secretagogue-induced release since peptide reuptake mechanisms have not been shown to exist in chromaffin cells in contrast with catecholamine reuptake (Kenigsberg and Trifaro 1980). Indeed it has been found that nicotine-stimulated secretion of Met-enkephalin is accompanied by an increase in Met-enkephalin biosynthesis, which occurs at a pretranslational locus (Waschek et al. 1987). Direct opening of voltage-dependent calcium channels by depolarization of the plasma membrane with 59mM K^+ produced an increase in $mRNA^{enk}$ levels similar to that caused by nicotine or acetylcholine. Moreover, release elicited by nicotine or 59mM K^+ was inhibited by D600, a voltage-dependent calcium channel antagonist, as was also the induction of $mRNA^{enk}$ produced by nicotine and potassium. Barium is able to elicit secretion from chromaffin cells even in medium containing reduced concentrations of calcium; however, barium-stimulated increase in $mRNA^{enk}$ levels was completely dependent on extracellular calcium, indicating that calcium entry into the cell is a prerequisite for enkephalin biosynthesis. Thus, calcium influx or calcium activation of an intracellular factor appears to couple enkephalin secretion and synthesis following cholinergic stimulation.

Calcium may act on Met-enkephalin biosynthesis by activating adenylate cyclase since the $mRNA^{enk}$ level was also increased by forskolin or other agents that elevate cAMP (Waschek et al. 1987). Alternatively, cholinergic stimulation produces both calcium influx and activation of adenylate cyclase in parallel and calcium and cAMP may thus act cooperatively to increase $mRNA^{enk}$.

b) Calcium and the regulation of chromogranin A biosynthesis

Chromogranin A (CGA), a large acidic protein, is the major polypeptide in chromaffin granules, comprising 40% of the total soluble proteins. Recent evidence indicates that CGA-derived peptides are probably involved in a negative feedback mechanism controlling secretion from chromaffin cells (Simon et al. 1988). We have examined the regulation of CGA biosynthesis in cultured chromaffin cells, using a radiolabelled aminoacid precursor and a specific antibody to immuno-precipitate CGA (Simon et al. 1989). Cholinergic stimulation provoked both the release of CGA and a compensatory increase in radiolabelled CGA, indicating that synthesis of CGA is coupled to its secretagogue-induced release, thereby maintaining intracellular CGA at a constant level. Direct depolarization with 59mM K^+ evoked a similar increase in radiolabelled CGA , an observation which indicates that CGA biosynthesis is dependent on an influx of extracellular calcium. In addition, the absence of secretagogue-induced CGA synthesis in calcium-free medium further implicates a second-messenger role of calcium in the regulation of CGA synthesis.

Agents that elevate cAMP (forskolin), which have been shown to increase $mRNA^{enk}$, had no effect on CGA synthesis. In contrast the phorbol ester TPA, a direct activator of protein kinase C, elicited an increase in CGA synthesis comparable to that provoked by nicotine or by high potassium. The role of protein kinase C in mediating calcium-dependent CGA synthesis is further supported by the observation that sphingosine, an inhibitor of protein kinase C, blocks both TPA- and nicotine-induced CGA synthesis. Moreover, long-term incubation of cells with phorbol ester which reduces protein kinase C activity also inhibits nicotine-evoked induction of CGA synthesis.

At this point it seems reasonable to conclude that CGA secretion and synthesis are driven by an influx of extracellular calcium following cholinergic stimulation. Calcium may increase CGA synthesis by activating protein kinase C (Fig. 1). The question of whether protein kinase C regulates CGA synthesis at a pretranslational locus is now awaiting further investigation.

CONCLUSIONS

It is now well established that calcium plays an essential role in stimulus-secretion coupling. On stimulation calcium enters the cytosol, provoking a cascade of events leading to the release of chromaffin granule contents. Here we report that the transient calcium induces the dissolution of the actin network in the subplasmalemmal space, thereby allowing secretory granules access to exocytotic sites. In addition to its role as a second messenger in the secretory process, calcium acts as a second messenger in coupling the synthesis of peptides and proteins with their secretagogue-induced release. Thus, calcium is the key messenger triggering and controlling a series of events termed the **stimulation-secretion-synthesis** coupling.

REFERENCES

Artalejo C.R., Bader M.F., Aunis D., and Garcia A.G., 1986, Biochim. Biophys. Res. Commun., 134, 1-7.
Artalejo C.R., Garcia A.G., and Aunis D., 1987, J. Biol. Chem., 262, 915-926.
Bader M.F., Thiersé D., Aunis D., Ahnert-Hilger G., and Gratzl M., 1986a, J. Biol. Chem., 261, 5777-5783.
Bader M.F., Trifaro J.M., Langley K.O., Thiersé D., and Aunis D., 1986b, J. Cell Biol., 102, 636-646.
Bernstein B.W., and Bamburg J.R., 1987, Neurochem. Res., 12, 929-935.
Burgoyne R.D., Cheek T.R., and Norman K.M., 1986, Nature, 319, 68-70.
Cheek T.R., and Burgoyne R.D., 1986, FEBS Lett., 207, 110-114.
Douglas W.W., 1968, Br. J. Pharmacol., 34, 451-474.
Gomperts B.D., and Fernandez J.M., 1985, Trends Biochem. Sci., 10, 414-417.
Grant N.J., Aunis D., and Bader M.F., 1987, Neuroscience, 23, 1143-1155.
Haycock J.W., Megileni J.A., Bennett W.F., and Waymire J.C., 1982, J. Biol. Chem., 257, 12641-12648.
Kao L.S., and Schneider A.S., 1986, J. Biol. Chem., 261, 4881-4888.
Katz B., 1969, "The Release of Neurotransmitter Substances", Liverpool University Press, Liverpool.
Kenigsberg R.L., and Trifaro J.M., 1980, Neuroscience, 5, 1547-1556.
Langley O.K., Perrin D., and Aunis D., 1986, J. Histochem. Cytochem., 34, 517-525.
Meligeni J.A., Haycock J.W., Bennett W.F., and Waymire J.C., 1982, J. Biol. Chem., 257, 12632-12640.
Perrin D., Langley O.K., and Aunis D., 1987, Nature, 326, 498-501.
Pfeiffer J.R., Seagrave J.C., Davis B.H., Deanin G.G., and Oliver J.M., 1985, J. Cell Biol., 101, 2145-2155.
Sarafian T., Aunis D., and Bader M.F., 1987, J. Biol. Chem., 262, 16671-16676.
Simon J.P., Bader M.F., and Aunis D. , 1988, Proc. Natl. Acad. Sci.USA, 85, 1712-1716.
Simon J.P., Bader M.F., and Aunis D., 1989, Biochem. J., 260, 915-922.
Sontag J.M., Aunis D., and Bader M.F., 1988, Eur. J. Cell Biol., 46, 316-326.
Waschek J.A., Pruss R.M. Siegel R.E., Eiden L.E., Bader M.F., and Aunis D., 1987, Ann. N.Y. Acad. Sci., 493, 308-323.
Wulf E., Deroben A., Bautz F.A., Faulstich H., and Whieland T., 1979, Proc. Natl. Acad. Sci. USA, 76, 4498-4502.

CA^{2+}-BINDING PROTEINS AS COMPONENTS OF THE CYTOSKELETON

Michael Schleicher, Ludwig Eichinger, Walter Witke and Angelika A. Noegel

Max-Planck-Institute for Biochemistry, 8033 Martinsried, FRG

One of the major filamentous systems of non-muscle cells is based on actin, a highly conserved protein of 42kD. The polymerization equilibrium between globular (G-) and filamentous (F-) actin determines the viscosity of the cytoplasm and is regulated by cytoplasmic salt conditions and by actin-binding proteins (for review see Stossel et al., 1985, Pollard and Cooper, 1986). Several of these actin-binding proteins are activated or inhibited by micromolar Ca^{2+}-concentrations either through direct binding to Ca^{2+} or indirectly through modulators such as calmodulin.

Investigations of the cytoskeleton in the amoeba Dictyostelium discoideum led to the description of several actin-binding proteins which belong to the distinct functional classes of capping, sequestering, crosslinking, severing proteins, or to the group of myosins and membrane associated proteins (for review see Noegel and Schleicher, 1989). We focus in the following report on two Ca^{2+}-regulated actin-binding proteins: α-Actinin and severin.

α-ACTININ

D. discoideum α-actinin was cloned from a genomic λgt11-library using monoclonal antibodies (Witke et al., 1986). The genome harbors a single α-actinin gene that codes for an mRNA of about 3.0kb. This RNA is present throughout the developmental cycle of D. discoideum. The deduced amino acid sequence of the complete cDNA (Noegel et al., 1987) revealed three structural features that together could be responsible for the Ca^{2+}-dependent F-actin crosslinking function of the homodimeric molecule.

a) The actin-binding domain is located at the N-terminus (Noegel et al., 1987). This region shows actin-binding activity and seems to be highly conserved. To date, similar regions are shared by α-actinins from other species (Arimura et al., 1989, Baron et al., 1987), by a second major crosslinking protein from D. discoideum (Noegel et al., 1989) and by dystrophin, the gene product from the Duchenne muscular dystrophy locus (Koenig et al., 1988). There is evidence that spectrin or filamin also might contain a similar actin-binding site.

b) Spectrin-like repeats most likely are responsible for the extended structure of the α-actinin molecule. They are rich in α-helix forming regions and a triple-helix model is discussed (Baron et al., 1987; Koenig et al., 1988).

c) α-Actinin from D. discoideum contains at the C-terminus two complete calcium-binding loops that show the characteristics of EF-hand structures (Noegel et al., 1987).

Fig. 1 aligns these regions with EF-hands from calmodulin, with related regions from muscle and non-muscle chicken α-actinin, with spectrin and with human dystrophin. It is remarkable that besides calmodulin only D. discoideum α-actinin and chicken brain spectrin contain the complete set of six liganding oxygens in the Ca^{2+}-binding loops. The corresponding domains in chicken fibroblast α-actinin, chicken skeletal muscle α-actinin and dystrophin seem to be degenerated to 3-5 liganding residues and it is

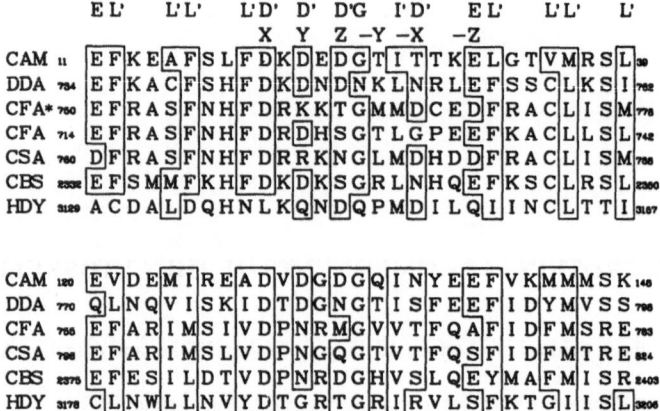

Fig. 1. Alignment of EF-hand domains from known and putative Ca^{2+}-binding proteins. The corresponding protein regions are aligned (boxed residues) according to the EF-hand model (first line, Kretsinger 1980 a/b), the octahedral vertices of the liganding oxygens are listed in the second line. The upper and lower panels show respectively calmodulin (CAM) from *Trypanosoma brucei gambiense* (Tschudi et al., 1985), *D. discoideum* α-actinin (DDA, Noegel et al., 1987, chicken fibroblast α-actinin (CFA*, Baron et al., 1987; CFA, Arimura et al., 1988), chicken skeletal muscle α-actinin (CSA, Arimura et al., 1988), chicken brain α-spectrin (CBS, Wasenius et al., 1989) and human dystrophin (HDY, Koenig et al., 1988). The positions of the corresponding regions are given in amino acid numbers.

questionable whether these structural motifs can bind Ca^{2+} at all. It should be noted, however, that α-actinin from rabbit alveolar macrophages has the potential of binding to 4 Ca^{2+} ions per homodimer with an affinity of 4×10^6 Mol⁻¹ (Bennett et al., 1984).

To investigate the involvement of the EF-hands in the function of D. discoideum α-actinin we constructed an α-actinin with a modification in the second Ca^{2+}-binding loop (Fig. 2). At position 2418 of the cDNA between a Cla I and an Asp 718 site a 42bp oligonucleotide was inserted which codes for a peptide of 13 amino acids from a Sendai virus protein and represents an epitope for a monoclonal antibody. The modified α-actinin was cloned into a D. discoideum transformation vector and introduced into an α-actinin deficient mutant (Schleicher et al., 1988; Wallraff et al., 1986). Cells expressing the modified gene were cloned, the gene product purified to homogeneity and assayed. The first data obtained with this engineered α-actinin indicate that in contrast to the native α-actinin the Ca^{2+}-binding and the affinity for F-actin are strongly reduced.

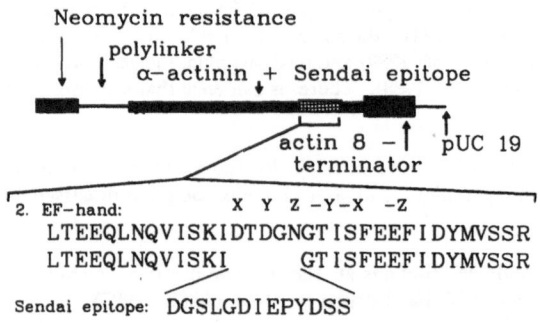

Fig. 2. Modification of the second EF-hand in *D. discoideum* α-actinin. A 42bp oligonucleotide was inserted into the α-actinin gene as reporter sequence encoding 13 amino acids from a Sendai virus protein. The construct was introduced into an α-actinin deficient mutant by homologous recombination using neomycin resistance as selection system.

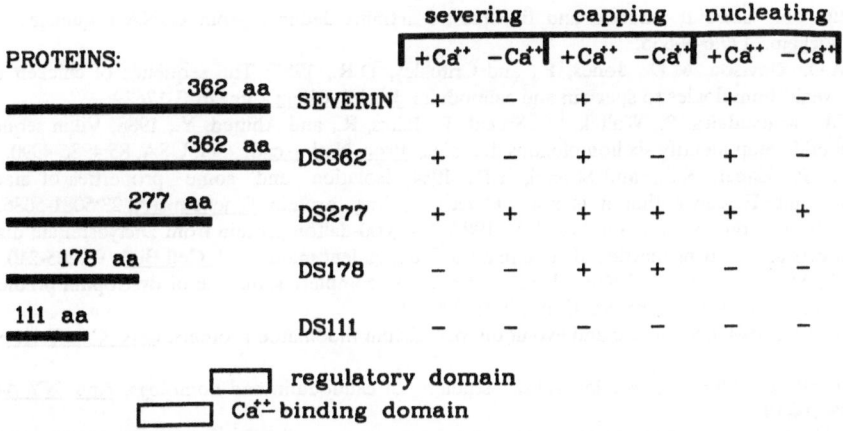

PROTEINS:		severing		capping		nucleating	
		+Ca⁺	−Ca⁺	+Ca⁺	−Ca⁺	+Ca⁺	−Ca⁺
362 aa	SEVERIN	+	−	+	−	+	−
362 aa	DS362	+	−	+	−	+	−
277 aa	DS277	+	+	+	+	+	+
178 aa	DS178	−	−	+	+	−	−
111 aa	DS111	−	−	−	−	−	−

☐ regulatory domain
☐ Ca²⁺ binding domain

Fig. 3. Domain mapping of severin, a Ca^{2+}-regulated F-actin fragmenting protein. A severin cDNA clone (DS362) encoding all 362 amino acids of the native severin was deleted from the 3'-end by exonuclease III. The complete clone and the truncated severin molecules (DS277, DS178, DS111) were expressed in *E. coli*, the proteins purified to homogeneity and assayed. The Ca^{2+}-dependency of the severing, capping and nucleating activities are listed. ^{45}Ca overlays showed that DS277 still bound Ca^{2+} but had lost its Ca^{2+}-dependent activity. This suggests a regulatory domain at the C-terminus that changes its conformation upon Ca^{2+}-binding further upstream.

SEVERIN

Severin is a 40kD protein that cuts actin filaments in a Ca^{2+}-dependent fashion (Brown et al., 1982); it remains bound at the fast growing ends of the newly formed short F-actin fragments, inhibits re-annealing and accelerates by its nucleation activity like other capping proteins the onset of actin polymerization. It is encoded by a single gene in D. discoideum and its protein sequence is very similar to other Ca^{2+}-regulated F-actin fragmenting proteins (André et al., 1988) like villin (Bazari et al., 1988), fragmin (Ampe and Vandekerckhove, 1987) or gelsolin (Kwiatkowski et al., 1986). The protein binds Ca^{2+} directly, but the sequence does not reveal an EF-hand structure or any other known consensus sequence that could contain the Ca^{2+}-binding site.

To determine the functional domains of the protein that could be responsible for the F-actin fragmenting, the F-actin capping, the nucleating and Ca^{2+}-binding activity we constructed truncated severin molecules. For this purpose the complete cDNA was deleted with exonuclease III from the 3'-end, the shortened coding regions were cloned into an ATG expression vector, the proteins expressed in E. coli, purified to homogeneity and their functions assayed by testing their impact on: i) the viscosity of F-actin solutions, ii) the kinetics of actin polymerization and iii) the Ca^{2+}-binding activity. Fig. 3 summarizes the domain structure of the severin molecule.

After removing 85 amino acids from the C-terminus, the protein lost its Ca^{2+}-dependency and was with regard to its severing, capping and nucleating functions always "on". However, it could be shown by ^{45}Ca overlay experiments that the truncated severin DS277 still bound Ca^{2+}. We conclude from these data that the Ca^{2+} binding site and the regulatory domain which confers the Ca^{2+}-dependent activity are distinct.

REFERENCES

Ampe, C., and Vandekerckhove, J., 1987, The F-actin capping proteins of Physarum polycephalum : Cap42(a) is very similar, if not identical, To fragmin and structurally and functionally very homologous to gelsolin. Cap42(b) is Physarum actin, EMBO J., 6:4149-4157.

André, E., Lottspeich, F., Schleicher, M., and Noegel, A., 1988, Severin, gelsolin, and villin share a homologous sequence in regions presumed to contain F-actin severing domains, J. Biol. Chem., 263:722-727.

Arimura, C., Suzuki, T., Yanagisawa, M., Imamura, M., Hamada, Y., and Masaki, T., 1988, Primary structure of skeletal muscle and fibroblast α-actinins deduced from cDNA sequences, Eur. J. Biochem., 177:649-655.

Baron, M.D., Davison, M.D., Jones, P., and Critchley, D.R., 1987, The sequence of chicken α-actinin reveals homologies to spectrin and calmodulin, J. Biol. Chem., 36:17623-17629.

Bazari, W.L., Matsudaira, P., Wallek, M., Smeal, T., Jakes, R., and Ahmed, Y., 1988, Villin sequence and peptide map identify six homologous domains, Proc. Natl. Acad. Sci. USA, 85:4986-4990.

Bennett, J. P., Zaner, K.S., and Stossel, T.P., 1984, Isolation and some properties of macrophage α-actinin: Evidence that it is not an actin gelling protein, Biochemistry, 23:5081-5086.

Brown, S., Yamamoto, K., and Spudich, J.A., 1982, A 40,000-dalton protein from Dictyostelium discoideum affects assembly properties of actin in a Ca^{2+}-dependent manner, J. Cell Biol., 93:205-210.

Koenig, M., Monaco, A.P., and Kunkel, L.M., 1988, The complete sequence of dystrophin predicts a rod-shaped cytoskeletal protein, Cell, 53:219-228.

Kretsinger, R.H., 1980a, Structure and evolution of calcium modulated proteins, CRC Crit. Rev. Biochem., 8:119-174.

Kretsinger, R.H., 1980b, Crystallographic studies of calmodulin and homologs, Ann. NY Acad. Sci., 356:14-19.

Kwiatkowski, D., Stossel, T.D., Orkin, S.H., Mole, J.E., Colton, H.R., and Yin, H.L., 1986, Plasma and cytoplasmic gelsolins are encoded by a single gene and contain a duplicated actin-binding domain, Nature, 323:455-458.

Noegel, A., Witke, W., and Schleicher, M., 1987, Calcium-sensitive non-muscle α-actinin contains EF-hand structures and highly conserved regions, FEBS Lett., 221:391-396.

Noegel, A.A., and Schleicher, M., 1989, The contractile system in non-muscle cells: Involvement of actin and actin-binding proteins, In "Stimulus-Response Coupling: The Role of Intracellular Calcium", J.R. Dedman and V.L. Smith, eds., Telford Press, in press.

Noegel, A.A., Rapp, S., Lottspeich, F., Schleicher, M., and Stewart, M., 1989, The Dictyostelium gelation factor shares a putative actin-binding site with α-actinins and dystrophin and also has a rod domain containing six 100-residue motifs that appear to have a cross-beta conformation, J. Cell Biol., 109:607-618.

Pollard, T.D., and Cooper, J.A., 1986, Actin and actin-binding proteins. A critical evaluation of mechanisms and functions, Ann. Rev. Biochem., 55:987-1035.

Schleicher, M., Noegel, A., Schwarz, T., Wallraff, E., Brink, M., Faix, J., Gerisch, G., and Isenberg, G., 1988, A Dictyostelium mutant with severe defects in α-actinin: its characterization using cDNA probes and monoclonal antibodies, J. Cell Sci., 90:59-71.

Stossel, T.P., Chaponnier, C., Ezzell, R.M., Hartwig, J.H., Janmey, P.A., Kwiatkowski, D.J., Lind, S.E., Smith, D.B., Southwick, F.S., Yin, H.L., and Zaner, K.S., 1985, Nonmuscle actin-binding proteins, Ann. Rev. Cell Biol., 1:353-402.

Tschudi, C., Young, A.S., Ruben, L., Patton, C.L., and Richards, F.F., 1985, Calmodulin genes in Trypanosomes are tandemly repeated and produce multiple mRNAs with a common 5' leader sequence, Proc. Natl. Acad. Sci. USA, 82:3998-4002.

Wallraff, E., Schleicher, M., Modersitzki, M., Rieger, D., Isenberg, G., and Gerisch, G., 1986, Selection of Dictyostelium mutants defective in cytoskeletal proteins: use of an antibody that binds to the ends of α-actinin rods, EMBO J., 5:61-67.

Wasenius, V.-M., Saraste, M., Salven, P., Erämaa, M., Holms, L., and Lehto, V.P., 1989, Primary structure of the brain a α-spectrin, J. Cell Biol., 108:79-93.

Witke, W., Schleicher, M., Lottspeich, F., and Noegel, A., 1986, Studies on the transcription, translation, and structure of α-actinin in Dictyostelium discoideum, J. Cell Biol., 103:969-975.

S-100 PROTEINS : RELATIONSHIPS WITH MEMBRANES AND THE CYTOSKELETON

Rosario Donato

Section of Anatomy, Department of Exper. Med. and Biochem. Sciences, Cas. Post. 81
06100 Perugia Succ. 3, Italy

INTRODUCTION

Proteins of the S-100 family, S-100a$_0$ ($\alpha\alpha$), (αß), and S-100b (ßß) belong to the superfamily of Ca^{2+}-binding proteins of the EF-hand type (for review see Donato, 1986). S-100 proteins are widely distributed in animal tissues, yet they are not ubiquitous. In the brain, $\sim$ 85% of S-100 is soluble in the cytoplasm and 15% of it is associated with membranes.

FUNCTIONAL ASPECTS OF S-100 PROTEINS

S-100 regulates the assembly-disassembly of microtubule protein (MTP) Ca^{2+}- and pH-dependently (Baudier et al., 1982; Endo and Hidaka, 1983; Donato, 1983, 1988), several kinase and phosphoprotein phosphatase activities both dependently and independently of Ca^{2+} (Patel and Marangos, 1982; Albert et al., 1984; Kuo et al., 1986), the phosphorylation of a number of proteins both dependently and independently of Ca^{2+} by interacting with the substrates rather than with the kinases (Baudier et al., 1987; Baudier and Cole, 1988; Hagiwara et al., 1988), and a brain aldolase activity (Zimmer and Van Eldik, 1986). S-100 inhibits the MTP assembly by interfering with the nucleation and the elongation processes (for references see Donato, 1986). It is suggested that S-100 binds to and sequesters unassembled tubulin (Donato, 1988). However, S-100 also binds to microtubule-associated tau proteins under nonreducing conditions (Baudier et al., 1987; Baudier and Cole, 1988). S-100 stimulates the disassembly of steady-state microtubules (MTs) Ca^{2+}- and pH-dependently (Baudier et al., 1982; Donato, 1983, 1988) by binding to the MT wall (Donato and Giambanco, 1989). Association of S-100 with MTs in vivo has been reported (Cocchia, 1981; Michetti and Cocchia, 1982; Uchida and Endo, 1988; Rambotti et al., 1988). S-100b is being secreted by glial cells and adipocytes (Shashoua et al., 1984; Van Eldik and Zimmer, 1987; Suzuki et al., 1984, 1987). S-100b behaves as a neurite extension factor when added to cultured neurons (Van Eldik et al., 1988).

RELATIONSHIPS OF S-100 PROTEINS WITH MEMBRANES

The S-100 family comprises additional members on the basis of sequence homology (for references see Van Eldik et al., 1988). Among these is p11, the light and regulatory chain of the cytoskeletal protein complex calpactin I. Bovine brain S-100 binds to and inhibits in the absence of Ca^{2+} the phosphorylation of the heavy chain (p36) of calpactin I (Hagiwara et al., 1988). It is possible that the S-100 binding to p36 is but a reflection of sequence homology between the α and ß subunits of S-100 and p11. However, it is quite possible that S-100 has physiological targets in membranes. It has been shown that the S-100 binds to specific acceptor sites in brain membranes and that Ca^{2+} regulates the S-100 binding to high, but not low, affinity sites on them (Donato, 1976). We have reinvestigated the association of S-100 with membranes. In the brain, membrane-associated S-100 is distributed in two pools, one resistant to EDTA plus 1.0 M NaCl and extractable with Triton X-100 (Donato et al., 1986), and one resistant to Triton X-100 and extractable with EDTA/KCl (Donato et al., 1989a). In the heart, no Triton X-100-soluble S-100a$_0$ is found, whereas a Triton X-100-resistant, EDTA/KCl-extractable fraction of S-100a$_0$ is recovered (Donato et al., 1989b). Thus in the brain S-100 both associates strongly with integral membrane proteins and/or the lipid bilayer, and binds to the Triton X-100-resistant material, i.e. to what is operationally defined as the cytoskeleton, whereas S-100a$_0$ only associates with the latter in the heart. Immunocytochemical data confirm that a fraction of S-100b in cultured Schwann cells is associated with Triton-cytoskeletons (in preparation). Several S-100-binding polypeptides have been detected in EDTA/KCl extracts of brain membranes pretreated with Triton X-100, by an overlay procedure (Donato et al., 1989a). The most prominent species are characterized by M$_r$ values ranging from 15 to 30 kDa and from 70 to 130 kDa. A 70 kDa doublet with ability to bind S-100 is also identified in the Triton X-100-extractable material from brain membranes. Our data suggest that in the brain S-100 interacts electrostatically with membrane skeleton proteins at very low Ca^{2+} concentrations; that transients of free Ca^{2+} stabilized these interactions by blocking S-100 in a "Ca^{2+}" conformation, thanks to which S-100 remains bound to the acceptors even after the free Ca^{2+} concentration had returned to low levels; and that the total Ca^{2+} concentration has to decrease to extremely low levels for membrane skeleton-associated S-100 to come again in solution.

We suggest that electric charges on S-100 play an important role in the interaction of S-100 with its targets, and that neutralization of positive charges on S-100 by negative charges on targets results in the enhancement of the Ca^{2+}-binding properties of S-100 subunits. A cationic site is found in the N-terminal half of individual α and ß S-100 subunits (residues 20-32), and an anionic site, including the Ca^{2+}-binding loop, is found in the C-terminal half (Isobe and Okuyama, 1978, 1981).

S-100 binds to negatively charged cardiolipin vesicles (CLVs) (Zolese et al., 1988). In the absence of Ca^{2+}, S-100 reduces the fluidity of the polar surface of the lipid bilayer. Also, CLVs induce a reduction (16%) in the α-helical content of S-100b, analogous to that produced by Ca^{2+}. In the presence of Ca^{2+}, S-100 enhances the Ca^{2+}-dependent reduction in the fluidity of the hydrophobic core of the lipid bilayer, and loses 52% of its α-helical content. In the absence of Ca^{2+}, S-100b promotes aggregation and fusion of CLVs, whereas in the presence of Ca^{2+} more of the S-100b binds to CLVs, but an S-100b-dependent inhibition of Ca^{2+}-induced aggregation and fusion of CLVs is registered (Donato et al., 1988). We suggest that in the absence of Ca^{2+}, S-100b binds to and cross-links adjacent negatively charged CLVs through its cationic sites, thereby promoting aggregation and fusion of CLVs, whereas in the presence of Ca^{2+} S-100b binds strongly and possibly penetrates partway into individual CLVs, thereby contrasting their Ca^{2+}-induced aggregation and fusion. We also suggest that the conformation of membrane-bound S-100b differs substantially from that of S-100b in solution and that, once S-100b had interact with membranes Ca^2-dependently, Ca^2 is no longer required for S-100b to remain bound to membranes.

It is possible that what has been observed with CLVs is but a reflection of an intrinsic property of S-100b to bind to and, hence, to cross-link two adjacent structures bearing clusters of negatively charged residues sterically disposed in an appropriate way. Thus S-100b could be involved in the regulation of the interaction of cytoskeletal proteins with the lipid bilayer. Alternatively, S-100b could play a role in the regulation of membrane structural organization, possibly in relation to membrane fusion and/or exocytosis.

Lastly, we have identified a molecular target of S-100 in striated muscle cell membranes. S-100b inhibits the basal (Mg^{2+}-activated) skeletal muscle adenylate cyclase (Fanò et al., 1988), whereas S-100a$_0$ strongly stimulates this activity (Fanò et al., 1989).

CONCLUSIONS

The view is emerging that S-100 is a family of multifunctional proteins whose dimeric nature would determine and condition the involvement of individual isoforms in a large number of cell activities.

ACKNOWLEDGEMENTS

Supported by M.P.I. and C.N.R. funds.

REFERENCES

Albert, K.A., Wu, W.C.-S., Nairn, A.C., and Greengard, P., 1984, Inhibition by calmodulin of calcium/phospholipid-dependent protein phosphorylation, Proc. Natl. Acad. Sci. USA, 81:3622.

Baudier, J., and Cole, R.D., 1988, Interactions between the microtubule-associated tau proteins and S-100b regulate tau phosphorylation by the Ca^{2+}/calmodulin-dependent protein kinase II, J. Biol. Chem., 263:5876.

Baudier, J., Briving, C., Deinum, J., Haglid, K., Sorskog, L., and Wallin, M., 1982, Effect of S-100 proteins and calmodulin on Ca^{2+}-induced disassembly of brain microtubule proteins, FEBS Lett., 161:235.

Baudier, J., Mochly-Rosen, D., Newton, A., Lee, S.-H., Koshland, D.E., jr., and Cole, R.D., 1987, Comparison of S-100b protein with calmodulin : interactions with melittin and microtubule-associated tau proteins and inhibition of phosphorylation of tau proteins by protein kinase C, Biochemistry, 26:2886.

Cocchia, D., 1981, Immunocytochemical localization of S-100 protein in the brain of adult rat. An ultrastructural study, Cell Tissue Res., 214:529.

Donato, R., 1976, Further studies on the specific interaction of S-100 protein with synaptosomal particulates, J. Neurochem., 27:439.

Donato, R., 1983, Effect of S-100 protein on assembly of brain microtubule proteins in vitro, FEBS Lett., 162:310.

Donato, R., 1986, S-100 proteins. Cell Calcium, 7:123.

Donato, R., 1988, Ca^{2+}-independent, pH-regulated effects of S-100 proteins on assembly-disassembly of brain microtubule protein in vitro, J. Biol. Chem., 263:106.

Donato, R., and Giambanco, I., 1989, Interaction between S-100 proteins and steady-state and taxol-stabilized microtubules in vitro. J. Neurochem., 52:1010.

Donato, R., Giambanco, I., Aisa, M.C., and Ceccarelli, P., 1989a, Identification of S-100 proteins and S-100-binding proteins in a detergent-resistant, EDTA/KCl-extractable fraction from bovine brain membranes, FEBS Lett., 247:31.

Donato, R., Giambanco, I., Aisa, M.C., Spreca, A., Rambotti, M.G., Ceccarelli, P., and Di Geronimo, G., 1989b, Cardiac S-100a$_0$ protein : purification by a simple procedure and related immunocytochemical and immunochemical studies, Cell Calcium, 10:81.

Donato, R., Giambanco, I., Aisa, M.C., Zolese, G., Amati, S., and Curatola, G., 1988, S-100b protein regulates bilayer-nonbilayer transitions in cardiolipin vesicles, Sixth Int. Symp. on Calcium Binding Proteins in Health and Disease, July 24-28, 1988, Nagoya, Japan, p. 135.

Donato, R., Prestagiovanni, B., and Zelano, G., 1986, Identity between cytoplasmic and membrane-bound S-100 proteins purified from bovine and rat brain, J. Neurochem., 46:1333.

Endo, T., and Hidaka, H., 1983, Effect of S-100 protein on microtubule assembly-disassembly, FEBS Lett., 161:235.

Fanò, G., Angelella, P., Mariggiò, D., Aisa, M.C., Giambanco, I., and Donato, R., 1989, S-100a$_0$ protein stimulates the basal (Mg^{2+}-activated) adenylate cyclase activity associated with skeletal muscle membranes, FEBS Lett., 248:9.

Fanò, G., Fulle, S., Della Torre, G., Giambanco, I., Aisa, M.C., Donato, R., and Calissano, P. 1988, S-100b protein regulates the activity of squeletal muscle adenylate cyclase in vitro. FEBS Lett, 240:177.

Hagiwara, M., Achiai, M., Owada, K., Tanaka, T., and Hidaka, H., 1988, Modulation of tyrosine phosphorylation of p36 and other substrates by the S-100 protein. J. Biol. Chem., 263:6438.

Isobe, T., and Okuyama, T., 1978, The amino-acid sequence of S-100 protein (PAPIb-protein) and its relationship to calcium-binding proteins, Eur. J. Biochem., 89:379.

Isobe, T., and Okuyama, T., 1981, The amino-acid sequence of the -subunit in bovine brain S-100 protein, Eur. J. Biochem., 116:79.

Kuo, W.-N., Blake, T., Cheema, I.R., Dominguez, J., Nicholson, J., Puente, K., Shells, P., and Lowery, J., 1986, Regulatory effects of S-100 protein and parvalbumin on protein kinases and phosphoprotein phosphatases from brain and skeletal muscles, Mol. Cell. Biochem., 71:19.

Michetti, F., and Cocchia, D., 1982, S-100-like immunoreactivity in a planarian. An immunochemical and immunocytochemical study, Cell Tissue Res., 223:575.

Patel, J., and Marangos, P.J., 1982, Modulation of brain protein phosphorylation by S-100 protein, Biochem. Biophys. Res. Commun., 109:1089.

Rambotti, M.G., Saccardi, C., Spreca, A., Aisa, M.G., Giambanco, I., and Donato, R., 1988, Immune localization of S-100ß antigen in ciliated and supporting cells of lamb olfactory mucosa, J. Cell Biol., 107:733a.

Shashoua, V.E., Hesse, J.W., and Moore, B.W., 1984, Proteins of the brain extracellular fluid : evidence for release of S-100 protein, J. Neurochem., 42:1536.

Suzuki, F., Kato, K., Kato, T., and Ogasawara, N., 1987, S-100 protein in clonal astroglioma cells in released by adenocorticotropic hormone and corticotropin-like intermediate-lobe peptide, J. Neurochem., 49:1557.

Suzuki, F., Kato, K., and Nakajima, T., 1984, Hormonal regulation of adipose S-100 protein release, J. Neurochem., 43:1336.

Uchida, T., and Endo, T., 1988, Immunoelectron microscopic demonstration of S-100b protein-like in centriole, cilia, and basal body, J. Histochem. Cytochem., 36:693.

Van Eldik, L.J., and Zimmer, D.B., 1987, Secretion of S-100 from rat C6 glioma cells, Brain Res., 436:367.

Van Eldik, L.J., Staecker, J.L., and Winningham-Major, F., 1988, Synthesis and expression of a gene coding for the calcium-modulated protein S-100 and designed for cassette-based, site-directed mutagenesis, J. Biol. Chem., 263:7830.

Zimmer, D.B., and Van Eldik, L.J., 1986, Identification of a molecular target for the calcium-modulated protein, S-100 : fructose-1,6-biphosphate aldolase, J. Biol. Chem., 261:11424.

Zolese, G., Tangorra, A., Curatola, G., Giambanco, I., and Donato, R., 1988, Interaction of S-100b protein with cardiolipin vesicles as monitored by electron spin resonance, pyrene fluorescence and circular dichroism, Cell Calcium, 9:149.

ONCOMODULIN IN NORMAL AND TRANSFORMED CELLS

John P. MacManus,[1] Linda M. Brewer,[1] and Denis Banville[2]

[1]Division of Biological Sciences, National Research Council, Ottawa K1A OR6, Canada
[2]BioTechnology Research Institute, National Research Council, Montreal H4P 2R2
Canada

The 3-dimensional crystal structure of oncomodulin from X-ray analysis reveals that it is quite similar to that of parvalbumin (F. Ahmed et al., in preparation). Oncomodulin has three domains composed of helix:metal-binding loop:helix arranged in a similar way to parvalbumin (Moews and Kretsinger 1975). This was not unexpected because it was known that oncomodulin and parvalbumin share 50% identical amino acid sequence, and an additional 30% conservative residue replacement (MacManus et al., 1987). Also both the circular dichroic and proton NMR spectra suggested the existence of great similarity of secondary structure (MacManus et al., 1984; Williams et al., 1987).

Despite this structural similarity between oncomodulin and parvalbumin, oncomodulin is more than a mere parvalbumin. Oncomodulin has a low affinity modulator-like calcium specific binding site. This was first proposed in 1984 after UV, circular dichroic and fluorescence spectroscopic studies on rat hepatoma oncomodulin (MacManus et al., 1984). This has been supported by lanthanide exchange studies, either terbium exchange monitored fluorometrically (Henzl et al., 1986), lutecium exchange monitored by NMR (Williams et al., 1987), or by europium exchange monitored by direct laser excitation of the metal (Henzl & Birnbaum, 1988). The final evidence for the CD site being the site prefering calcium over magnesium has been obtained by site directed mutagenesis of bacterially expressed oncomodulin (MacManus et al., 1989).

Oncomodulin has a unique distribution for a calcium-binding protein of the calmodulin superfamily. It has never been detected in normal adult tissues. During prenatal development oncomodulin is expressed very early (Figure 1). Following fertilization in the rat, at the morula stage, oncomodulin can be located immunohistochemically in all cells (Figure 1A). At the blastula stage, oncomodulin mRNA can be located in all cell types by in situ hybridization (Figure 1B & C).

By this stage of development, the totipotent cells of the morula have undergone the first differentiation into trophectoderm and inner cell mass (Rossant,1986). By in situ hybridization and immunohistochemistry, both of these cell types appear to express oncomodulin in cultured blastocysts (Brewer et al., 1989). Following implantation in the rat, oncomodulin appears in the ectoplacental cone (Figure 1D). After this time as the fetus develops, oncomodulin remains restricted to the extraembryonic structures, the rat placental trophoblast (Brewer & MacManus 1985), the cytotrophoblast of the shell and villi of the human (Brewer & MacManus 1987), and in both the yolk sacs and amnion of the rat (Brewer & MacManus 1985). Oncomodulin has also been detected in trophoblast from bat, cat, mouse and sheep (MacManus et al., 1988). The mRNA for oncomodulin can be quantitated, and it appears that the control of expression in rat placenta is transcriptional (Gillen et al., 1988). The fetal tissues have always been free of detectable oncomodulin by any quantitative or qualitative method, as are tissues of the neonate and adult rat. Thus oncomodulin appears soon, if not immediately, following fertilization. As differentiation proceeds, oncomodulin remains in the inner cell mass, but following further differentiation, the oncomodulin gene is selectively turned off in cells destined to be fetal. However, the gene remains on in cells which are extraembryonic.

This distinctive distribution throughout development suggests a role for oncomodulin during the pre-implantation phase, but this remains undefined. Heterologous expression of oncomodulin in Rat-1 or baby hamster kidney cells has been achieved, but failed to define a role for oncomodulin (Mes-Masson, personal communication).

Oncomodulin is capable of calcium-dependent enzyme modulation in vitro (MacManus et al., 1987, 1988). It was recently demonstrated that a dimer of oncomodulin formed through Cys-18 of the protein was as potent a stimulator of calcineurin as calmodulin (Mutus et al., 1989). No dimer has ever been detected in vivo. The presence of an interactive surface, assembled or not, on oncomodulin whets the appetite to know what is the genuine oncomodulin target.

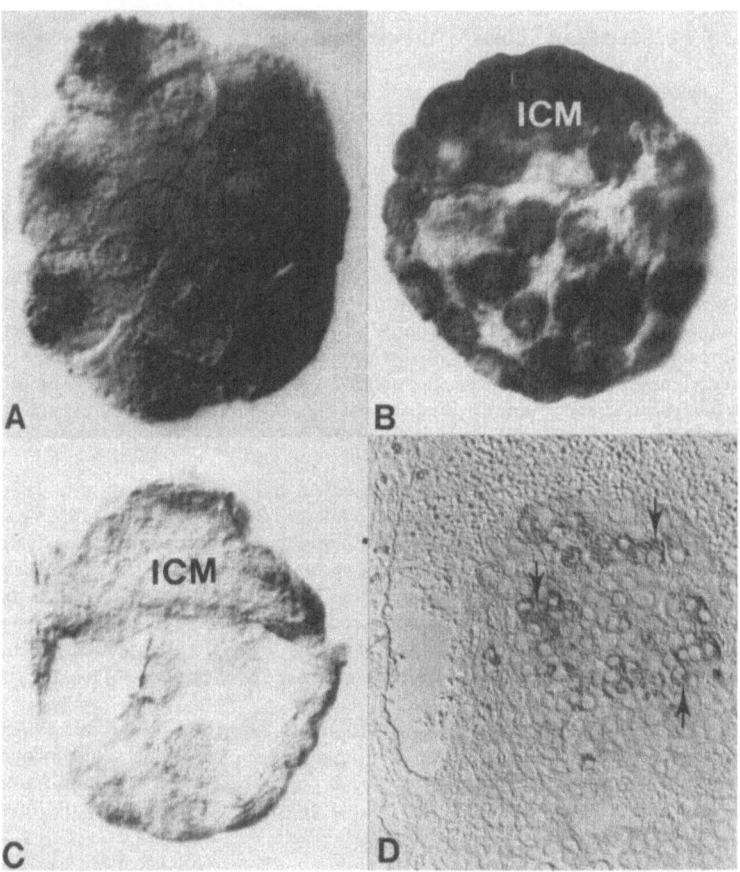

Fig. 1 . Localisation of oncomodulin mRNA and protein during early development. A: Morula (day 4), immuno-staining in all cells, *1380. B: Blastula (day 5), *in situ* hybridization showing oncomodulin mRNA in all cells, *830. C: Control for B using sense probe. D: Positive immuno-staining in ectoplacental cone (day 9), *435

The oncomodulin gene is activated in adult mammals following neoplastic transformation (MacManus et al.,1987, 1988). During chemically induced (by diethylnitrosamine) hepatocarcinogenesis, oncomodulin was detected in liver foci after three months by immunostaining (Bernaert et al., 1989). About 80% of foci were positive by six months. The proportion of positive neoplastic nodules increased from 50% at five months to about 70% by nine months. The majority of the resulting carcinomas at ten to twenty months remained oncomodulin positive. This early appearance suggests that oncomodulin does not correlate with full malignant transformation, since most foci are not regarded as developing into frank carcinomas.

In another model of chemical carcinogenesis, oncomodulin could be detected very early in methylnitro-nitrosoguanidine (MNNG) or procarbazine induced sarcomas (Sommer & Heizmann 1989). The presence of oncomodulin in the lesions was correlated with the ability to form colonies in soft agar. However, no correlation between the presence of oncomodulin and the ability to grow in soft agar was seen with a series of Morris hepatoma cell lines (Brewer et al., 1985). Despite these contradictory findings, the important observation derived from studies such as these is that oncomodulin is present in the majority of tumors. However the amount of protein present in the various tumors does vary enormously from one species to another (MacManus et al., 1987, 1988).

The structure of the rat oncomodulin gene has been determined (Banville & Boie, 1989). The rat gene sequence is 9kb long and contains 5 exons. The coding portion of the gene is contained in exons 2-5. The nucleotide sequence of these exons was found to be identical to that determined from the cDNA (Gillen et al., 1987). Comparison with other genes of the calmodulin family clearly show that the oncomodulin gene is related, and fits into the envisioned picture of the evolution of these proteins (Perret et al., 1989). The oncomodulin gene is most like the parvalbumin gene (Berchtold et al., 1987). The coding region is interrupted by the 4 introns at homologous positions in both genes suggesting that they had an ancestor in common. However, the first exon of the gene for rat oncomodulin is totally different from that for parvalbumin. Comparison of this sequence with those present in GENBANK revealed >75% identity with the last 70 nucleotides of the long terminal repeat (LTR) structure of an intracisternal A particle (IAP) from Syrian hamster (Ono et al., 1985).

The cellular DNA of all mammals contains repeated sequences related to retroviral genomes. IAPs are one category of these encapsidated non-infectious endogenous retroviral entities, which occur in multiple copies (100-1000) in rodent genomes (Hojman & Peries, 1986; Kuff & Lueders 1988). They reside in cisternae after budding from the endoplasmic reticulum, and share the properties of LTRs in their DNA, reverse transcriptase and capsid proteins with the B and C type endogenous proviruses. The presence of LTRs allows the IAPs to be insertional mutagens in a similar way to infectious retroviruses. Contiguous insertion of IAPs has led to activation of several genes (Horwitz et al.,1984; Ymer et al.,1985). In a similar way to viral LTR replication, unequal crossing over during IAP duplication can yield a solo LTR, eg in a gamma-actin pseudogene (Man et al., 1987). It is such an unusual solo LTR which constitutes the rat oncomodulin promoter region. The solo LTR is delimited by a 3bp inverted repeat, and a 6bp (GGTAGG) target site duplication can be seen.

The acquisition of this retroviral promoter by the oncomodulin gene is not unique to the Buffalo rat strain from which the genomic library was prepared. Indeed, oncomodulin cDNAs isolated from a different library (a Leydig cell tumor from a Fischer rat) were shown to have the exact same sequence. However, this LTR structure is absent from the mouse and human oncomodulin genes (Banville, unpublished). The rat gene is unique in being the only known mammalian gene controlled by a promoter of retroviral origin. LTRs from a mouse IAP have been shown to be the target for transactivation by oncogene products such as those from the myc gene, or the adenovirus E1A gene. The presence of such a strong promoter in the rat oncomodulin gene is probably the reason for the high levels of expression found in rat tumors compared to those in other species such as mice and humans. It is of interest to note that the pattern of appearance during early development and later in adult neoplasms is common both to IAPs and to oncomodulin (Hojman & Peries, 1986; Kuff & Lueders 1988).

Although the transposition of a LTR sequence within the rat oncomodulin gene can best be viewed as an accident of evolution, it is remarkable that the pattern of expression of oncomodulin during early development, and later in neoplasms of the adult is comparable in other mammals (MacManus et al., 1987, 1988). This can be taken as an indication of the strong need for this gene product in both prenatal development and perhaps in the process of carcinogenesis.

REFERENCES

Banville, D., and Boie, Y., 1989, J. Mol. Biol., 207:481.
Berchtold, M. W., Epstein, P., Beaudet, A. L., Payne, M. E., Heizmann, C. W., and Means, A. R., 1987, J. Biol. Chem., 262:8696.
Bernaert, D., Brewer, L. M., MacManus, J. P., and Galand, P., 1989, Internat. J. Cancer, 43:719.
Brewer, L. M., Boynton, A. L., and MacManus, J. P., 1985, J. Nutr. Growth Cancer, 2:25.
Brewer, L. M., Gillen, M. F., and MacManus, J. P., 1989, Placenta, 10:359.
Brewer, L. M., and MacManus, J. P., 1985, Dev. Biol., 112:49.

Brewer, L. M., and MacManus, J. P., 1987, Placenta, 8:351.

Gillen, M. F., Banville, D., Rutledge, R. G., Narang, S., Seligy, V., Whitfield, J. F., and MacManus, J. P., 1987, J. Biol. Chem., 262:5308.

Gillen, M. F., Brewer, L. M., and MacManus, J. P., 1988, Cancer Lett., 40:151.

Henzl, M.T., and Birnbaum, E. R., 1988, J. Biol. Chem.,263:10674.

Henzl, M. T., Hapak, R. C., and Birnbaum, E. R., 1986, Biochim. Biophys. Acta, 872:16.

Hojman, F., and Peries, J., 1986, Ann. Inst. Pasteur/Virol., 137E:3.

Horowitz, M., Luria, S., Rechavi, G., and Givol, D., 1984, EMBO J., I3:2937.

Kuff,E.L., and Lueders, K.K., 1988, Adv. Cancer Res., 51:183.

MacManus, J. P., Brewer, L. M., and Gillen, M. F., 1987 In: "The Role of Calcium in Biological Systems", Vol. IV L.J. Anghileri, ed., pp. 1-19, CRC Press, Boca Raton.

MacManus, J. P., Brewer, L. M., and Gillen, M. F., 1988, In: "Calcium and Calcium-Binding Proteins", C. Gerday, R. Gilles, & L. Bolis, eds., pp. 128-138, Springer-Verlag, Berlin.

MacManus, J. P., Hutnik, C. M. L., Sykes, B. D., Szabo, A. G., Williams, T. C., and Banville, D., 1989, J. Biol. Chem., 264:3470.

MacManus, J. P., Szabo, A. G., and Williams, R. E., 1984, Biochem. J., 220:261.

Man, Y. M., Delius, H., and Leader, D. P., 1987, Nucl. Acids Res., 15:3291.

Moews, P. C., and Kretsinger, R. H., 1975, J. Mol. Biol., 91:201.

Mutus, B., Palmer, E. J., and MacManus, J. P., 1988, Biochemistry, 27:5615.

Ono, M., Toh, H., Miyata, T., and Awaya, T., 1985, J. Virol., 55:387.

Perret, C., Lomri, N., and Thomasset, M., 1988, J. Mol. Evol., 27:351.

Rossant, J., 1986, In: "Experimental Approaches to Mammalian Embryonic Development", J. Rossant, & R.A. Pedersen, eds., pp. 97-120, Cambridge University Press, Cambridge.

Sommer, E. W., and Heizmann, C. W., 1989, Cancer Res., 49:899.

Williams, T.C., Corson, D.C., Sykes, B.D., and MacManus, J. P., 1987, J. Biol. Chem., 262:6248.

Ymer, S., Tucker, W.Q.J., Sanderson, C. J., Hapel, A. J., Campbell, H.D., and Young, I.G., 1985, Nature, 317:255.

PERTURBATION OF THE CALMODULIN SYSTEM IN

TRANSFORMED CELLS

Linda J. Van Eldik,[1,2] Warren E. Zimmer,[1,3,*] Steven W. Barger,[2] and D. Martin Watterson[1,3]

[1]Department of Pharmacology, Vanderbilt University, Nashville, Tn 37232, USA

[2]Department of Cell Biology, Vanderbilt University, Nashville, Tn 37232, USA

[3]Howard Hughes Medical Institute, Vanderbilt Univ., Nashville, TN 37232 USA
[*]current address: Dept. of Structural and Cellular Biology, Univ. of South Alabama Medical School, Mobile, AL 36688 USA

INTRODUCTION

In eukaryotic cells, calcium acts as an intracellular signal transducer primarily through its interaction with a class of calcium binding proteins, of which calmodulin (CaM) is the most highly conserved, phylogenetically ubiquitous member (for reviews, see Van Eldik et al., 1982; Van Eldik and Roberts, 1988; Cohen and Klee, 1988). CaM transduces a calcium signal into a biological response by its ability to regulate the activity of other proteins in a calcium dependent manner. Because CaM is probably the most widely distributed mediator of intracellular calcium signals, fundamental insights can be derived from an enhanced knowlege about the genetic encoding, biosynthetic assembly and regulation of CaM-modulated calcium response pathways. Regardless, in order to understand fully the roles of CaM in the eukaryotic cell and obtain insight into how CaM-modulated pathways can respond differentially to calcium signals, it is necessary to be able to describe in some detail all of the CaM pathways for at least one biological system. This has not been done yet for any biological system. Chicken embryo fibroblasts (CEF) represent one biological system for which a relatively extensive body of information has been described. Also, CEF transformed by Rous sarcoma virus exhibit a number of phenotypic alterations that are potentially mediated by CaM and CaM binding proteins, and perturbations in CaM regulation have been described for normal and transformed CEF. In a more general sense, because perturbations of CaM pathways occur in many kinds of virus-transformed cells, knowledge of how alterations in CaM expression are coupled to oncogene expression may yield insight into how CaM-regulated calcium response pathways are involved in mechanisms of oncogenic transformation.

Our previous studies showed that in the normal CEF/v-src transformed CEF system, there is a perturbation of CaM pathways. To summarize, we showed that: 1) CaM levels are 2- to 4-fold higher in transformed CEF than in normal CEF (Watterson et al., 1976; Van Eldik and Burgess, 1983), 2) there is no major subcellular redistribution of CaM upon transformation (Van Eldik and Burgess, 1983), 3) the increase in CaM is not seen if non-transforming viruses are used (Watterson et al., 1976), 4) the CaM molecule itself is not altered upon transformation (Van Eldik and Watterson, 1979), 5) the elevated levels of CaM are due to a selective increase in the rate of synthesis of CaM (Zendegui et al., 1984), and 6) the molecular mechanism is an increase in CaM-specific mRNA in transformed CEF (Zendegui et al., 1984). These studies were extended to demonstrate the presence of multiple classes of CaM binding proteins in CEF and to identify several of these binding proteins (Burgess et al., 1984). Initial analyses of the changes in levels of CaM binding proteins demonstrated that myosin light chain kinase (MLCK) activity and immunoreactive levels are 2- to 3-fold lower in transformed CEF than in normal CEF (Van Eldik et al., 1984). These observations, coupled with pilot studies of CaM and MLCK levels in proliferating CEF, raise the possibility of coordinate regulation of CaM and MLCK.

Our current studies are addressing the overall question of the relationship between cellular phenotype and perturbation of the CaM system; i.e., how are CaM pathways regulated? In order to address this question, it is necessary to define in detail the spatial and temporal relationships between CaM and CaM binding proteins, including definition of the number, concentration, distribution, and regulation of CaM binding proteins. We report here some of our more recent studies on the CaM system in CEF, with particular emphasis on elucidation of the basic features of the regulation of the CaM-MLCK signal transduction complex. We are focusing on the non-muscle/smooth muscle MLCK because it is the simplest and best characterized of the CaM regulated protein kinases and its phosphotransferase activity is directly related to its cellular role, the initiation of contraction. The long term goal is to understand how a calcium signal transduction complex is encoded and assembled in eukaryotic cells, and how the expression of the components of such a complex is modulated in virus-transformed cells.

METHODS

Cell Culture

Normal CEF and CEF transformed with the PrC strain of Rous sarcoma virus were prepared as previously described (Van Eldik and Burgess, 1983). The temperature-sensitive NY68 mutant of Rous sarcoma virus was obtained from Dr. H. Hanafusa, Rockefeller University. Experiments utilizing this temperature-sensitive mutant were done by preparing secondary CEF and infecting them with the virus mutant. After 4 days at 36°, cultures were trypsinized and seeded into 100 mm dishes at 4×10^6 cells/dish. After 8 to 16 hrs at 36°, cultures were shifted to 42° and kept in media containing 1% (v/v) calf serum for 72 hrs. At this point (designated the 0 hr time point), some of the cultures were shifted to the permissive temperature (36°) to initiate transformation, and some of the cultures were kept at the non-permissive temperature (42°). Cells were harvested at various times after temperature shift, and analyzed for uptake of ^{3}H-deoxyglucose (Knauer and Smith, 1979) or for CaM levels (Watterson et al., 1980).

Immunoprecipitation of CaM and CaM binding protein complexes

CEF were labeled for 22 hrs with 0.5 mCi of ^{35}S-methionine in 1.5 ml of methionine-free F10 (GIBCO) containing 1% calf serum. For analysis of labeled proteins by 2-d gels, cell extracts were prepared by rinsing dishes 3 times with phosphate buffered saline, and then scraping cells in SDSBME-DNARNAse buffer (Protein Databases Inc, Cold Spring Harbor NY), and freezing in liquid N_2. For immunoprecipitations, cell extracts were prepared by rinsing dishes 3 times with TBS (150 mM NaCl, 50 mM Tris-HCl, pH 7.5, 1 mM EDTA), and scraping cells in TBS. Cells were pelleted at 500xg and cell pellets were resuspended in TBS containing 1% Triton X-100 and 1% sodium deoxycholate. After centrifuging at 10,000xg for 30 min, supernatants were preabsorbed with normal rabbit serum and Protein A-Sepharose (Pharmacia) for 20 min on ice. The mixture was centrifuged in the microfuge at 15,000xg for 2 min, and 200 μl of the supernatant was used for incubation with antibody. The supernatant was incubated with 20 μl of anti-CaM serum or normal rabbit serum for 1 hr on ice. Protein A-Sepharose was then added for 30 min on ice with occasional mixing, and the tubes were then centrifuged in the microfuge at 15,000xg for 2 min. Pellets were washed 3 times with TBS containing 1% Triton X-100, 1% sodium deoxycholate, 0.1% sodium dodecyl sulfate, and then once with TBS. The final pellet was resuspended in 50 μl of SDSBME and boiled 5' to release the immune complexes. After centrifugation, the supernatant was frozen until analysis by 2-d gels. The 2-d gels were prepared and analyzed by Protein Databases Inc as described (Garrels, 1989).

Isolation of Chicken Calmodulin cDNA and Genomic Clones

A chicken gizzard cDNA library in the plasmid vector pBR322 was obtained from E. Lazarides (California Institute Technology). The library was screened with a synthetic oligonucleotide probe (CCaM.1) corresponding to the nucleotides coding for amino acids 140-148 of vertebrate calmodulin. The nucleotide sequence of the CCaM.1 probe is 3' TGAAGCGACAGTAGTAGACATGTTTGAGAA 5'. The CCaM.1 probe was synthesized on an Applied Biosystems 380A synthesizer and purified by gel electrophoresis as previously described (Roberts et al., 1987). Filters were hybridized with ^{32}P-labeled CCaM.1 essentially as previously described (Roberts et al., 1987), except that: a) hybridization buffer was 0.9 M NaCl, 0.09 M sodium citrate, 0.01 M sodium phosphate, pH 7.5, 0.5% sarcosyl, 0.2% ficoll, 0.2% polyvinylpyrrolidone, and 100 μg/ml yeast tRNA; b) filters were hybridized at 37° for 24 hrs; and c) stringent washes were done in 0.9 M NaCl, 0.09 M sodium citrate, 0.5% sodium dodecyl sulfate at 70° for 1 hr, then at 80° for 2 x 15 min before drying and exposing the filters.

For isolation of CaM genomic clones, a chicken library was prepared in the lambda-phage vector EMBL4 and the non-amplified library was screened with the CaM cDNA clone 9H8 by standard methods (Zimmer et al., 1988). Four overlapping genomic clones were isolated. Clones were characterized by restriction enzyme mapping and partial DNA sequence analysis by methods previously described (Van Eldik et al., 1988; Zimmer et al., 1988).

Analysis of RNA Levels

For analysis of CaM mRNA levels, total RNA was prepared from normal and transformed CEF as described (Zimmer et al., 1988). Equal amounts of RNA from normal and transformed CEF were denatured, subjected to electrophoresis, and hybridized to a 750 bp NcoI-XbaI fragment of the 9H8 CaM cDNA clone using methods similar to those previously described (Zimmer et al., 1988). For analysis of MLCK mRNA levels, poly(A)$^+$ RNA was prepared from normal and transformed CEF as described (Leof et al., 1986). Equal amounts of RNA from normal and transformed CEF were denatured, subjected to electrophoresis in a formaldehyde/agarose gel as described (Davis et al., 1986a), and transferred to GeneScreen Plus membranes (DuPont). Blots were pre-hybridized in 50% formamide, 10% dextran sulfate, 1% sodium dodecyl sulfate, 1 M NaCl, 100 μg/ml salmon sperm DNA for 2 hr at 42ú, then hybridized overnight at 42° in the same buffer containing ^{32}P-labeled MLCK cDNA probe. In these experiments, a cDNA clone for MLCK was isolated from a CEF cDNA library (Lukas et al., 1988; Watterson, manuscript in preparation) and an 1100 bp PstI fragment was used as a probe. After hybridization, blots were washed in 2 changes (5 min each) of 2X SSC (0.3 M NaCl, 0.03 M sodium citrate, pH 7.0) at room temperature, followed by 2 changes (30 min each) of 2X SSC, 1% sodium dodecyl sulfate at 42ú, and finally by 2 changes (30 min each) of 0.1X SSC at room temperature.

RESULTS

Detection of CaM-Associated Proteins by Immunoprecipitation

We have previously prepared a variety of anti-CaM antibodies that react with different regions of the CaM molecule, and that react selectively with CaM in immunoblot analyses of whole cell extracts (Van Eldik and Watterson, 1981; Van Eldik et al., 1983a; Van Eldik et al., 1983b; Van Eldik and Lukas, 1987). We have shown that in addition to binding monomeric CaM, some of these site-specific antibodies bind to CaM when it is part of a supramolecular complex such as phosphorylase kinase (Burgess et al., 1983). These data suggest that our antibodies may be useful for isolation of CaM as a complex with particular target proteins, and we have initiated studies to immunoprecipitate CaM from CEF as an attempt to isolate and characterize any associated binding proteins. For these experiments, CEF were labeled with ^{35}S-methionine and cell extracts prepared as described in Methods. The extracts were incubated with either anti-CaM

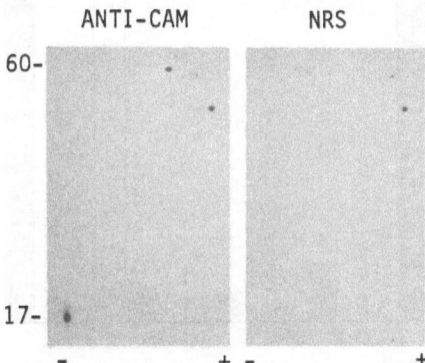

Fig. 1 . Detection of CaM-associated Proteins. ^{35}S-labeled CEF extracts were immunoprecipitated with either anti-CaM antibodies (anti-CaM) or normal rabbit serum (NRS). Immunoprecipitated material was subjected to 2-dimensional gel electrophoresis, and autoradiography was done. The acidic end of the isoelectric focusing dimension is on the left of each panel (-) and the basic end is on the right (+). The SDS gel electrophoresis dimension is from top to bottom. Two proteins that precipitate with anti-CaM but not with NRS (CaM at 17,000 MW (17) and a protein at 60,000 MW (60)) are marked.

antibody or normal rabbit serum, and the immune complexes were then precipitated with Protein A-Sepharose and analyzed by 2-dimensional gel electrophoresis. Figure 1 shows the results obtained with one of our anti-CaM antisera (rabbit #465). We found that CaM was precipitated with the anti-CaM antiserum, but not with the normal rabbit serum control. We also found that there are other proteins that precipitate along with the CaM. Some of the proteins also precipitate with normal rabbit serum (such as proteins that co-migrate with actin and tubulin). However, there are certain proteins that are precipitated selectively with the anti-CaM antibody and not with the normal rabbit serum. An example is the protein that migrates at apparent molecular weight 60,000 in the anti-CaM panel.

In experiments with other anti-CaM antibodies, we have found six protein spots that are either missing or very low in the normal rabbit serum controls compared to the anti-CaM immunoprecipitations, suggesting that these proteins may be CaM binding proteins. Alignment of these six proteins with proteins in 2-d gels of ^{35}S-labeled CEF cell lysates has allowed a tentative identification of their position in relationship to other CEF proteins. We then compared the relative density of these proteins in ^{35}S-labeled extracts from normal and transformed CEF; the result from one gel analysis is shown in Figure 2. It is interesting to note that most of the six proteins decrease in transformed CEF, although some are relatively unchanged and one increases slightly (Figure 2A). When the total PPM units in the six gel spots are summed and compared to the density of the CaM spot, it is evident that the combined levels of the six CaM-associated proteins are higher than the levels of CaM alone in both normal and transformed CEF (Figure 2B). These data should be interpreted with caution, since spot intensity represents radioactivity associated with ^{35}S-labeled protein and not absolute protein levels, and since the CaM binding ability of these proteins has not been confirmed. However, the results are consistent with previous data (Burgess et al., 1984; Van Eldik et al., 1984) suggesting that there is not a large excess of CaM in the cell compared to the CaM binding proteins.

Interrelationship between CaM, CaM binding proteins, and transformation

A growing body of evidence suggests that there is quantitative regulation of the levels of CaM and CaM binding proteins, implying that the balance of the expression of CaM and its target proteins may be important in cellular homeostasis, and raising the possibility of a coupling between the production of CaM and CaM binding proteins. As an initial approach to this question, we have analyzed a model system for studying the perturbed expression of the CaM system as a function of transformation. This system is CEF infected with a Rous sarcoma virus mutant (NY68) that is temperature sensitive for transformation.

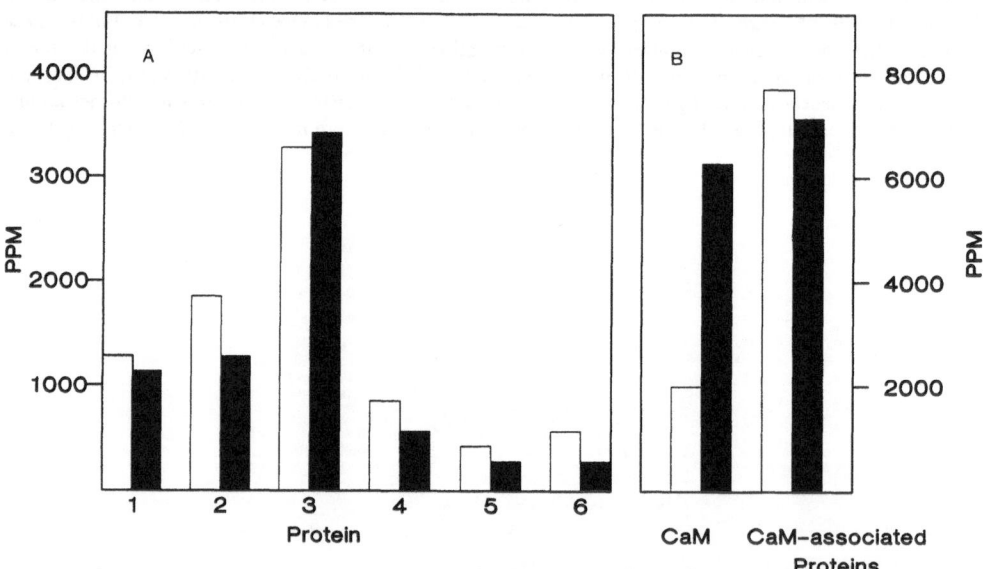

Fig. 2. Comparison of CaM and six CaM-associated proteins in normal and transformed CEF. The radioactivity present in CaM and six CaM-associated proteins (detected by immunoprecipitations as in Fig. 1) was quantitated in whole cell lysates of ^{35}S-labeled normal (open bars) and transformed (solid bars) CEF. The height of each bar represents the intensity of the protein as parts per million (PPM) of the total radioactivity applied to the gel. Panel A shows the PPM in each of the six proteins; panel B shows the PPM in CaM compared to the total PPM in the six proteins combined.

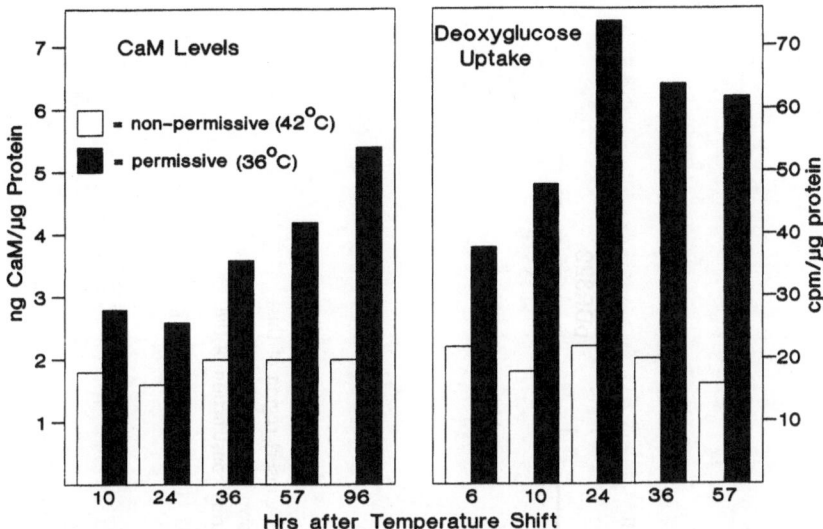

Fig. 3 . CaM Levels and Deoxyglucose Uptake in CEF Infected with Temperature-Sensitive Virus Mutants. The CaM levels and deoxyglucose uptake were measured in cells at various times after shift to permissive temperature or in control cells that remained at non-permissive temperature.

At the non-permissive temperature (42°), the cells are phenotypically normal. However, upon shift to permissive temperature (36°), the cells become phenotypically transformed by multiple criteria (changes in morphology, transport, growth in soft agar). Figure 3 shows an initial analysis of the changes in CaM levels after shift of the temperature-sensitive CEF to permissive temperature. We found that the increase in CaM levels occurred at a later time than the increase in deoxyglucose uptake (another criterion of the transformed phenotype). Because different criteria of transformation appear at different times after virus infection and because of the complexities in the definition of when transformation occurs in cells, it is difficult to correlate the changes in CaM levels with molecular mechanisms of transformation. However, the data in Figure 3 suggest that the increase in CaM occurs as a consequence of the transformed state, and not vice versa. In addition, these data show that utilization of cells that are temperature-sensitive for transformation will provide a useful biological tool for analyzing the potential interrelationships between expression of CaM and CaM binding proteins.

Analysis of CaM and MLCK Expression in CEF

To examine the basic features of the regulation of the CaM-MLCK complex in CEF, and analyze the perturbations of this CaM pathway that occur in transformed CEF, we have been developing the molecular tools necessary to analyze the expression of the genes for CaM and MLCK. A chicken calmodulin cDNA clone (9H8) that contains the entire coding region and some 3' and 5' untranslated regions was isolated. A portion of the restriction enzyme map is shown in Figure 4. The restriction mapping done to date on this clone agrees with that previously reported for a chicken CaM cDNA (Putkey et al., 1983).

This CaM cDNA was used as a probe to screen a chicken genomic library. Several overlapping genomic clones were isolated, and these are being characterized by detailed restriction mapping and selected DNA sequence analysis. A portion of a composite restriction map of our chicken CaM gene is shown in Figure 5. Our map and limited DNA sequence analyses are in general agreement with the revised CaM gene map (Simmen et al., 1987) and additional 5' flanking DNA sequence (Epstein et al., 1989).

We have utilized the chicken CaM 9H8 cDNA clone to test the relative abundance of CaM mRNA in normal and transformed CEF. Equal amounts of total RNA were subjected to electrophoresis and blotted to a membrane. Blots were hybridized to the CaM cDNA probe, and the results are shown in Figure 6. One major species of mRNA migrating at approximately 1.8 Kb that hybridizes to the CaM cDNA probe was detected in RNA preparations from both normal and transformed CEF. Quantitation of the intensity of the bands showed that there was approximately 2.5- to 3-fold more density associated with the mRNA band

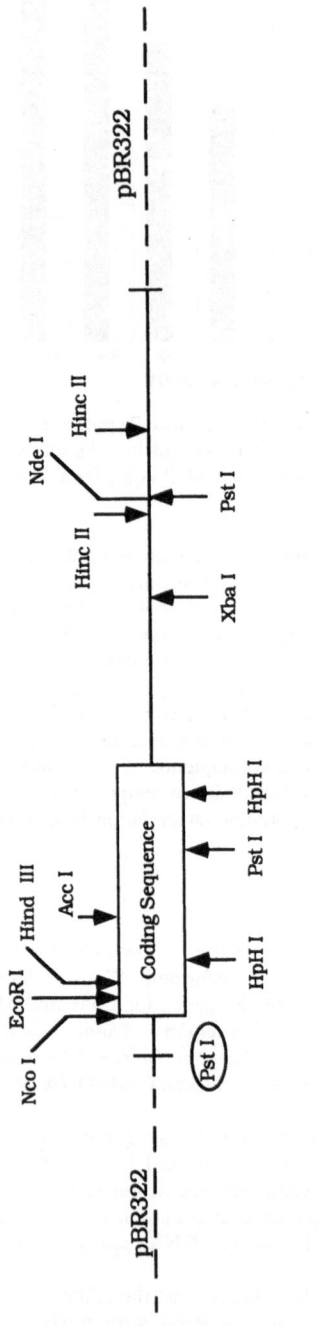

Fig. 4. Selected restriction enzyme sites on the CaM cDNA clone 9H8. The cDNA insert region of this clone is denoted by the solid lines and the pBR322 vector sequences are shown by the broken lines. The circled PstI restriction enzyme site represents a cloning site added during construction of the cDNA library.

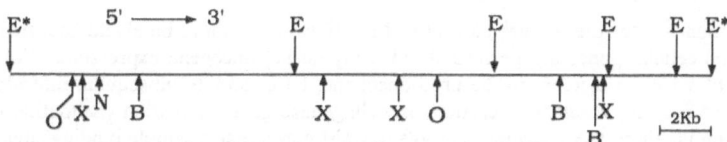

Fig. 5 . Partial restriction enzyme map of the chicken CaM gene locus as determined from analyses of overlapping genomic clones. B = BamHI, E = EcoRI, N = NcoI, O = XhoI, X = XbaI, E* = EcoRI restriction enzyme sites from the EMBL4 vector adjacent to the cloning sites.

in transformed CEF than in normal CEF. These data are in agreement with our previous Northern analysis (Zendegui et al., 1984) using an eel CaM cDNA probe, and demonstrate that transformed CEF contain higher levels of CaM-specific mRNA than normal CEF.

Our previous studies (Van Eldik et al., 1984) had shown that MLCK protein levels were decreased in transformed CEF compared to normal CEF. In order to examine the mechanism responsible for the changes in MLCK levels, we determined the relative abundance of MLCK mRNA in normal and transformed CEF. Northern analysis was performed on poly(A)$^+$ RNA preparations from normal and transformed CEF by using a fragment of a CEF MLCK cDNA clone as a probe (Figure 7). The probe utilized in this experiment has been shown to hybridize exclusively with the MLCK mRNA. We found that the MLCK probe hybridized to a single RNA species of approximately 5.5 Kb in the fibroblast RNA samples, and that there was approximately 2-fold less MLCK-specific mRNA in transformed CEF compared to normal CEF. These data agree with our previous observations of a 2- to 3-fold decrease in MLCK protein in transformed CEF.

DISCUSSION

In this report, we have presented data that: 1) v-src expression results in an increase in CaM mRNA and protein, 2) the perturbation of CaM is a late event in the transformation process, 3) there does not appear to be an extreme excess of CaM over CaM binding proteins in the cell, and 4) v-src expression results in a decrease in MLCK mRNA and protein, and may selectively alter the levels of other CaM binding proteins. Altogether, the data indicate how CaM pathways are perturbed in transformed cells. Although

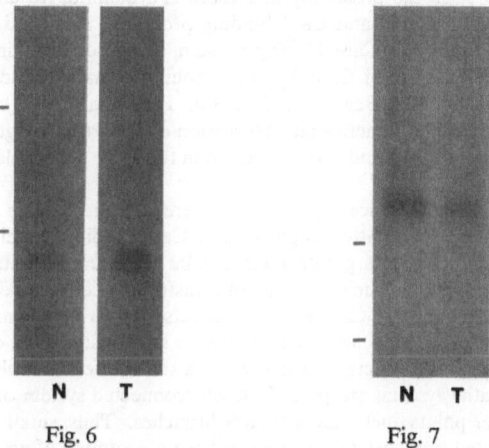

Fig. 6   Fig. 7

Fig. 6 . CaM mRNA Levels in Normal and Transformed CEF. Equal amounts (20 μg) of total RNA from normal (N) and transformed (T) CEF were subjected to electrophoresis, and hybridized to a ^{32}P-labeled chicken CaM cDNA probe as described in Methods. The position of 28S and 18S RNA standards is marked.

Fig. 7 . MLCK mRNA Levels in Normal and Transformed CEF. Equal amounts (3 μg) of poly(A)$^+$ RNA from normal (N) and transformed (T) CEF were subjected to electrophoresis, and hybridized to a chicken MLCK cDNA probe as described in Methods. The position of the 28S and 18S RNA standards is marked.

the molecular mechanisms for the perturbation of CaM pathways remain to be elucidated, there are precedents where normal cellular genes are perturbed indirectly during oncogene expression. For example, in cells transformed by v-src, an increase in the protooncogene, c-fos, and its subsequent interaction with the transacting factor, AP-1/jun, result in alterations in collagenase gene expression (Schonthal et al., 1988). Related to this example, there are 5' sequence motifs in CaM genes that resemble binding sites for transacting factors, such as AP-1/jun (Zimmer et al., 1988; and unpublished observations). Thus, the data here and in previous reports, as well as the precedents of other genes that are perturbed in transformed cells, raise the question of whether perturbation of CaM and CaM binding proteins after v-src expression may occur through mechanisms involving c-fos or c-fos/jun regulation, and suggest logical directions for a mechanistic analysis of transcriptional regulation of CaM and CaM binding proteins.

One interesting feature about the biosynthetic regulation of CaM and CaM binding proteins that has emerged indirectly from studies over the past ten years is that there does not appear to be an extreme excess of CaM in the cell as had been previously assumed in the earlier stages of research in this field. In the 1970s, CaM had been identified as an activator of only a few enzymes. Thus, from this perspective, it was logical to assume that there was a much greater concentration of CaM in the cell compared to its target proteins. However, as we learn more about CaM expression and as additional CaM binding proteins are identified, it is becoming clear that there is not a large excess of CaM in the cell. Although the levels of all the CaM binding proteins have not been quantitated for any cell type, our previous studies of CEF (Burgess et al., 1984; Van Eldik et al., 1984) and the data reported here strongly suggest that the concentration of CaM would be only slightly higher than, if not equal to, the sum of the amount of known CaM binding proteins. Studies of the effects of CaM gene inactivation in yeast (Davis et al., 1986b; Takeda and Yamamoto, 1987) are also consistent with the possibility that the CaM concentration in a cell does not greatly exceed the sum of the concentration of CaM binding proteins. In general, it may be that the CaM concentration in the cell approximates the sum of the concentrations of all the CaM binding proteins in that cell.

Related to the above possibility are observations that raise the question of whether the biosynthesis of CaM and CaM binding proteins is coupled. One example is our previous work with normal and virus-transformed cells (Van Eldik and Burgess, 1983; Zendegui et al., 1984; Burgess et al., 1984), where we showed that in v-src transformed cells, when there is a perturbation in CaM, there are also perturbations in certain CaM binding proteins, including MLCK. In this report, we have extended those previous findings to show that there are perturbations in certain CaM-associated proteins detected by immunoprecipitation with anti-CaM antibodies. In addition, we have shown here that there is a decrease in MLCK mRNA levels in transformed CEF. These data raise the possibility that there is a quantitative and possibly coordinate, or coupled, regulation of the levels of CaM and CaM binding proteins. A second example is in certain inherited disorders, where a decrease in a CaM binding protein, phosphorylase kinase, is accompanied by a concomitant decrease in the tissue level of CaM by that amount normally found in a complex with phosphorylase kinase (Cohen and Klee, 1988; Bender et al., 1988). Finally, a related example are recent studies (Zimmer et al., 1988) of the molecular genetics and expression of CaM that suggest a possible biosynthetic coupling between the production of CaM and proteins found in the same subcellular structure.

Overall, a growing body of evidence suggests that there is a quantitative regulation of the levels of CaM and a qualitative, as well as quantitative, regulation of CaM binding proteins. This implies that the balance of the expression of CaM and its target proteins may be important in cellular homeostasis. Further, a 2- to 4-fold change in CaM production, such as occurs in transformed cells, could have profound effects on calcium signal transduction events because CaM exerts its effects early in signal amplification pathways. It is worth noting that a number of precedents are available where small changes in effector molecules can be amplified into biologically significant effects. In many cases these effector molecules are acting at early steps in signal amplification pathways that are part of an interconnected system of individual pathways with negative and positive cross-over points, including feedback branches. Thus, small perturbations early in one branch of such an interconnected homeostatic system can have a profound effect which can be restored to basal level by the inherent dampening in the system. Some of the more extensively characterized examples in regulatory biology come from the fields of endocrinology and metabolism. For example, 2- to 3-fold changes in insulin, glucagon, or catecholamines result in dramatic changes in metabolism (Cherrington and Vranic, 1986). Quantitative oscillations or variations are an inherent feature of the pathways in which these molecules are involved, yet further perturbations can result in a disease state.

How the production of CaM and CaM binding proteins is regulated at the molecular level, and how this regulation might be altered with changes in the physiological state of the cell are not known. Clearly, the overall pattern of distribution and expression of CaM binding proteins under a selective set of conditions or in specific cell types could influence the particular CaM pathways that operate in response to calcium fluxes in a given tissue or cell. Additional studies addressing these questions are required if we are to have a sound understanding of how calcium signal transduction complexes are synthesized and regulated in the eukaryotic cell. The knowledge and results gained from this research will provide a firm foundation on which to base future studies, will help give a framework for interpretation of studies of other calcium signal transduction complexes, and may yield insight into how perturbation of calcium and protein kinase mediated pathways might be involved in disease states.

ACKNOWLEDGEMENTS

These studies were supported in part by National Science Foundation grant DCB8302912 (LVE), American Cancer Society institutional research grant IN-25V (LVE), National Cancer Institute predoctoral training grant CA09592 (SWB/LVE), National Institutes of Health grant GM33481 (DMW), and funds from the Howard Hughes Medical Institute (LVE).

REFERENCES

Bender, P.K., Dedman, J.R., and Emerson, C.P., 1988, The abundance of calmodulin mRNAs is regulated in phosphorylase kinase-deficient skeletal muscle. J. Biol. Chem., 263:9733.

Burgess, W.H., Schleicher, M., Van Eldik, L.J., and Watterson, D.M., 1983, Comparative studies of calmodulin, in: "Calcium and Cell Function," W.Y. Cheung, ed., Academic Press, N.Y., p. 209.

Burgess, W.H., Watterson, D.M., and Van Eldik, L.J., 1984, Identification of calmodulin binding proteins in chicken embryo fibroblasts. J. Cell Biol., 99:550.

Cherrington, A.D., and Vranic, M., 1986, in: "Hormonal Regulation of Gluconeogenesis," N. Kraus and Friedmann, eds., CRC Press, FL.

Cohen, P., and Klee, C.B., eds., 1988, "Molecular Aspects of Cellular Regulation - Calmodulin," Elsevier, Amsterdam, Vol. 5.

Davis, L.G., Dibner, M.D., and Battey, J.F., 1986a, "Basic Methods in Molecular Biology," Elsevier, N.Y.

Davis, T.N., Urdea, M.S., Masiarz, F.R., and Thorner, J., 1986b, Isolation of the yeast calmodulin gene: calmodulin is an essential protein. Cell, 47:423.

Epstein, P.N., Christenson, M.A., and Means, A.R., 1989, Chicken calmodulin promoter activity in proliferating and differentiated cells. Mol. Endocrinol., 3:193.

Garrels, J.I., 1989, The QUEST system for quantitative analysis of two-dimensional gels. J. Biol. Chem., 264:5269.

Knauer, D.J., and Smith, G.L., 1979, Regulation of the proliferative response in Rous sarcoma virus transformed chicken embryo fibroblasts by serum and multiplication-stimulating activity (MSA). J. Cell. Physiol., 100:311.

Leof, E.B., Proper, J.A., Gaustin, A.S., Shipley, G.D., DiCorlto, P.E., and Moses, H.L., 1986, Induction of c-sis mRNA and activity similar to platelet-derived growth factor by transforming growth factor ß: a proposed model for indirect mitogenesis involving autocrine activity. Proc. Natl. Acad. Sci. USA, 83:2453.

Lukas, T.J., Haiech, J., Lau, W., Craig, T.A., Zimmer, W.E., Shattuck, R.L., Shoemaker, M.O., and Watterson, D.M., 1988, Calmodulin and calmodulin-regulated protein kinases as transducers of intracellular calcium signals. Cold Spring Harbor Symposia on Quantitative Biology, Vol. 53, p. 185.

Putkey, J.A., Tsui, K.F., Tanaka, T., Lagace, L., Stein, J.P., Lai, E.C., and Means, A.R., 1983, Chicken calmodulin genes. A species comparison of cDNA sequences and isolation of a genomic clone. J. Biol. Chem., 258:11864.

Roberts, D.M., Zimmer, W.E., and Watterson, D.M., 1987, The use of synthetic oligodeoxyribonucleotides in the examination of calmodulin gene and protein structure and function. Methods in Enzymology, 139:290.

Schonthal, A., Herrlich, P., Rahmsdorf, H.J., and Ponta, H., 1988, Requirement for fos gene expression in the transcriptional activation of collagenase by other oncogenes and phorbol esters. Cell, 54:325.

Simmen, R.C.M., Tanaka, T., Ts'ui, K.F., Putkey, J.A., Scott, M.J., Lai, E.C., and Means, A.R., 1987, The structural organization of the chicken calmodulin gene: a correction. J. Biol. Chem., 262:4928.

Takeda, T., and Yamamoto, M., 1987, Analysis and in vivo disruption of the gene coding for calmodulin in Schizosaccharomyces pombe. Proc. Natl. Acad. Sci. USA, 84:3580.

Van Eldik, L.J., and Watterson, D.M., 1979, Characterization of a calcium modulated protein from transformed chicken fibroblasts. J. Biol. Chem., 254:10250.

Van Eldik, L.J., and Watterson, D.M., 1981, Reproducible production of antiserum against vertebrate calmodulin and determination of the immunoreactive site. J. Biol. Chem., 256:4205.

Van Eldik, L.J., Zendegui, J.G., Marshak, D.R., and Watterson, D.M., 1982, Calcium binding proteins and the molecular basis of calcium action. Intl. Rev. Cytol., 77:1.

Van Eldik, L.J., and Burgess, W.H., 1983, Analytical subcellular distribution of calmodulin and calmodulin binding proteins in normal and virus-transformed fibroblasts. J. Biol. Chem., 258:4539.

Van Eldik, L.J., Watterson, D.M., Fok, K.-F., and Erickson, B.W., 1983a, Elucidation of a minimal immunoreactive site of vertebrate calmodulin. Arch. Biochem. Biophys., 227:522.

Van Eldik, L.J., Fok, K.-F., Erickson, B.W., and Watterson, D.M., 1983b, Engineering of site-directed antisera against vertebrate calmodulin by using synthetic peptide immunogens containing an immunoreactive site. Proc. Natl. Acad. Sci. USA, 80:6775.

Van Eldik, L.J., Watterson, D.M., and Burgess, W.H., 1984, Immunoreactive levels of myosin light chain kinase in normal and virus-transformed chicken embryo fibroblasts. Mol. Cell. Biol., 4:2224.

Van Eldik, L.J., and Lukas, T.J., 1987, Site-directed antibodies to vertebrate and plant calmodulins. Methods in Enzymology, 139:393.

Van Eldik, L.J., and Roberts, D.M., 1988, Calcium modulated proteins in pathophysiology, in: "Calcium Binding Proteins," M.P. Thompson, ed., CRC Press, Boca Raton, FL, Vol. II, p.59.

Van Eldik, L.J., Staecker, J.L., and Winningham-Major, F., 1988, Synthesis and expression of a gene coding for the calcium modulated protein S100ß and designed for cassette-based, site-directed mutagenesis. J. Biol. Chem., 263:7830.

Watterson, D.M., Van Eldik, L.J., Smith, R.E., and Vanaman, T.C., 1976, Calcium-dependent regulatory protein of cyclic nucleotide metabolism in normal and transformed chicken embryo fibroblasts. Proc. Natl. Acad. Sci. USA, 73:2711.

Watterson, D.M., Iverson, D.B., and Van Eldik, L.J., 1980, Rapid separation and quantitation of 3':5'-cyclic nucleotides and 5'-nucleotides in phosphodiesterase reaction mixtures using high-performance liquid chromatography. J. Biochem. Biophys. Methods, 2:139.

Zendegui, J.G., Zielinski, R.E., Watterson, D.M., and Van Eldik, L.J., 1984, Biosynthesis of calmodulin in normal and virus transformed chicken embryo fibroblasts. Mol. Cell. Biol., 4:883.

Zimmer, W.E., Schloss, J.A., Silflow, C.D., Youngblom, J., and Watterson, D.M., 1988, Structural organization, DNA sequence, and expression of the calmodulin gene. J. Biol. Chem., 263:19370.

HIGH LEVELS OF ONCOMODULIN AND CALMODULIN EXPRESSION IN THE LOG PHASE OF CELL GROWTH IN A CHEMICALLY TRANSFORMED RAT FIBROBLAST CELL LINE

Janaki K. Blum, Ernst W. Sommer, Marianne C. Berger and Martin W. Berchtold

Institute of Pharmacology and Biochemistry, University of Zürich-Irchel, CH-8057 Zürich and Institute of Toxicology, ETH-Zürich, CH-8603 Schwerzenbach, Switzerland

INTRODUCTION

A variety of cellular processes are under the regulatory control of calcium. Alterations in fluxes and steady state levels of intracellular calcium appear to be responsible for evoking these events. Calcium homeostasis has been reported to be altered during tumorigenesis as well as upon transformation of cells in vitro by chemicals or viruses[1]. Neither the underlying molecular mechanisms related to calcium regulation in normal and tumor cells, nor the fact that transformed cells do not respond properly to normal regulatory controls are as yet well understood. Increased intracellular calcium levels could be partially responsible for a permanent activation of DNA synthesis and/or an increased mobility (invasive behaviour) of tumor cells.

The expression of intracellular calcium-binding proteins (CaBPs) such as calmodulin (CaM), which is found in all eukaryotic cells examined, or S-100 proteins is increased in many transformed cells as compared to their normal counterparts. In contrast, the CaBP oncomodulin (OM), which belongs to the CaM superfamily, is expressed in various tumor cells but not in normal animal or human cells[2]. The lack of suitable cell lines producing significant amounts of OM has been a limiting factor in the study of the regulation of this protein and the consequences of its expression. By exposure of rat fibroblasts to N-methyl-N'-nitro-N-nitrosoguanidine (MNNG) in vivo and/or in vitro, a chemically transformed cell line (T14c) which produces high levels of OM was obtained.

Our aim was to compare the morphology and growth characteristics of transformed T14c cells and normal fibroblasts as well as to examine the expression of CaM and OM in relation to cell growth.

CHEMICAL TRANSFORMATION OF RAT FIBROBLASTS

A rapidly proliferating granulation tissue was exposed to MNNG in vivo or in vitro[3]. As an end point of transformation, the soft agar assay was chosen and tumorigenicity verified by tumor growth in nude mice. A chemically transformed fibroblast cell line (T14c) was established by isolation of a soft agar colony and expansion of the cells in culture. Both normal and T14c cells were cultured in DMEM supplemented with 10% fetal calf serum (FCS) and gentamycin in an atmosphere of 8% CO_2 and 5% O_2 at 36°C. Cell monolayers were detached by trypsin treatment and counted in a coulter counter. In log phase growth, normal and T14c fibroblasts exhibit distinct morphologies. Whereas normal cells adhere quite strongly to the culture dish and exhibit characteristic cytoskeletal structures, the neoplastic T14c phenotype shows a more elongated, spindle-like morphology as demonstrated in Fig. 1. Untransformed cells also occupy a larger area of the dish compared to transformed cells. Therefore, confluence is reached at much lower cell numbers in normal cells.

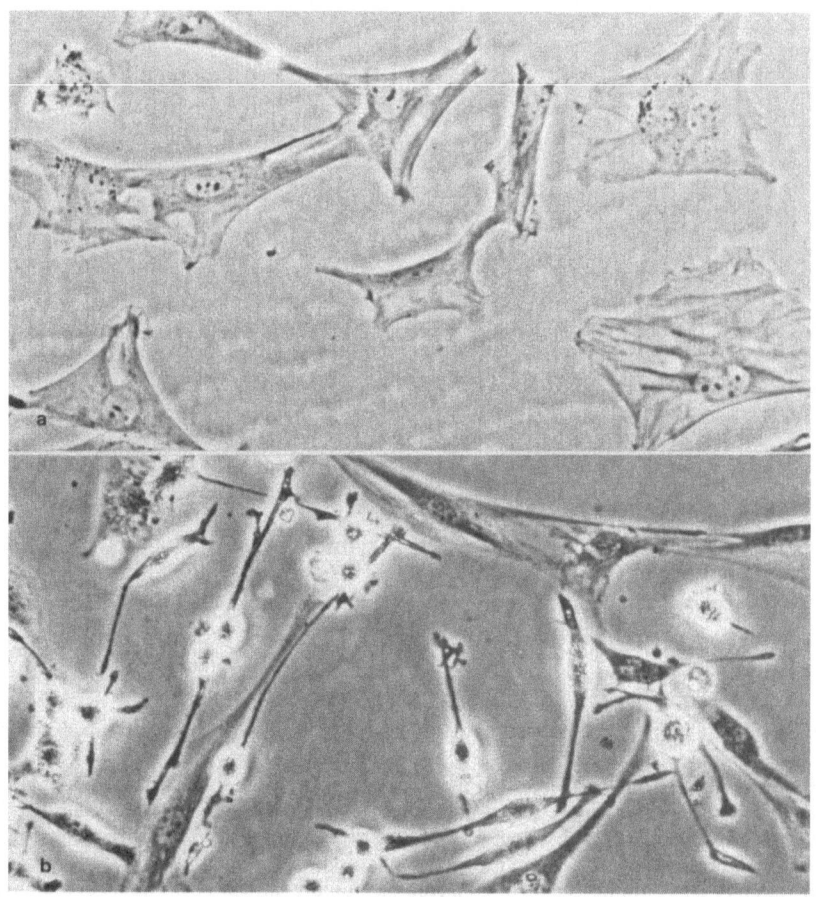

Fig. 1 . Normal rat fibroblasts (a) and chemically transformed fibroblasts derived from a soft agar colony T14c (b) in log phase growth. Cells were seeded at an initial density of 10^6 on 140 mm diameter plastic culture dishes. The neoplastic phenotype reveals a spindle-shaped morphology while some cells that are in mitotic phase have a rounded form.

GROWTH CURVES

The doubling times for both cell types in culture was about 24 hours. However, the doubling time for normal cells appear to be strongly dependent on the number of cell passages. Later passages exhibit a slower rate of exponential growth. Therefore, for growth curves, early cell passages (1 to 4) of normal fibroblasts were used.

When dishes were seeded with 10^6 cells in 20 ml medium, normal cells reached a contact inhibited state after 36 to 48 hours, whereas T14c cells grew in log phase for about 120 to 144 hours (Fig. 2).

NORTHERN BLOT ANALYSES

To study the distribution and abundance of OM transcripts in comparison to CaM mRNA in the rat, RNA from different sources was isolated and analyzed by Northern blots. A specific cDNA probe for OM was obtained from a rat placenta λ gt11 library screened with synthetic oligonucleotides[4] corresponding to the 5' leader sequence of rat OM mRNA[5]. CaM cDNA probe CAM 22[6] was obtained from J. Putkey, University of Texas, Houston. Fig. 3 shows that whereas T14c and Morris Hepatoma RNA gave strong hybridisation signals on Northern blots, neither normal fibroblasts with or without air pouch treatment, nor normal rat brain, nor control tRNA were oncomodulin positive. In both T14c and Hepatoma RNA, a single

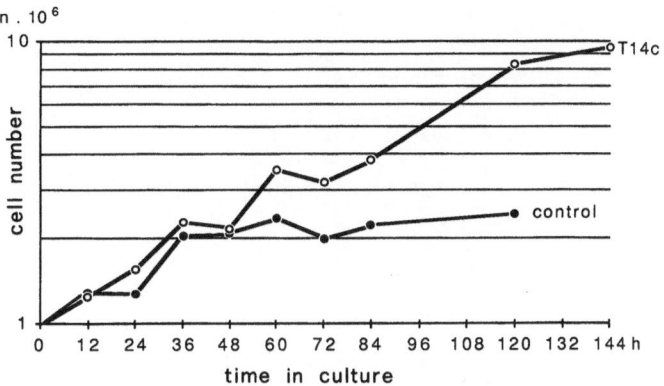

Fig. 2 . Comparison of growth curves for normal control fibroblasts (closed circles) and for the transformed T14c cell line (open circles) reveals differences in their lengths of exponential growth. Both cell lines were seeded at 10^6 cells at time 0. A medium change was performed every 48 hours.

transcript of approximately 600 - 700 nucleotides (as measured by staining of the internal standards of 18S and 28S ribosomal RNA with acridine orange) were detected. For a semiquantitative estimate of the relative levels of OM transcripts, dot blot analysis was performed. Densitometric staining indicated that OM steady state levels were 2 - 4 fold higher in Morris Hepatoma as compared to T14c cells.

Northern and dot blot analyses were also carried out to investigate the relative abundance of OM and CaM mRNA at different stages of cell confluence and in three different rat fibroblast-derived, chemically (T10, T14c) or virally (ras 1) transformed cell lines. In addition, the influence of cell passage number on the expression of mRNAs in one cell line (T10) was examined. T14c RNA contains more than 10 fold the OM mRNA levels as compared to T10, which is comparable to the amounts found in ras 1 virally transformed fibroblasts. There were variations in CaM and OM mRNA levels in RNA from the same cell line (T10) harvested at subconfluence from different passage numbers (Figs. 4, C and D, lanes 1-4). Ratios between OM and CaM signals were roughly constant.

It appears that CaM was expressed at equal or higher levels as compared to OM in T10 and ras 1 cells in contrast to T14c cells, where much stronger OM signals were found, though it must be emphasized that signal intensities on Northern and dot blots cannot be directly compared, owing to different lengths and species origin of the probes.

To test the dependence of OM and CaM expression, RNA from cells at 50% and 100% confluence was hybridized to OM (Fig. 4A) and CaM (Fig. 4B) cDNA probes. In T14c cells, levels of transcripts for both CaM and OM were significantly higher in the log phase of cell growth as compared to confluent stages. Three *bona fide* CaM genes in the rat genome encode for the same protein[7]. Using a chicken calmodulin probe, we found four signals on Northern blots with rat RNA. The major transcript is markedly reduced when cells enter the plateau phase of growth (Fig. 4B). Since the three other signals were much fainter, it is not possible to estimate if the corresponding transcripts are regulated in a similar fashion to the major CaM transcript.

A semiquantitative analysis using dot blot hybridization revealed that the reduction in signal intensities for both transcripts were in the range of 40 - 60% (Figs. 4C and 4D, lanes 5 and 6). A similar situation was found for both OM and CaM transcripts in the ras 1 cell line (Fig. 4C and 4D, lanes 7 and 8.).

CONCLUSIONS

The high level of OM expression in T14c cells opens up new possibilities in the study of the regulation of OM synthesis. Though the level of both CaM and OM transcripts were progressively reduced with increasing time in culture, it is not yet possible to correlate these changes with events in the cell cycle. It is known that the progression of cells through the G1 phase of the cell cycle is critically dependent upon CaM expression[8]. It has been reported that OM can substitute for CaM in stimulating DNA synthesis in Ca^{2+}-deprived cells[9]. In addition, it was shown that OM could replace CaM in its activation of heart phosphodiesterase[10]. These results indicate that OM could play a role similar to CaM with respect to cell

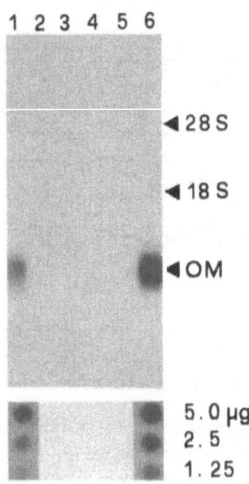

Fig. 3 . Northern and dot blots of different rat cell cultures and tissues hybridized to OM and CaM cDNA. 10 μg of total RNA was electrophoresed on a 1.4% agarose, 6% formaldehyde gel followed by blotting onto a nylon membrane. *Lane 1:* T14c; *lane 2:* fibroblasts, without pouch treatment; *lane 3:* fibroblasts, pouch activated; *lane 4:* rat brain; *lane 5:* tRNA; *lane 6:* Morris Hepatoma. Dot blots containing RNA in dilutions of 5, 2.5, 1.25 μg are shown for semiquantitative comparison. Hybridisation was performed with 2 - 5 x 10^6 cpm/ml of random oligolabelled OM cDNA with a specific activity of 1-2 x 10^9 cpm/μg. Kodak XAR films were exposed for 4 days with intensifying screens.

cycle progression and tumor cell proliferation. So far, no experimental data concerning cell phase specific expression of OM is available. Therefore we plan to examine the possible cell cycle association of OM expression using the transformed granuloma pouch cell (T14c) system as a model.

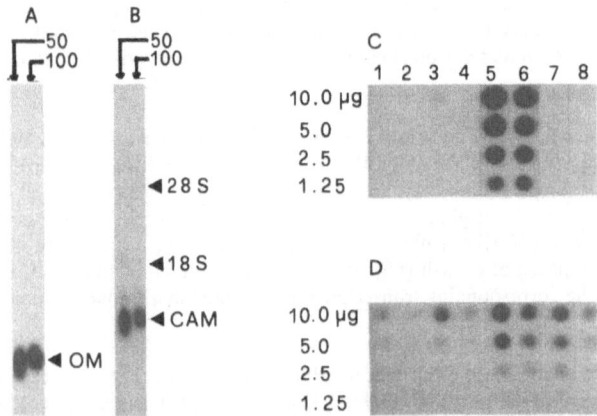

Fig. 4 . Northern and dot blots of transformed cells at different degrees of confluence or passage number hybridized to OM and CaM cDNA. 30 μg of total RNA was treated as in Fig. 3. Autoradiographic exposure was 24 hours for A and 72 hours for B, C, and D. A: OM cDNA hybridized to T14c RNA; B: CaM cDNA hybridized to T14c RNA; C and D: *lane 1*, T10 passage 3; *lane 2*, T10 passage 7; *lane 3*, T10 passage 8; *lane 4*, T10 passage 9; *lane 5*, T14c 50% confluence; *lane 6*, T14c 100% confluence; *lane 7*, ras 1 50% confluence; *lane 8*, ras 1 100% confluence.

ACKNOWLEDGEMENTS

The Swiss National Science Foundation (grant 3.634-0.87), the Swiss cancer league (FOR 337.87.2

and FOR 406.89.1) and the Kanton Zürich League against Cancer are acknowledged for financial support. The ras 1 cell line was kindly provided by Dr. I.D. Diamantis, FMI, Basle.

REFERENCES

1. J.F. Whitfield, A.L. Boynton, J.P. MacManus, M. Sikorska and B.K. Tsang, The regulation of cell proliferation by Ca^{2+} and cAMP. Mol.Cell Biochem. 27:155 (1979)

2. J.P. MacManus, L.M. Brewer and J.F. Whitfield, The widely distributed tumor protein, oncomodulin is a normal constituent of human and rodent placentas. Cancer Lett. 21:309 (1985)

3. E.W. Sommer and C.W. Heizmann, Expression of the tumor-specific and calcium-binding protein oncomodulin during chemical transformation of rat fibroblasts. Cancer Res. 49:899 (1989)

4. C.S. Furter, C.W. Heizmann and M.W. Berchtold, Isolation of a genomic oncomodulin clone and full length placental cDNA from rat. Experientia 45:A21 (1989)

5. M.F. Gillen, D.Banville, R.G. Rutledge, S. Narang, V.L. Seligy, J.F. Whitfield, and J.P. MacManus, A complete complementary DNA for the oncodevelopmental calcium-binding protein, oncomodulin. J.Biol.Chem. 262:5308 (1987)

6. J.A. Putkey, K.F. Ts'ui, T. Tanaka, L. Lagacé, J.P. Stein, E.C. Lai, and A.R. Means, Chicken calmodulin genes. A species comparison of cDNA sequences and isolation of a genomic clone. J.Biol.Chem. 258:11864 (1983)

7. H. Nojima and H. Sokabe, Structure of calmodulin genes in the rat genome. In: "Ca^{2+} signalling", H. Hidaka, ed., Plenum Press, in press (1989)

8. C.D. Rasmussen, and A.R. Means, Calmodulin is involved in regulation of cell proliferation. EMBO J. 6:3961 (1987)

9. A.L. Boynton, J.P. MacManus and J.F. Whitfield, Stimulation of liver cell DNA synthesis by oncomodulin, a MW 11,500 calcium-binding protein from hepatoma. Exp.Cell.Res. 138: 454 (1982)

10. B. Mutus, N. Karrupiah, and J.P. MacManus, The differential stimulation of brain and heart cyclic-AMP phosphodiesterase by oncomodulin. Biochem. Biophys. Res. Commun. 131:500 (1985)

CALMODULIN AND CALBINDIN IN PANCREATIC ISLET CELLS

W.J. Malaisse, F. Blachier, R. Pochet*, B. Manuel y Keenoy and A. Sener

Laboratory of Experimental Medicine, Brussels Free University, Brussels, Belgium

*Laboratoire d' Histologie, Faculté de Médecine, Université Libre de Bruxelles
2 rue Evers, B-1000, Bruxelles, Belgique

ABSTRACT

The process of insulin release evoked by D-glucose and other nutrient secretagogues is triggered by an increase in cytosolic Ca^{2+} activity. However, some other insulinotropic agents may stimulate insulin release at a close-to-basal concentration of cytosolic ionized calcium. The control of cytosolic Ca^{2+} concentration depends not solely on the rate of Ca^{2+} entry into the cell through voltage-sensitive channels and Ca^{2+} exit via Na^+-Ca^{2+} countertransport or active Ca^{2+} pumping, but also on the subcellular distribution of Ca^{2+}, as dependent, for instance, on both Ca^{2+}-ATPase activity and inositol 1,4,5-triphosphate-sensitive release in microsomes and calcium accumulation in mitochondria. Calmodulin and calbindin were both identified in pancreatic islet cells. Activation of adenylate cyclase by calcium-calmodulin may account for the increased production of cyclic AMP in islets stimulated by nutrient secretagogues. Calbindin is present in both normal and tumoral islet cells, and might participate to the alteration of islet function encountered in vitamin D-deprived or repleted rats. However, no target enzyme for calbindin was yet identified in islet cells. Independently of the role of calcium-binding regulatory proteins, the mitochondrial accumulation of calcium may account in part at least, for the preferential stimulation of mitochondrial oxidative events in the process of nutrient-stimulated insulin release.

INTRODUCTION

The ion Ca^{2+} plays a critical role in the process of insulin release from the pancreatic islet B-cell[1]. Thus, several secretagogues provoke insulin secretion by causing a rise in the cytosolic concentration of ionized Ca^{2+}. Some insulinotropic agents, however, affect insulin release independently of any obvious change in cytosolic Ca^{2+} activity. For instance, tumor-promoting phorbol esters may cause activation of protein kinase C and subsequent stimulation of insulin secretion at apparently close-to-basal cytosolic Ca^{2+} concentration. In other words, Ca^{2+} should be considered only as one, but not the sole, factor involved in the coupling between the identification of insulin secretagogues by the B-cell and the eventual activation of the microtubular-microfilamentous effector system for insulin release. The present report aims at summarizing recent acquisitions concerning the regulation of Ca^{2+} fluxes and the regulatory roles of Ca^{2+} in the pancreatic B-cell, with emphasis on the possible participation of the calcium-binding regulatory proteins calmodulin and calbindin in these processes.

CALMODULIN

Calmodulin was identified in islet cells ten years ago[2]. Its concentration is close to 0.1 pmol/islet or 30-50 μM. Further work also contributed to the identification of calmodulin-binding proteins in insulin-producing cells, as reviewed elsewhere[3]. In acellular material prepared from islet cells, calmodulin causes a Ca^{2+}-dependent activation of adenylate cyclase, cyclic AMP phosphodiesterase, plasma membrane-associated Ca-ATPase and protein kinase (see ref. 3 for review). The activation of adenylate cyclase by calcium-calmodulin may account for the finding that several insulin secretagogues known to increase cytosolic Ca^{2+} activity also stimulate cyclic AMP generation in intact islet cells[2]. The latter effect is indeed abolished in the absence of extracellular Ca^{2+} (see ref. 4).

CALBINDIN

Vitamin D and islet function

Studies on the presence of calbindin in islet cells were first undertaken in the light of findings indicating that insulin release may be affected by vitamin D depletion and repletion. Thus, vitamin D deprivation decreases and vitamin D treatment increases the insulin content of the whole pancreas or isolated islets and the secretory response of the islets to D-glucose[5]. The changes in insulin release remain significant when the hormonal output is expressed relative to the insulin content of the islets. Moreover, the time course for changes in biological variables during vitamin D deprivation and treatment suggests that the alteration of pancreatic B-cell function cannot be solely accounted for by concomitant anomalies in either plasma calcium or glucose concentration. It was proposed, therefore, that vitamin D deprivation or repletion may affect in a direct but delayed manner the function of insulin-producing cells[5].

The view that the B-cell may represent a target for a direct action of the active metabolites of vitamin D gained further support from a study of cationic fluxes in islets prepared from control, vitamin D-deprived and vitamin D-repleted rats[6]. Briefly, it was first observed that vitamin D deprivation decreases and vitamin D repletion restores the secretory response not only to D-glucose but also to a rise in extracellular Ca^{2+} concentration. Second, the vitamin D-induced changes in insulin release were found to coincide with an unaltered K^+ conductance, as judged from the fractional outflow rate of $^{86}Rb^+$ from prelabelled islets, in both resting and stimulated islets. This finding argues against any major alteration in nutrient catabolism. Last, the perturbation of ^{45}Ca handling induced by vitamin D deprivation or repletion also differed from that found in starved rats and were compatible with an altered regulation of Ca^{2+} uptake by and/or release from some intracellular binding or sequestration site[6].

Presence of calbindin in islet cells

Calbindin was recently identified in both normal and tumoral islet cells[7,8]. In rat as distinct from chick islets[9], the calbindin content, relative to cell volume, appears higher in non-B than B islet cells (Table 1). In tumoral islet cells of the RINm5F line, heat-resistant proteins of higher molecular weight than calbindin-D 28K also reacted with a polyclonal antibody raised against chick duodenal calbindin. The calbindin content of RINm5F cells was little affected when they were cultured in a medium containing 1,25-dihydroxyvitamin D_3-depleted calf serum.

Functional role of calbindin

To our knowledge, no target system for calbindin was so far identified in islet cells. The cytosolic concentration of Ca^{2+} is thought to affect the activity of a number of enzymes (e.g. protein kinase C or phospholipase C) involved in the functional response of the B-cell to D-glucose or other insulin secretagogues. For instance, the interference of Ca^{2+} with phospholipase C activity was documented in isolated cell membrane[11] and permeabilized islets[12].

Activation of phospholipase C by Ca^{2+} could also be documented in RINm5F cells first preincubated for 120 min in the presence of 2.8 mM D-glucose and myo-[2-^{3}H] inositol (4-6 μM) and then sonicated and incubated for 10 min at 37°C in a phosphate buffer (1.0 mM; pH 7.4) containing NaCl (20 mM), KCl (100 mM), $MgSO_4$ (5 mM), LiCl (5 mM), $NaHCO_3$ (equilibrated against ambient air), bovine serum albumin (20 mg/ml) and, as required EGTA (1.0 mM) and/or Ca^{2+} (0.1 μM to 1.0 mM). At the end of the incubation period, the samples (10^6 cells/60 μl) were treated for extraction of tritiated inositol phosphates and lipids, as described elsewhere[13]. Under these experimental conditions, Ca^{2+} caused a concentration-related stimulation of tritiated inositol phosphates production up to a several fold increase in reaction velocity relative to the paired basal value recorded in the presence of EGTA. In this system, neither calmodulin nor calbindin, both tested at a concentration of 8 μM, affected significantly the Ca^{2+}-stimulated production of tritiated inositol phosphates (Table 2). A possible contribution of endogenous calcium-binding regulatory proteins should, however, not be ruled out. In these experiments, calbindin-D 28K was prepared from rat cerebellum. Briefly, the rat cerebellum was extracted with a teflon-glass homogenizer in four volumes of a Tris-HCl buffer (15 mM, pH 7.4) containing NaCl (0.12 M), KCl (3 mM) and aprotinine (0.02 U/ml).The homogenate was centrifuged for 60 min at 100,000 g, and the resulting supernatant heated for 10 min at 65°C. After a further centrifugation for 15 min at 10,000 g, the supernatant was lyophilized and passed through an affinity column prepared with chick intestinal calbindin antibody (generously donated by D.E.M. Lawson, Cambridge, U.K.) coupled to CNBr Sepharose (Pharmacia, Uppsala, Sweden) as previously described.[14] The amount of purified protein was determined using the Lowry procedure, and visual-

Table 1 . Western blot scanning densitometry of the calbindin immunoreactive band

Cell type	Cell number x 10^{-3}	A^b	Calbindin relative area (per cell)[a]	
			B^b	B^b
Islets (10)	20.0^c	1.61 (0.08)		
B-cells	11.5	N.D.^{e}(?)	1.45 (0.13)	
	30.0			1.83 (0.06)
	47.0	2.60 (0.06)	2.76 (0.06)	
	120.0			3.40 (0.03)
Non-B-cells	33.5	3.06 (0.09)	4.55 (0.14)	
	52.5			6.12 (0.12)
	133.0	14.42 (0.11)	12.14 (0.09)	
	210.0			11.81 (0.06)

Non-B/B cell ratio

- per cell^d		1.90 (1.81)	1.39 (1.23)	1.96 (1.94)
- per fl^d		7.34 (6.97)	5.38 (4.74)	7.56 (7.50)

[a] The area of the calbindin band is expressed in arbitrary units (in each experiment, the mean area for each non-B-cell, as derived from the readings collected in both concentrated and diluted extracts, was taken as 0.10); the value quoted in parentheses refers to the area per individual cell.

[b] Results are derived, as indicated, from the experiments illustrated either in Fig. 3A or B in ref. 7.

[c] Results derived from 10 islets; each islet was assumed to contain 2×10^3 cells.

[d] The non-B/B-cell ratio for calbindin content is expressed either per cell or per fl, taking into account a mean volume of 776 fl and 201 fl for B- and non-B-cells, respectively[10]; the ratios were derived either from the integrated values for calbindin area and cell numbers (obtained by summation of the data collected in both concentrated and diluted extracts) or from the mean of available results per cell (in parentheses).

[e] N.D. : not determined because of poor density.

ized in SDS-PAGE by silver staining (Biorad, Richmond, USA). After loading the electrophoresis gel with 2 μg of protein eluted from the anticalbindin column and further dialyzed, a single band with an apparent molecular weight of 28 KDa was detected.

In the light of the results collected in islets removed from vitamin D-depleted or vitamin D-repleted rats, investigations were also undertaken to assess the possible effect of calbindin upon the net uptake of

Table 2 . Effect of Ca^{2+}, calmodulin and calbidin upon phospholipase C activity in RINm5F cell homogenates

EGTA	Ca^{2+}	Calmodulin	Calbindin	[^{3}H]Inositol phosphates
1.0 mM	-	-	-	100 ± 2 (3)^a
-	1.0 mM	-	-	162 ± 2 (3)
-	1.0 mM	8.0 μM	-	163 ± 3 (3)
-	1.0 mM	-	8.0 μM	154 ± 3 (3)

[a] Mean results ($\pm$ SEM) are expressed relative to the basal readings recorded in the presence of EGTA. Such a basal value was little affected over the 10 min incubation period.

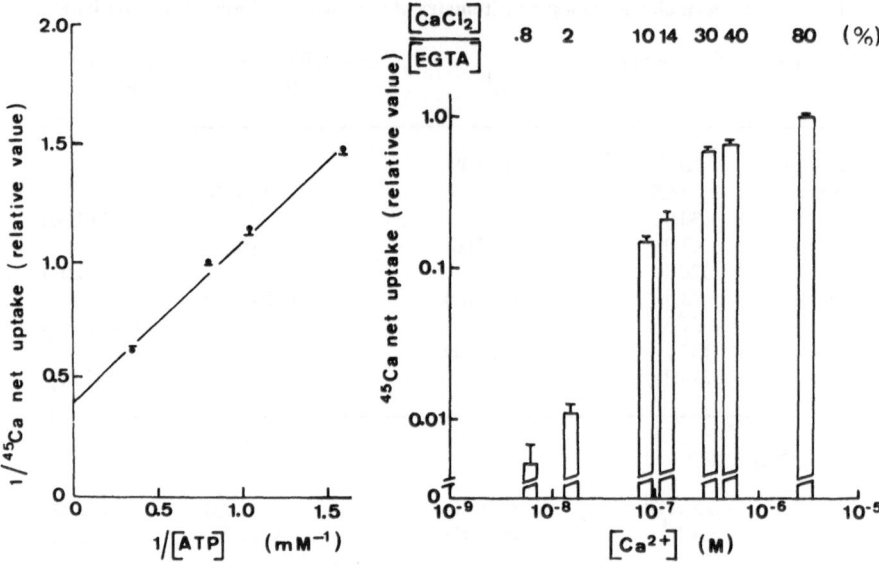

Fig. 1. Net uptake of ^{45}Ca by RINm5F microsomes incubated for 30 min at 37°C. Left panel : double reciprocal plot as a function of ATP concentration, all results being expressed relative to the value found in the presence of 1.3 mM ATP; mean value ($\pm$ SEM) are derived from 3 individual experiments. Right panel : effect of increasing [CaCl$_2$]/[EGTA] ratios (EGTA concentration : 0.38 - 0.50 mM) in the presence of 1.3 mM ATP; data are plotted in logarithmic scales as a function of the estimated Ca^{2+} concentration, mean values ($\pm$ SEM) refer to 3 individual experiments and are expressed relative to the mean reading recorded at the highest Ca^{2+} concentration.

^{45}Ca^{2+} by a microsome-rich subcellular fraction isolated from RINm5F cells. This fraction was isolated by centrifugation for 60 min at 160,000 g and 4°C of material prepared by sonication (1 x 20 s) of RINm5F cells (approximately 10 x 10^6 cells/ml) in a Hepes-KOH buffer (5.0 mM, pH 7.4) containing mannitol (220 mM), sucrose (70 mM) and EDTA (1.0 mM). After centrifugation for 10 min at 770 g to remove nuclei and cell debris, a mitochondria-rich pellet was first obtained over 10 min centrifugation at 12,000 g and 4°C. The corresponding supernatant was then used to isolate the microsomal fraction. In a typical experiment, microsomes (34 μg protein per sample) were incubated for 30 min at 37°C in 0.1 ml of a Hepes-KOH buffer (10 mM, pH 7.1) containing sucrose (125 mM), KCl (40 mM), MgCl$_2$ (5 mM), ammonium oxalate (10 mM), EGTA (0.38 mM) and, as required, ATP (1.3 mM) and CaCl$_2$ (0.11 to 0.45 mM) mixed with a tracer amount of ^{45}CaCl$_2$. The reaction was halted by addition of 3.0 ml of an iced solution of sucrose (300 mM) in a Hepes-KOH buffer (5.0 mM, pH 7.0). The sample was then immediately passed through a Millipore filter (HAWP, 0.45 μm pore size) with the aid of a vacuum pump. The tube was washed twice with 3.0 ml of the same iced solution, which was then also passed through the filter, the full procedure being completed within 20 s. Each filter was then placed in a counting vial and dried in an oven at 110°C. The radioactive content of the counting vial was eventually examined by liquid scintillation, using 10 ml of Lumagel (Lumac, Landgraaf, The Netherlands). In the absence of calbindin, the net uptake of ^{45}Ca^{2+} (CaCl$_2$ concentration : 0.45 mM) amounted to 11.3 $\pm$ 0.7 and 113.1 $\pm$ 2.5 nmol/mg protein in the absence and presence of ATP, respectively. The ATP-dependent net uptake of ^{45}Ca by the microsomal fraction was markedly inhibited by either a lowering of temperature to 10°C or the addition of vanadate (2.0 mM) to the assay medium, whilst being little affected by ruthenium red (10 μM). It also depended on the Ca^{2+} concentration of the medium. For instance, in the presence of only 0.11 mM CaCl$_2$, the net uptake of ^{45}Ca^{2+} was decreased to 65.9 $\pm$ 1.3 nmol/mg protein. Investigations on the effect of calbindin in this system were hampered by the likely binding of Ca^{2+} by this regulatory protein. This binding phenomenon may well account for the fact that, at the low CaCl$_2$ concentration mentioned above, calbindin (2.2 μM) decreased ^{45}Ca net uptake from 65.9 $\pm$ 1.3 to 24.0 $\pm$ 0.4 nmol/mg protein. Indeed, at a much higher Ca^{2+} concentration (CaCl$_2$ 0.45 mM), calbindin failed to inhibit ^{45}Ca net uptake. These results are reminiscent of those obtained by Ghijsen et al[15]. in basolateral membrane vesicles isolated from rat duodenum.

Further work is obviously required to fully assess the possible effect of calbindin upon this and other Ca^{2+}-sensitive processes in islet cells.

MICROSOMAL HANDLING OF Ca^{2+}

In addition to the classical role of cytosolic Ca^{2+} as a trigger for insulin release and as possibly mediated by Ca^{2+}-binding protein(s), other regulatory roles of the divalent cation in islet cells should not be overlooked. For instance, within the range of concentrations between 10^{-8} and 10^{-5} M, Ca^{2+} causes a dramatic ATP-dependent and vanadate-sensitive stimulation of Ca^{2+} uptake by islet cell microsomes (see above), with an apparent Ka close to 0.2 μM (Fig.1). Since the ATP-dependency of such a process yields a Km close to 1.2-1.7 mM, it appears that both cytosolic Ca^{2+} and ATP, within their physiological range of concentrations, may regulate in a synarchistic manner the uptake of Ca^{2+} by these organelles. The latter process, however, apparently accounts for only a minor fraction of the total amount of ATP consumed by intact islet cells.

MITOCHONDRIAL HANDLING OF Ca^{2+}

It should be realized that, independently of the regulatory roles attributed to calmodulin and calbindin in either the movements of Ca^{2+} or enzymatic response to Ca^{2+} in islet cells, this ion is likely to play itself a major role in the stimulus-secretion coupling process for insulin release. It has long been known both that mitochondria may participate in the gross control of Ca^{2+} cytosolic activity in islet cells[16] and that, in these cells, nutrient secretagogues increase the calcium content of mitochondria, at least as judged by radioisotopic criteria[17,18]. We now wish to propose that the mitochondrial accumulation of Ca^{2+}, which could conceivably be attributable to a change in mitochondrial redox state[19], may participate in the feedback control of metabolic events by cationic factors in nutrient-stimulated B-cells.

This proposal is based on the following considerations. In normal islets, D-glucose causes a

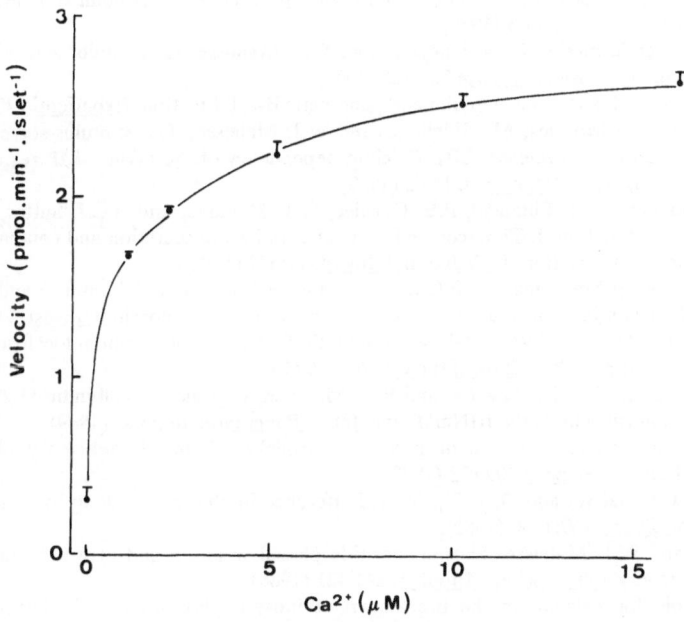

Fig. 2. Effect of increasing concentrations of Ca^{2+} (EGTA 0.5 mM or EGTA and $CaCl_2$ both 10μM, 40 μM, 0.2 mM, 0.8 mM and 2.0 mM) upon the activity of 2-ketoglutarate dehydrogenase in islet homogenates incubated at 37°C in a imidazole-HCl buffer (50 mM, pH 7.2) containing [1-^{14}C]2-ketoglutarate (0.5 mM), NAD$^+$ (1.0 mM) and coenzyme A (0.5 mM). Mean values ($\pm$ SEM) are derived from triplicate measurements in each of 3 individual experiments.

preferential stimulation of mitochondrial oxidative events, such as aerobic glycolysis (as coupled with the transfer of reducing equivalents into the mitochondria at the intervention of the glycerol phosphate shuttle; see ref. 21), pyruvate decarboxylation or oxidation of acetyl residues in the Krebs cycle, relative to the total rate of D-glucose utilization[20]. This unusual situation[21,22], represents a concentration- and time-related phenomenon[23]. It is severely impaired by environmental manipulations aiming at preventing the activation by D-glucose of ATP-consuming functional processes such as proinsulin biosynthesis or Ca^{2+} active pumping[23]. Such is the case, for instance, in the absence of extracellular Ca^{2+}. It is conceivable therefore that, in the normal process of insulin release evoked by D-glucose (or L-leucine; see ref. 23), the mitochondrial accumulation of Ca^{2+} accounts, in part at least, for the preferential stimulation of mitochondrial oxidative events. This is compatible with the Ca^{2+}-dependency of several mitochondrial dehydrogenases, such as 2-ketoglutarate dehydrogenase (Fig.2).

CONCLUDING REMARKS

Although the participation of Ca^{2+} in the process of insulin release was first disclosed more than two decades ago, it is obvious that our knowledge on the role of this cation in the secretory process is still dependent on a constant inflow of novel information. The identification of calcium-binding regulatory proteins in islet cells provided a new conceptual framework to account for such phenomena as the activation of adenylate cyclase by calmodulin in islet cells stimulated by secretagogues causing an increase in cytosolic Ca^{2+} concentration or the direct but delayed control of B-cell secretory activity by vitamin D, as possibly mediated by calbindin. It should not be overlooked, however, that the remodelling of Ca^{2+} fluxes in islet cells stimulated by suitable secretagogues may, independently of any intervention of calcium-binding cytosolic proteins, exert a major regulatory role in the reciprocal coupling between metabolic and cationic events. In this respect, the present report introduces the concept that the mitochondrial accumulation of calcium may represent an essential determinant of the preferential stimulation of mitochondrial oxidative events by nutrient secretagogues.

REFERENCES

1. W.J. Malaisse, Cellular calcium : secretion of hormones, in: Calcium in Human Biology, B.E.C. Nordin, ed., Springer-Verlag, London (1988).
2. I. Valverde, A. Vandermeers, R. Anjaneyulu, and W.J. Malaisse, Calmodulin activation of adenylate cyclase in pancreatic islets, Science 206:225 (1979).
3. I. Valverde, and W.J. Malaisse, Calmodulin and pancreatic B-cell function, Experientia 40:1061 (1984).
4. I. Valverde, P. Garcia-Morales, M. Ghiglione, and W.J. Malaisse, The stimulus-secretion coupling of glucose-induced insulin release. LIII. Calcium-dependency of the cyclic AMP response to nutrient secretagogues, Horm. Metab. Res. 15:62 (1983).
5. H. Labriji-Mestaghanmi, B. Billaudel, P.E. Garnier, W.J. Malaisse, and B.C.J. Sutter, Vitamin D and pancreatic islet function. I. Time-course for changes in insulin secretion and content during vitamin D deprivation and repletion, J. Endocrinol. Invest. 11:577 (1988).
6. B. Billaudel, H. Labriji-Mestaghanmi, B.C.J. Sutter, and W.J. Malaisse, Vitamin D and pancreatic islet function. II. Dynamics of insulin release and cationic fluxes, J. Endocrinol. Invest. 11:585 (1988).
7. R. Pochet, D.G. Pipeleers, and W.J. Malaisse, Calbindin D-27kDa : preferential localization in non-B islet cells of the rat pancreas, Biol. of the Cell 61:155 (1987).
8. R. Pochet, F. Blachier, D.E.M. Lawson, and W.J. Malaisse, Presence of calbindin-D 28K in endocrine pancreatic tumoral cells of the RINm5F line. Int. J. Pancreatol. in press (1989).
9. A.W. Norman, B.J. Frankel, A.M. Heldt, and G.M. Grodsky, Vitamin D deficiency inhibits pancreatic secretion of insulin, Science 209:823 (1980).
10. F.K. Gorus, W.J. Malaisse, and D.G. Pipeleers, Difference in glucose handling by pancreatic A and B cells, J. Biol. Chem. 259:1196 (1981).
11. M.E. Dunlop, and W.J. Malaisse, Phosphoinositide phosphorylation and hydrolysis in pancreatic islet cells membrane, Arch. Biochem. Biophys. 244:421 (1986).
12. L. Best, A role for calcium in the breakdown of inositol phospholipids in intact and digitonin-permeabilized pancreatic islets, Biochem. J. 238:773 (1986).
13. F. Blachier, and W.J. Malaisse, Possible role of a GTP-binding protein in the activation of phospholipase C by carbamylcholine in tumoral insulin producing cells, Res. Commun. Chem. Pathol. Pharmacol. 58:237 (1987).
14. B. Pasteels, N. Miki, S. Hatakenaka, and R. Pochet, Immunohistochemical cross-reactivity and electrophoretic comigration between calbindin D-27kDa and visinin, Brain Res. 412:107 (1987).

15. W.E.J.M. Ghijsen, C.H. Van Os, C.W. Heizmann, and H. Murer, Regulation of duodenal Ca^{2+} pump by calmodulin and vitamin D-dependent Ca^{2+}-binding protein, <u>Am. J. Physiol.</u> 251:G223 (1986).

16. M. Prentki, D. Janjic, and C.B. Wollheim, The regulation of extramitochondrial steady-state free Ca^{2+} concentration by rat insulinoma mitochondria, <u>J. Biol. Chem.</u> 258:7597 (1983).

17. T. Andersson, C. Betsholtz, and B. Hellman, Granular calcium exchange in glucose-stimulated pancreatic B-cells, <u>Biomed. Res.</u> 3:29 (1982).

18. T. Andersson, P.-O Berggren, E. Gylfe and B. Hellman, Amounts and distribution of intracellular magnesium and calcium in pancreatic B-cells, <u>Acta Physiol. Scand.</u> 114:235 (1982).

19. W.J. Malaisse, and A. Sener, The redox potential, <u>in:</u> Energetics of Secretion Response, J.W.N. Akkerman, ed., CRC Press, Boca Raton (1988).

20. A. Sener, and W.J. Malaisse, Stimulation by D-glucose of mitochondrial oxidative events in islet cells, <u>Biochem. J.</u> 246:89 (1987).

21. A. Sener, J. Rasschaert, D. Zähner and, W.J. Malaisse, Hexose metabolism in pancreatic islets. Stimulation by D-glucose of [2^{-3}H] glycerol detritiation, <u>Int. J. Biochem</u> 20:595 (1988).

22. A. Sener, F. Blachier, and W.J. Malaisse, Crabtree effect in tumoral pancreatic islet cells, <u>J. Biol. Chem.</u> 263:1904 (1988).

23. W.J. Malaisse, and A. Sener, Hexose metabolism in pancreatic islets. Feedback control of D-glucose oxidation by functional events, <u>Biochim. Biophys. Acta</u> 971:246 (1988).

STRUCTURAL AND FUNCTIONAL CHARACTERIZATION OF PROTEIN I

(p36₂p11₂) and II (p32) - CALCIUM/PHOSPHOLIPID BINDING PROTEINS

WITH HOMOLOGIES TO LIPOCORTIN I

Volker Gerke

Max-Planck-Institute for Biophysical Chemistry, Goettingen, FRG

INTRODUCTION

Certain Ca^{2+}-binding proteins, which interact with phospholipids in a Ca^{2+}-dependent manner and are discussed to be involved in the control of membrane fusion events, belong to a newly characterized multigene family (for review see Klee, 1988; Crumpton et al., 1988). Members of this family show a high degree of sequence homology and share a common structural feature : a segment of 70 to 80 amino acids in length, which is repeated either four- or eight-fold (depending on the molecular weight of the individual member). These repeated motifs exhibit homologies with each other, not only within one particular protein, but also between different members of the multigene family. Mild proteolytic cleavage defines two regions in the Ca^{2+}/lipid-binding proteins : a protease-resistent core, which comprises the repeat motifs, and a short N-terminal tail, which is sensitive to protease and variable in sequence and lenght. While the Ca^{2+}- and phospholipid-binding sites reside in the core, the N-terminal tail often bears tyrosine and/or serine/threonine phosphorylation sites (Glenney and Tack, 1985; De et al., 1986; Gould et al., 1986; Johnsson et al., 1986; Weber et al., 1987). Two major cellular substrates of tyrosine-specific protein kinases belong to this family (for review see Brugge, 1986). p35 is phophorylated by the EGF receptor/kinase. It is identical to lipocortin I, originally described as a steroid-induced inhibitor of phospholipase A_2. p36 (also called calpactin I heavy chain, or lipocortin II) is the major cytoplasmic substrate of src-related viral tyrosine kinases. In contrast to other members of this family, p36 forms a tetrametric complex (protein I) with a dimer of a unique p11 polypeptide (Erikson et al., 1984; Gerke and Weber, 1984, 1985a). Sequence analysis revealed that p11 belongs to the S-100 family of Ca^{2+}-binding proteins (Gerke and Weber, 1985b; Glenney and Tack, 1985; Hexham et al., 1986). The intact protein I complex (p36₂p11₂) was originally isolated from intestinal epithelial cells by exploiting its Ca^{2+}-dependent interaction with cytoskeletal and/or membrane structures. In addition, a different Ca^{2+}/lipid-binding protein (protein II) was also found to reside within these cells and was purified following a similar protocol (Gerke and Weber, 1984).

Here, some biochemical and structural characteristics of the two intestinal Ca^{2+}/lipid-binding proteins, p36 and protein II, are described. In addition, the expression of both mRNAs is compared in different tissues and cell lines.

RESULTS AND DISCUSSION

Biochemical Characterization of Proteins I and II

Intestinal brush border membrane vesicles prepared in the presence of Ca^{2+} contain substantial amounts of proteins I and II. Both are specifically extracted by the addition of EGTA and can be purified by a series of conventional chromatographic steps (Gerke and Weber, 1984). The purified proteins have been characterized in several laboratories (Gerke and Weber, 1984, 1985a; Glenney, 1986). Their properties are listed in Table 1. Although both proteins have several properties in common, only protein I (as well as its p36 subunit) interacts in a Ca^{2+}-dependent manner with cytoskeletal elements. While the biological significance of the F-actin binding is not yet understood, the interaction of protein I (and p36) with non-erythroid-spectrin could reflect the situation within the living cell. Immunofluorescence as well as immuno-electron microscopy revealed that both proteins show a very similar, if not identical, distribution within cultured fibroblasts, and are localized in the submembraneous network (Gerke and Weber, 1984, and references therein; Semich, Gerke, Robenek, Weber, submitted).

Phosphorylation of p36 by the EGF Receptor/kinase

The parameters influencing tyrosine phosphorylation of p36 were studied using the EGF receptor/kinase to phosphorylate p36 *in vitro*. As shown in Fig.1, this phosphorylation is EGF-dependent and tyrosine-specific. Interestingly, the presence of Ca^{2+} significantly increases the phosphate incorporation into p36. Since autophosphorylation of the EGF receptor/kinase is not affected by Ca^{2+}, the observed Ca^{2+}-effect is most likely based on the well documented Ca^{2+}-induced conformational changes in the p36 substrate (Gerke and Weber, 1985a).

Phosphorylation of p36 by pp60[src] and serine/threonine-specific protein kinases occurs in the N-terminal tail region of the molecule (Glenney and Tack, 1985; Gould et al., 1986; Johnsson et al., 1986). Since this tail also harbors the binding site for p11 (Glenney, 1986; Johnsson et al., 1988), it was of interest to investigate whether p11 binding would affect p36 phosphorylation by the EGF receptor/kinase. Figure 1 (panel D) shows that p11 binding interferes with the phosphorylation. Monomeric p36 serves as a much better substrate, compared to p36 associated with p11 in the heterotetrameric protein I complex. Densitometric scanning revealed that the difference is in the order of 5 to 10 fold. This result could explain the finding that EGTA extracts from intestinal epithelium contain no significant phosphorylated substrates of the EGF receptor/kinase (De et al., 1986). Since protein I is by far the most prominent p36 species in these extracts (Gerke and Weber, 1985a), phosphorylation of this complexed p36 by the EGF receptor/kinase can be expected to be rather poor. Exact delineation of the p11 binding site revealed that it is formed by the first 12 residues of p36 (Johnsson et al., 1988). Since the phosphorylated Tyr-23 is not part of this binding site, it seems likely that p11 binding renders this residue less accessible for the kinase by steric hindrance, or by forcing the p36 molecule into a different conformation.

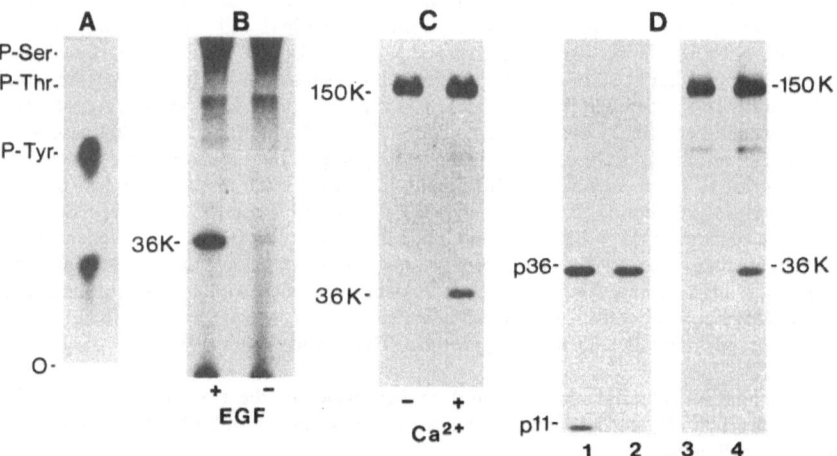

Fig. 1 . Phosphorylation of p36 and protein I by the EGF receptor/kinase. Purified p36 (panels A, B, C and panel D, lanes 2, 4) or protein I (panel D, lanes 1, 3) were reacted with A-431 cell membrane vesicles containing an enzymatically active 150 K fragment of the EGF receptor/kinase as described (Fava and Cohen, 1984). Products of the total reaction (panels C, D) or immunoprecipitates using a polyclonal p36 antibody (panel B) were separated in SDS-gels and phosphorylated proteins were visualized by autoradiography. Phosphoamino acid analysis of the phosphorylated p36 is shown in panel A. Reactions performed in the absence or presence of EGF and Ca^{2+} are seen in panels B and C, respectively. Panel D compares the phosphorylation of monomeric p36 (lanes 2, 4) and protein I ($p36_2 p11_2$; lanes 1, 3). Reactions were carried out with identical amounts of p36 polypeptide. Lanes 1 and 2 show the Coomassie stain, lanes 3 and 4 the corresponding autoradiogram.

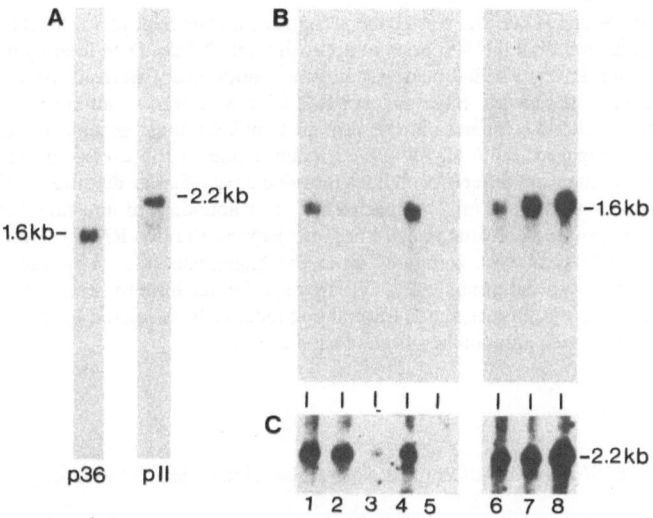

Fig. 2 . p36 and protein II mRNA size and tissue distribution. Total RNA from different cell lines and tissues was analyzed by Northern blotting using oligonucleotide probes specific for p36 (corresponding to amino acids 289-299 of murine p36; Saris et al, 1986) and protein II (amino acids 215-225 of porcine protein II; Weber et al, 1987). Panel A reveals the transcript size and shows the specificity of the probes on RNA from MDBK cells. Panels B and C show identical blots hybridized with either the p36 (B) or the protein II (C) probe. RNA samples were from mouse 3T3 cells (lane 6), human A-431 cells (lane 7), bovine MDBK cells (lane 8), and the following rat tissues 1, lung; 2, spleen; 3, kidney; 4, intestine; 5, brain. Lanes 1 and 2 of panel C contain 5 times the amount of RNA as compared to all other lanes.

Distribution of p36 and Protein II in Different Cell Lines and Tissues

Several studies have shown that different Ca^{2+}/lipid-binding proteins often coexist in different cell types (for review see Klee, 1988). In order to extend these observations and obtain unambiguous data on the protein II expression (in direct comparison to p36), I used oligonucleotide probes to analyze the synthesis of protein II, and p36, mRNA in Northern blots. Figure 2 (panel A) shows the specificity of the

Table 1 . Biochemical Properties of Intestinal Calcium/lipid-binding Proteins

	Protein I	Protein II
Native M_r	85000 ± 3000	32000 ± 2000
Subunit M_r (in SDS-PAGE)	36000	32000
	11000	
Subunit composition	$p36_2$ $p11_2$	monomer
Stokes radius	41 ± 1 Å	21 ± 1 Å
Isoelectric point	$7.4 + 7.2$	6.5
Ca^{2+}-binding, apparent K_d	$\approx 10^{-4}$ M	4.5×10^{-5} M
Ca^{2+}-binding in the presence of phospholipid, app. K_d	4.5×10^{-6} M	
Ca^{2+}-dependent lipid-binding	PS (phosphatidylserine) PI (phosphatidylinositol)	PS
Ca^{2+}-dependent binding to cytoskeletal elements	F-actin non-erythroid-spectrin	- -
Phospholipase A_2 inhibition	+	+

137

oligo nucleotides employed. While the p36 probe recognizes a transcript of 1.6 kb (Saris et al., 1986), the protein II probe hybridizes with a RNA species of approximately 2.2 kb. The identity of this RNA with the mRNA encoding protein II was verified using an oligonucleotide probe derived from a different region of the protein II sequence (not shown). Since protein II is slightly shorter in amino acid sequence than p36, additional 5' or 3' untranslated sequences in the protein II mRNA must account for the longer protein II transcript (2.2 kb as compared to 1.6 kb for p36). Examination of the expression of p36 and protein II mRNAs in various rat tissues and different cell lines revealed a very similar distribution (Fig. 2, panels B and C). Among the tissues tested, both mRNA species are most abundant in intestine, but virtually absent in brain. Several murine, bovine, and human cell lines also express the two RNAs. Here, the relative abundance of the two mRNAs cannot be compared since the oligonucleotides show varying degrees of complementary to transcripts from different species. Future experiments have to reveal whether coexpression of p36 and protein II in certain cell types, e.g. intestinal epithelial cells, is related to their biological functions, which could include the participation in membrane fusion events.

REFERENCES

Brugge, J., 1986, The p35/p36 substrates of protein-tyrosine kinases as inhibitors of phospholipase A_2, Cell, 46:149.

Crumpton, M.R., Moss, S.E., and Crumpton, M.J., 1988, Diversity in the lipocortin/calpactin family, Cell, 55:1.

De, B.K., Misono, K.S., Lukas, T.J., Mroczkowski, B., and Cohen, S., 1986, A calcium-dependent 35-kilodalton substrate for epidermal growth factor receptor/kinase isolated from normal tissue, J. Biol. Chem., 261:13784.

Erikson, E., Tomasiewicz, H.G., and Erikson, R.L., 1984, Biochemical characterization of a 34-kilodalton substrate for pp60[src] and an associated 6-kilodalton protein, Mol. Cell. Biol., 4:77.

Fava, R.A., and Cohen, S., 1984, Isolation of a calcium-dependent 35-kilodalton substrate for the epidermal growth factor receptor/kinase from A-431 cells, J. Biol. Chem., 259:2636.

Gerke, V., and Weber, K., 1984, Identity of p36K phosphorylated upon Rous sarcoma virus transformation with a protein purified from brush borders : calcium-dependent binding to non-erythroid spectrin and F-actin, EMBO J., 3:227.

Gerke, V., and Weber, K., 1985a, Calcium-dependent conformational changes in the 36-kDa subunit of intestinal protein I related to the 36-kDa target of Rous sarcoma virus tyrosine kinase, J. Biol. Chem., 260:1688.

Gerke, V., and Weber, K., 1985b, The regulatory chain in the p36-kd substrate complex of viral tyrosine-specific protein kinases is related in sequence to the S-100 protein of glial cells, EMBO J., 4:2917.

Glenney, J.R., Jr., 1986, Phospholipid-dependent Ca^{2+}-binding by the 36-kDa tyrosine kinase substrate (calpactin) and its 33-kDa core, J. Biol. Chem., 261:7247.

Glenney, J.R., Jr., and Tack, B.F., 1985, Aminoterminal sequences of p36 and associated p10 : identification of the site of tyrosine phosphorylation and homology with S-100, Proc. Natl. Acad. Sci. USA, 87:7884.

Gould, K.L., Woodgett, J.R., Isacke, C.M., and Hunter, T., 1986, The protein-tyrosine kinase substrate, p36, is also a substrate for protein kinase C in vivo and in vitro, Mol. Cell. Biol., 6:2738.

Hexham, J.M., Totty, N.F., Waterfield, M.D., and Crumpton, M.J., 1986, Homology between the subunits of S-100 and a 10kDa polypeptide associated with p36 of pig lymphocytes, Biophys. Biochem. Res. Comm., 134:248.

Johnsson, N., Van Nguyen, P., Soeling, H.-D., and Weber, K., 1986, Functionally distinct serine phosphorylation sites of p36, the cellular substrate of retroviral protein kinase; differential inhibition of reassociated with p11, EMBO J., 5:3455.

Johnsson, N., Marriott, G., and Weber, K., 1988, p36, the major cellular substrate of src tyrosine protein kinase binds to its p11 subunit via a short amino-terminal amphiphatic helix, EMBO J., 7:2435.

Saris, C.J.M., Tack, B.F., Kristensen, T., Glenney, J.R., and Hunter, T., 1986, The cDNA sequence for the protein-tyrosine kinase substrate p36 (calpactin heavy chain) reveals a multidomain protein with internal repeats, Cell, 46:201.

Weber, K., Johnsson, N., Plessmann, U., Nguyen Van, P., Soeling, H.-D., Ampe, C., and Vandekerckhove, J., 1987, The amino acid sequence of protein II and its phophorylation site for protein kinase C; the domain structure Ca^{2+}-modulated lipid binding proteins, EMBO J., 6:1599.

POSSIBLE ROLE FOR TWO CALCIUM-BINDING PROTEINS OF THE S-100 FAMILY, CO-EXPRESSED IN GRANULOCYTES AND CERTAIN EPITHELIA

Veronica van Heyningen and Julia Dorin

MRC Human Genetics Unit, Western General Hospital, Edinburgh, EH4 2XU, UK

INTRODUCTION

Calgranulins A and B - originally designated cystic fibrosis antigen (CFAG) - were first recognized as components of an anonymous serum protein found at elevated level in cystic fibrosis (CF) homozygotes and heterozygotes (van Heyningen et al., 1985; Hayward et al., 1987). Polyclonal and monoclonal antibodies recognizing this protein were used in tissue localization, chromosomal mapping, and purification of the heteropolymeric protein (van Heyningen et al., 1985; Dorin et al., 1987). Partial N-terminal aminoacid analysis revealed one polypeptide. Multiple redundant synthetic oligonucleotides corresponding to part of this sequence were used to isolate a clone encoding this gene from a chronic myeloid leukemia (CML) cDNA library (Dorin et al., 1987). The deduced aminoacid sequence of the gene product showed homology to: S-100 a and b, intestinal calcium-binding protein, calcyclin, and to the 11K calpactin light chain (Dorin et al.,1987) as well as to p9Ka which has been associated with breast cancer (Barraclough et al., 1987; Murphy et al., 1988). Clones encoding part of a second polypeptide with a blocked N-terminus (Michal Novak personal communication; Odink et al., 1987) were isolated subsequently from the same CML cDNA library, using the published sequence information of Odink et al. (1987). This second subunit, calgranulin B, also shows homology to the S-100 family, with approximately the same degree of similarity to calgranulin A as to other members of the S-100 family.

Chromosomal localization of the "cystic fibrosis antigen" gene by following expressed protein in granulocyte-derived somatic cell hybrids (van Heyningen et al., 1985), was confirmed on chromosome 1 (region q12-q21) for both calgranulins by DNA blot analysis with the cDNA probes (Dorin et al., 1987; Dorin et al., in preparation). As soon as the CF disease locus was assigned to chromosome 7 (Knowlton et al., 1985; White et al., 1985; Wainwright et al., 1985), it became clear that the involvement of the calgranulins in the etiology of CF is secondary. Elucidation of the biochemical/physiological role of these calcium-binding proteins should help in the unravelling of CF gene function.

APPROACHES TO ELUCIDATING FUNCTION

One of the problems frequently posed by the molecular biological approach to identifying gene products is that precise function is difficult to assign. These problems are not insurmountable as there are usually several pointers to likely biochemical role for the protein under study. Our approach falls into the following categories:

1. Homology studies : In the case of the calgranulins the homology relationships gave the immediate clue that these molecules may be functionally modulated by calcium. The binding of calcium can be directly demonstrated.

2. Tissue distribution : Knowing at least some of the tissues in which CF-associated physiological abnormalities have been demonstrated, we have carried out an extensive tissue search to determine where the calgranulins are expressed and have considered to what extent these sites reflect the CF defect.

3. Subcellular molecular associations : We are currently exploring the subcellular localization of the calgranulins using the techniques of immunohistochemistry and immunoprecipitation. Another way of analyzing molecular association is the use of labelled purified protein in protein blotting and in microinjection into whole cells.

4. Functional comparisons among members of the gene family : understanding of the in vivo role of members of the S-100 family is emerging only slowly. The time is not yet ripe to decide whether any generalizations and extrapolations will be possible for this group, despite the fact that several of them, though not S-100b, map to the same region of chromosome 1 in man.

HOMOLOGY STUDIES

Calgranulins A and B should bind calcium as judged by their sequence similarity to other members of the S-100 family which in turn belong to the superfamily of EF-hand proteins (for reviews see Kretsinger, 1987; Perret et al.,1988). We have demonstrated (unpublished data) the binding of $^{45}Ca^{2+}$ to purified calgranulins in a nitrocellulose binding assay in which native rather than SDS-denatured protein is used.

TISSUE DISTRIBUTION

Our panel of monoclonal antibodies directed to CFAG (Hayward et al., 1986) has been used to determine the tissue distribution of calgranulins A and B. Antibodies CF145 and CF557, directed to different epitopes on the native CFAG protein, were most widely used. CF145 recognizes the 11K calgranulin A subunit in both protein blot analysis (Wilkinson et al., 1988) and when this cDNA clone is expressed in a bacterial system (unpublished data) CF557 binds strongly in protein blot analysis to the 14K calgranulin B subunit and weakly to calgranulin A (Wilkinson et al., 1988); its binding to bacterially expressed calgranulin A is very weak or absent. Nevertheless the two antibodies revealed completely parallel expression profiles in all tissues in the immunohistochemical search (Wilkinson et al., 1988) including in extensive double labelling analysis (D. B. Jones, personal communication). In sandwich ELISA assays to quantitate CFAG in serum samples (Hayward et al., 1987), our two epitope assay must detect only AB heterodimers or their multiples. Protein blot analysis of SDS-PAGE separated components reveals that if the cell lysate is neither boiled nor reduced, then most of the calgranulin reactivity is in an approximately 45K band, to which the two antibodies bind synergistically. This 45K protein is probably an A_2B_2 heterotetramer and the most likely form of the serum component which is probably a mislocalized product of peripheral granulocytes in CF homozygotes and heterozygotes, suggesting that these cells express the CF defect.

In addition to their presence in all mature granulocytes and in some monocytes and macrophages, the calgranulins are expressed in the normal mucous epithelia of tongue, oesophagus and buccal tissues (Wilkinson et al., 1988). Normal and CF pancreas, lung and skin are negative (except for occasional positive hair follicle cells). Hyperproliferative skin epithelial cells, as well as frankly malignant squamous cell carcinomas of skin, lung and buccal origin, express both calgranulins strongly. RNA blot analysis confirms that this shared reactivity is not due to immunological cross-reactivity but to true identity of expressed genes (Wilkinson et al., 1988). A very similar tissue distribution has been described for the L1 myelomonocytic antigen (Brandtzaeg et al., 1988), which is probably identical to the calgranulins.

How does the observed tissue distribution fit the clinical picture in CF ?

The major abnormalities in cystic fibrosis are those of exocrine secretion. The first symptom is usually failure to thrive because of exocrine pancreatic insufficiency which can be corrected by dietary supplements of pancreatic enzymes. Definitive diagnosis is normally achieved by the demonstration of elevated sweat chloride - another exocrine abnormality. The median age of survival has been extended to the early twenties mainly by improvements in antibiotic therapy to combat the progressive obstructive lung disease which is the predominant cause of mortality. The onset of lung involvement, with recurrent infection accompanied by the production of thick mucus (also ascribed to abnormal exocrine function), is normally in later childhood. Could this late onset lung abnormality be a clinical manifestation of aberrant granulocyte function ?

Altered regulation of apical membrane chloride channels in CF homozygote-derived tissues has been demonstrated in cultured cells from airway epithelia (Schoumacher et al., 1987; Li et al., 1988), sweat glands (Bijman & Fromter, 1986), and most recently from small intestine and colon (Berschneider et al., 1988). It has been shown that in CF cells chloride channel responsiveness to beta-adrenergic stimuli is absent although the signal elicits normal rises in cyclic AMP levels. Normal but not CF channels can respond to phosphorylation signals from protein kinase A. Using the CF145 and CF557 antibodies, Hoogeveen and colleagues have found calgranulin expression in cultured epithelial cells with physiologically demonstrable chloride channels of the CF type (personal communication).

The postulated involvement of the calgranulins in the physiological abnormality described in CF epithelia suggests that these calcium binding proteins, together with the product of the CF gene, participate in the signal transduction pathway involved in control of chloride channel opening.

SUBCELLULAR MOLECULAR ASSOCIATIONS

There is strong circumstantial evidence (e.g. Neer and Clapham, 1988) that members of a signal transduction pathway must interact at least transiently in order to pass on a specific signal. This in turn suggests that by manipulating the physiological variables, it may be possible to "catch" interacting members of the pathway for example by controlled immunoprecipitation or by gentle crosslinking. This is the approach we are attempting in various available cell types where we know calgranulins are expressed and which may be implicated as affected cell types in CF. Thus, a buccal squamous cell carcinoma line which expresses calgranulins and may be an implicated cell type in the the basic CF defect (McPherson et al., 1987), is being analyzed. Granulocytes, from the peripheral circulation and the inducible HL60 cell line are other possibilities in which manipulation is possible and where the subcellular molecular associations of the calgranulins may reveal the product of the CF gene, which should of course map to chromosome 7.

Another approach is to study subcellular associations by immunofluorescence histochemistry. In this way members of the calpactin family have been shown to associate with cytoskeletal elements (Owens et al., 1984; Osborn et al., 1988). The p11 calpactin light chain, which is part of the S-100 family, also shows such association (Osborn et al., 1988). Similarly, we have demonstrated a fine reticular pattern of calgranulin staining in the buccal squamous epithelial cell line (Wilkinson et al., 1988).

FUNCTIONAL COMPARISONS BETWEEN MEMBERS OF THE S-100 FAMILY

No clear function has been assigned to any proteins with homology to the calgranulins. p11 which shows calcium-dependent association with calpactin 1 is one possible model (Glenney et al., 1986). This molecular interaction with an actin-binding member of the lipocortin family poses the question whether other S-100 family members might associate with other lipocortins which belong to yet another multigene family of calcium-dependent phospholipid binding proteins some of which may be phospholipase A2 inhibitors and hence involved in signal transduction.

The in vitro interaction of S-100 proteins with brain microtubule protein is suggestive of function (Donato, 1988).

Calcyclin was isolated as a cell cycle variable message (Calabretta et al., 1986) and subsequently growth factor regulation of the 5' upstream sequences of the genomic clone has been demonstrated (Ghezzo et al., 1988). We have tentative evidence that calgranulin expression in the buccal cell line TR146, is cell density-dependent. Induction of calgranulins in certain cultured epithelia may not be a rare event: both subunits are expressed at high level in cultured keratinocytes and in cultured shallow skin biopsy specimens strong antibody staining is elicited within 24 hours. This expression is not necessarily coupled to proliferation, as in psoriasis patients treated with methotrexate doses which completely suppress DNA synthesis, calgranulin expression persists (unpublished results).

Consideration of such biological phenomena are all possible avenues to unravelling the function of members of the same gene family. It is of interest in this context that several S-100 family members map to the same small region of the chromosome 1 long arm (Dorin et al., in preparation) and may have co-evolved to be under similar molecular control.

CONCLUSIONS

The problem of assigning biological function to genes which are cloned in a search for genetic abnormalities, will arise with increasing frequency. It is important to evolve biological, biochemical and genetic techniques to solve these puzzles.

ACKNOWLEDGEMENTS

We thank the Medical Research Council and the Cystic Fibrosis Research Trust for generous support.

REFERENCES

Barraclough, R., Savin, J., Dube, S. K., and Rudland, P. S., 1987, Molecular cloning and sequence of the gene for p9Ka a cultured myoepithelial cell protein with strong homology to S-100, a calcium-binding protein. J. Mol. Biol., 198: 13-20.

Berschneider, H. M., Knowles, M. R., Azizkhan, R. G., Boucher, R. C., Tobey, N. A., Orlando, R. C., and Powell, D. W., 1988, Altered intestinal chloride transport in cystic fibrosis. FASEB J., 2: 2625-2629.

Bijman, J., and Fromter, E., 1986, Direct demonstration of high transepithelial chloride-conductance in normal human sweat duct which is absent in cystic fibrosis. Pflugers Arch., 407: S123-S127.

Brandtzaeg, P., Jones, D. B., Flavell, D. J., and Fagerhol, M. K., 1988, Mac 387 antibody and the detection of formalin resistant myelomonocytic L1 antigen. J. Clin. Pathol., 41: 963-970.

Calabretta, B., Battini, R., Kaczmarek, L., De Riel, J. K., and Baserga, R., 1986, Molecular cloning of the cDNA for a growth factor-inducible gene with strong homology to S-100, a calcium-binding protein. J. Biol. Chem., 261: 12628-12632.

Donato, R., 1988, Calcium-independent, pH-regulated effects of S-100 proteins on assembly-disassembly of brain microtubule protein in vitro. J. Biol. Chem., 263: 106-110.

Dorin, J. R., Novak, M., Hill, R. E., Brock, D. J. H., Secher, D. S., and van Heyningen, V., 1987, A clue to the basic defect from cloning the CF antigen gene. Nature, 326: 614-617.

Ghezzo, F., Lauret, E., Ferrari, S., and Baserga, R., 1988, Growth factor regulation of the promoter for calcyclin, a growth-regulated gene. J. Biol. Chem., 263: 4758-4763.

Glenney, J. R., Boudreau, M., Galyean, R., Hunter, T., and Tack, B., 1986, Association of the S-100-related calpactin light chain with the NH2-terminal tail of the 36-kDa heavy chain. J. Biol. Chem., 261: 10485-10488.

Hayward, C., Chung, S., Brock, D. J. H., and van Heyningen, V., 1986, Monoclonal antibodies to cystic fibrosis antigen. J. Immunol. Meth., 91: 117-122.

Hayward, C., Glass, S., van Heyningen, V., and Brock, D. J. H., 1987, Serum concentrations of a granulocyte-derived calcium-binding protein in cystic fibrosis patients and heterozygotes. Clin. Chim. Acta, 170: 45-56.

Knowlton, R. G., Cohen-Haguenauer, O., Van Cong, N., F: zal, J., Brown, V. A., Barker, D., Braman, J. C., Schumm, J. W., Tsui, L. C., Buchwald, M., and Donis-Keller, H., 1985, A polymorphic marker linked to cystic fibrosis is located on chromosome 7. Nature, 318: 380-382.

Kretsinger, R. H., 1987, Calcium coordination and the calmodulin fold: divergent versus convergent evolution. Cold Spring Harbor Symposia on Quantitative Biology, 52: 499-510.

Li, M., McCann, J. D., Liedtke, C. M., Nairn, A. C., Greengard, P., and Welsh, M. J., 1988, Cyclic AMP-dependent protein kinase opens chloride channels in normal but not cystic fibrosis airway epithelium. Nature, 331: 358-360.

Lin, P., and Gruenstein, E., 1987, Identification of a defective cAMP-stimulated chloride channel in cystic fibrosis fibroblasts. J. Biol. Chem., 262: 15345-15347.

McPherson, M. A., Tiligada, E., Bradbury, N. A., and Goodchild, M. C., 1987, Altered calmodulin activity in buccal epithelial cells from cystic fibrosis patients. Clin. Chim. Acta, 170: 135-142.

Murphy, L. C., Murphy, L. J., Tsuyuki, D., Duckworth, L., and Shiu, R. P. C., 1988, Cloning and characterization of a cDNA encoding a highly conserved, putative calcium binding protein, identified by an anti-prolactin receptor antiserum. J. Biol. Chem., 263: 2397-2401.

Neer, E. J., and Clapham, D. E., 1988, Roles of G protein subunits in transmembrane signalling. Nature London, 333-129.

Odink, K., Cerletti, N., Bruggen, J., Clerc, R. G., Tarcsay, L., Zwadlo, G., Gerhards, G., Schlegel, R., and Sorg, C., 1987, Two calcium-binding proteins in infiltrate macrophages of rheumatoid arthritis. Nature, 330: 80-82.

Osborn, M., Johnsson, N., Wehland, J., and Weber, K., 1988, The submembranous location of p11 and its interaction with the p36 substrate of pp60 src kinase in situ. Expl Cell Res., 175: 81-96.

Owens, R. J., Gallagher, C. J., and Crumpton, M. J., 1984, Cellular distribution of p68, a new calcium-binding protein from lymphocytes. EMBO J., 3: 945-952.

Perret, C., Lomri, N., and Thomasset, M., 1988, Evolution of the EF-hand calcium-binding protein family: evidence for exon shuffling and intron insertion. J. Mol. Evol., 27: 351-364.

Schoumacher, R. A., Shoemaker, R. L., Halm, D. R., Tallant, E. A., Wallace, R. W., and Frizzell, R. A., 1987, Phosphorylation fails to activate chloride channels from cystic fibrosis airway cells. Nature London, 330-752.

van Heyningen, V., Hayward, C., Fletcher, J., and McAuley, C., 1985, Tissue localization and chromosomal assignment of a serum protein that tracks the cystic fibrosis gene. Nature, 315: 513-515.

Wainwright, B. J., Scambler, P. J., Schmidtke, J., Watson, E. A., Law, H. Y., Farrall, M., Cooke, H. J., Eiberg, H., and Williamson, R., 1985, Localization of cystic fibrosis locus to chromosome 7cen-q22. Nature, 318: 384-385.

White, R., Woodward, S., Leppert, M., O'Connel, P., Nakamura, Y., Hoff, M., Herbst, J., Lalouel, J. M., Dean, M., and Vande Woude, G., 1985, A closely linked genetic marker for cystic fibrosis. Nature, 318: 382-384.

Wilkinson, M. M., Busuttil, A., Hayward, C., Brock, D. J. H., Dorin, J. R., and van Heyningen, V., 1988, Expression pattern of two related cystic fibrosis-associated calcium-binding proteins in normal and abnormal tissues. J. Cell Sci., 91: 221-230.

143

PHOSPHORYLATION OF THE CALCIUM-BINDING PROTEIN, p68, IN THE

SUBMEMBRANOUS CYTOSKELETON OF HUMAN PLACENTAL

SYNCYTIOTROPHOBLAST

Paul Kenton, Peter M. Johnson and Paul D. Webb

Dept. of Immunology, University of Liverpool, PO Box 147, Liverpool, UK

INTRODUCTION

Human placental syncytiotrophoblast expresses cell surface receptors for a number of growth factors including those for epidermal growth factor (EGF) (Richards, 1983), platelet-derived growth factor (PDGF) (Goustin, 1985) and insulin (Lai, 1981). mRNA transcripts coding for the receptor for macrophage colony stimulating factor (M-CSF), have been identified in fetal syncytiotrophoblast (Hoshina, 1985). Cell-surface receptors for transferrin (Loh, 1980) and for the Fc region of human immunoglobulin G (IgG) (Johnson, 1981) are also expressed.

The EGF, PDGF, M-CSF and insulin receptors are transmembrane proteins possessing intrinsic tyrosyl kinase activity (Yarden, 1988). Activation of these receptor kinases is an early event following ligand binding and may be central to the transduction of growth factor signals. An important key to the understanding of these events is the identification of the cytosolic substrates of these receptor kinases. A group of membrane-associated Ca^{2+}-binding proteins have been identified as substrates of such receptor kinases. These include p35 (lipocortin I), p36 (lipocortin II) and p32 (endonexin). p35 is a substrate of both the EGF receptor tyrosyl kinase (De, 1986) and the Ca^{2+}/phospholipid-dependent serine/threonine kinase, protein kinase C (PKC) (Nishizuka, 1984; Khanna, 1987). Similarly, p36 is a substrate of PKC (Khanna, 1987) and the tyrosyl kinase pp60[v-src] (Radke, 1979). These are also substrates for other kinases involved in signal transduction, e.g. cAMP-dependent kinase (PKA) (Pepinsky, 1986). The protein p68 (protein III) is closely related to p35 and p36 with respect to amino acid sequence, antiserum crossreactivity and binding of phospholipid and Ca^{2+} (Owens, 1984; Crompton, 1988).

EVIDENCE FOR p68 IN ISOLATED SYNCYTIOTROPHOBLAST MEMBRANES VESICLES

Webb and Mahadevan (1987) found that extraction of syncytiotrophoblast plasma membrane vesicles with Triton X-100 and EDTA yielded a detergent-insoluble pellet lacking protein bands at 68, 36 and 34 kDa whereas these bands were present after preparing the detergent-insoluble pellet in the presence of 1mM $CaCl_2$. These proteins were recognised by rabbit antisera to human lymphocyte cytoskeletal p68, p36 and p32 (a gift of Dr. M.J. Crumpton, ICRF, London (Owens, 1984; Crompton, 1988)). In addition, the 36 kDa band cross-reacted with a rabbit antiserum to recombinant lipocortin II (a gift of Dr. B. Pepinsky, Biogen, Cambridge, MA, USA. (Huang, 1986)). A rabbit antiserum to recombinant lipocortin I (from Dr. B. Pepinsky) reacted with a series of bands in the 37-34 kDa range. However neither of the antisera raised against recombinant lipocortins reacted with the detergent-insoluble fraction from syncytiotrophoblast plasma membrane vesicles prepared in the presence of EDTA. Thus, syncytiotrophoblast plasma membrane vesicles express proteins, recognised immunologically using antisera raised against p36, p32 and p68 as well as those raised against recombinant lipocortins I and II, which associate with the submembranous cytoskeleton in a Ca^{2+}-dependent manner.

Webb and Mahadevan (1987) also noted that the 68 kDa protein was phosphorylated in unstimulated syncytiotrophoblast membrane vesicles. Phosphoamino acid analysis indicated that serine was the principal residue phosphorylated. Incubation of the membrane vesicles with EGF led to an increase in phosphotyrosine content. A similar phosphorylation pattern was observed for the 36 kDa protein, with phosphotyrosine increasing after EGF treatment. By contrast the 34 kDa protein was not phosphorylated before or after treatment with EGF.

In summary the reactivity of these proteins with antisera raised against p32, p36, p68 and recombinant lipocortins, their Ca^{2+}-dependent association with the submembranous cytoskeleton and their phosphorylation patterns, all indicate that the trophoblastic 68 kDa band is p68, the 36 kDa band is p36 (lipocortin II) and the 34 kDa protein is endonexin (p32).

THE PHOSPHORYLATION OF p68

The observation that phosphotyrosine is increased in p68 and p36 after treatment with EGF prompted us to study the action of agents capable of modulating protein phosphorylation. Little is known about the phosphorylation of p68 and it is on this protein that these studies have been centred.

The phosphorylation of p68 in isolated syncytiotrophoblast plasma membrane vesicles (prepared from term placental villous tissue in the presence of $CaCl_2$) was assessed following treatment of vesicles with a growth factor (or other agent) for 10 mins on ice. [γ-^{32}P] ATP was then added, together with an excess of cold ATP and the divalent cations Mn^{2+}, Mg^{2+} and Ca^{2+}. The membrane proteins were separated on SDS-PAGE and gel autoradiography or Cerenkov counting of the excised p68 bands was subsequently performed. In unstimulated syncytiotrophoblast membrane vesicles p68 underwent rapid cycling of phosphorylation/dephosphorylation with 80-85% of p68 being processed within 10 minutes (Kenton, 1989). The phosphporylation of p68 was independent of Mn^{2+}, Mg^{2+} or Ca^{2+}, unlike the phosphorylation of p36 which showed an absolute requirement for Mn^{2+}. The Mn^{2+} requirement for p36 phosphorylation could be mimicked by diacylglycerol and Ca^{2+} in combination.

Figure 1 shows the results treatment of syncytiotrophoblast membrane vesicles with recombinant human M-CSF (a gift from Dr. P. Ralph, Cetus, CA, USA.). M-CSF induced a dose-dependent reduction in p68 phosphorylation. Similar results were obtained with both insulin and PDGF. The rank order of ability to inhibit p68 phosphorylation was M-CSF > insulin > PDGF. By contrast, EGF showed no effect on overall p68 phosphorylation. Transferrin enhanced p68 phosphorylation.

The effect of PKC activators on p68 phosphorylation was also examined. The phorbol esters 12-O-tetradecanoyl phorbol-13-acetate (TPA) and phorbol-12,13-dibutyrate (PdBu), but not phorbol-12,13-

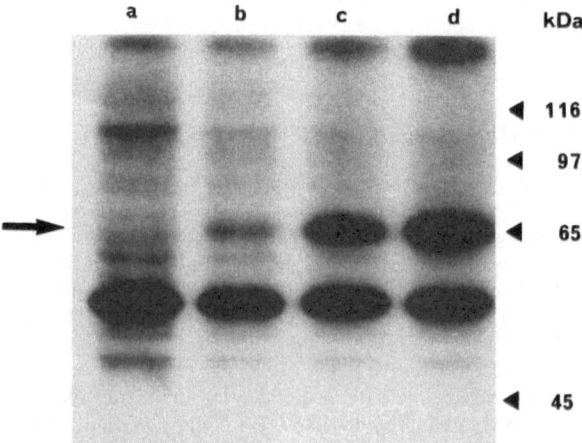

Fig. 1 . Autoradiograph of ^{32}P incorporation into p68 (arrowed) following treatment of syncytiotrophoblast membrane vesicles with macrophage colony stimulating factor. Units added: (a) 10, (b) 2, (c) 0.4, (d) 0, i.e. control buffer.

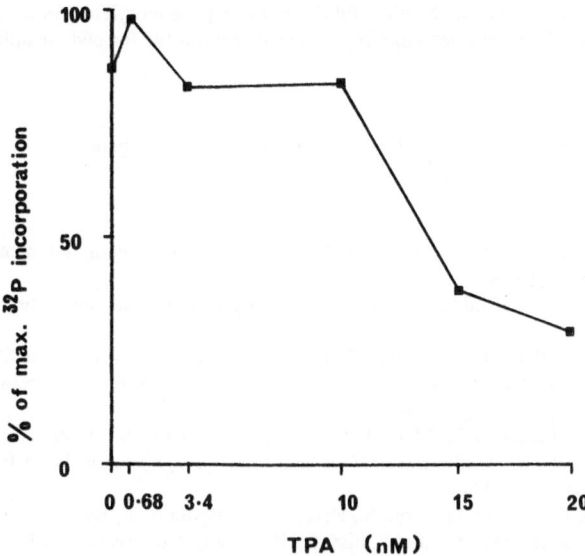

Fig. 2. Dose response curve of ^{32}P incorporation into p68 following treatment of syncytiotrophoblast membrane vesicles with the phorbol ester, TPA.

didecanoate (PdDn), activate PKC (Ashendel, 1985). Hydrolysis of membrane phosphotidylinositol to diacylglycerol by phosphatidylinositol-specific phospholipase C (PI-PLC, Webb, 1988) may also result in activation of PKC (Nishizuka, 1984). TPA induced a dose-dependent decrease in p68 phosphorylation (fig. 2). The same effect was achieved with PdBu whereas PdDn had no effect on p68 phosphorylation. Similarly PI-PLC showed a dose-dependent decrease in p68 phosphorylation but no effect was noted with non-specific phospholipase C (Webb, 1988). These results suggested that p68 is not a direct substrate of PKC.

The phosphoamino acid composition of p68 was investigated (Webb, 1987) after treatment of isolated syncytiotrophoblast plasma membranes with growth factors, transferrin or PKC activators. Serine was the major phosphorylated component and was the only phosphoamino acid identified in p68 from membrane vesicles treated with TPA or in the control. Tyrosine was phosphorylated upon treatment of membrane vesicles with EGF. Both transferrin and M-CSF increased the phosphotyrosine component, despite an overall reduction in p68 phosphorylation induced by M-CSF. These results suggested that p68 is a substrate of tyrosyl kinases.

Cholera toxin (CT) stimulates and Pertussis toxin (PT) prevents inhibition of cAMP accumulation respectively by interacting with the guanine nucleotide binding proteins (G proteins) coupling receptors to adenylate cyclase (Vaughan, 1986). p68 phosphorylation was inhibited by both CT and PT. Similarly addition of cAMP inhibited p68 phosphorylation. These results suggested that p68 is not a direct substrate of PKA.

In an attempt to identify specific pathways for the inhibition of p68 phosphorylation, a number of ligands were applied to syncytiotrophoblast membrane vesicles after treatment with CT or PT. Both toxins were synergistic with PDGF and insulin, indicating separate pathways of action. EGF, however, completely prevented the CT- or PT-induced reduction in p68 phosphorylation, suggesting an effect on the cAMP-generating system which bypasses G proteins.

Isolated syncytiotrophoblast membrane vesicles were also treated with several phosphatase inhibitors prior to carrying out the phosphorylation assay. Neither tetramisole (van Belle, 1972) nor trifluoperazine (Stewart, 1983) prevented the M-CSF-induced decrease in p68 phosphorylation.

These results have indicated that the Ca^{2+}-binding protein, p68, is a potential substrate for both serine and tyrosyl kinases. p68 is phosphorylated in human syncytiotrophoblast plasma membrane vesicles by a kinase other than that phosphorylating p36 in the same system. p68 phosphorylation is down-regulated by insulin, M-CSF, PDGF, activators of PKC and cAMP. EGF does not directly effect overall p68 phosphorylation whilst transferrin increases p68 phosphorylation. None of these ligands affecting p68 phosphorylation

147

appear to act via the cAMP pathway. Neither PKC nor PKA phosphorylate p68 in this membrane system. Current studies are directed towards determining the agent responsible for p68 phosphorylation.

ACKNOWLEDGEMENTS

This research work was supported by the Cancer Research Campaign.

REFERENCES.

Ashendel, C.L., 1985, The phorbol ester receptor: a phospholipid-regulated protein kinase, Biochim. Biophys. Acta, 822:219-242.

van Belle, H., 1972, Kinetics and inhibition of alkaline phosphatases from canine tissues, Biochim. Biophys. Acta, 289:158-168.

Crompton, M.R., Owens, R.J., Totty, N.F., Moss, S.E., Waterfield, M.D., and Crumpton, M.J., 1988, Primary structure of the human, membrane-associated Ca^{2+}-binding protein p68: a novel member of a protein family, EMBO J., 7:21-27.

De, B.K., Misono, K.S., Lukas, T.J., Mroczkowski, B., and Cohen, S., 1986, A calcium-dependent 35-kilodalton substrate for epidermal growth factor receptor/kinase isolated from normal tissue, J. Biol. Chem., 261:13784-13792.

Goustin, A.S., Betsholtz, C., Pfeifer-Ohlsson, S., Persson, H., Rydnert, J., Bywater, M., Holmgren, G., Heldin, C.-H., Westermark, B., and Ohlsson, R., 1985, Coexpression of the sis and myc proto-oncogenes in developing human placenta suggests autocrine growth control of trophoblast growth, Cell, 41:301-312.

Hoshina, M., Nishio, A., Bo, M., Boime, I., and Mochizuka, M., 1985, The Expression of oncogene fms in human chorionic tissue, Acta Obst. Gynec. Jpn., 12:2791-2798.

Huang, K-S., Wallner, B.P., Mattaliano, R.J., Tizard, R., Burne, C., Frey, A., Hession, C., McGray, P., Sinclair, L.K., Chow, E.P., Browning, J.L., Ramachandran, K.L., Tang, J., Smart, J.E., and Pepinsky, R.B., 1986, Two human 35 kd inhibitors of phospholipase A2 are related to substrates of $pp60^{v-src}$ and of the epidermal growth factor receptor/kinase, Cell, 46:191-199.

Johnson, P.M., and Brown, P.J., 1981, Fc receptor in the human placenta, Placenta, 2:355-370.

Kenton, P., Johnson, P.M., and Webb, P.D., 1989, Biochim. Biophys. Acta, in press.

Khanna, N.C., Tokuda, M., and Waisman, D.M., 1987, Purification of three forms of lipocortin from bovine lung, Cell Calcium, 8:217-228.

Lai, W.H., Guyda, H.J., Branchaud, C.L., and Goodyer, C.G., 1985, Insulin-induced receptor regulation in early gestation and term human placental cell cultures, Placenta, 6:505-517.

Loh, T.T., Higuchi, D.A., van Bockxmeer, F.M., Smith, C.M., and Brown, E.B., 1980, Transferrin receptor on the human placental microvillous membrane, J. Clin. Invest., 65:1182-1191.

Nishizuka, Y., 1984, The role of protein kinase C in cell surface signal transduction and tumour promotion, Nature, 308:693-698.

Owens, R.J., and Crumpton, M.J., 1984, Isolation and characterization of a novel $68000-M_r$ Ca^{2+}-binding protein of lymphocyte plasma membrane, Biochem. J., 219:309-316.

Owens, R.J., Gallagher, C.J., and Crumpton, M.J., 1984, Cellular distribution of p68, a new calcium-binding protein from lymphocytes, EMBO J., 3:945-952.

Pepinsky, R.B., and Sinclair, L.K., 1986, Epidermal growth factor-dependent phosphorylation of lipocortin, Nature, 321:81-84.

Radke, K., Gilmore, T., and Martin, G.S., 1980, Transformation by Rous sarcoma virus: a cellular substrate for transformation-specific protein phosphorylation contains phosphotyrosine, Cell, 21:821-828.

Richards, R.C., Beardmore, J.M., Brown, P.J., Molloy, C.M., and Johnson, P.J., 1983, Epidermal Growth Factor receptor on isolated human placental syncytiotrophoblast plasma membrane, Placenta, 4:133-138.

Stewart, A.A., Ingebritsen, T.S., and Cohen, P., 1983, The protein phosphatases involved in cellular regulation, Eur. J. Biochem., 132:289-295.

Vaughan, M., and Moss, J., 1986, Guanyl nucleotide-binding proteins and regulation of cAMP metabolism, in: "Mechanisms of Insulin Action," Belfrage, P., Donner, J., and Stralfors, P., eds., Elsevier, Amsterdam.

Webb, P.D., and Mahadevan, L.C., 1987, Calcium-dependent binding proteins associated with human placental syncytiotrophoblast microvillous cytoskeleton, Biochim. Biophys. Acta, 916:288-297.

Webb, P.D., and Todd, J., 1988, Plasma membrane attachment of human placental-type alkaline phosphatase to phosphatidylinositol, Eur. J. Biochem., 172:647-652.

Yarden, Y., and Ullrich, A., 1988, Molecular analysis of signal transduction by growth factors, Biochemistry, 27:3313-3119.

CALCYCLIN-LIKE PROTEIN FROM EHRLICH ASCITES TUMOUR CELLS - Ca^{2+}-BINDING PROPERTIES, DISTRIBUTION AND TARGET PROTEIN[1]

Jacek Kuźnicki and Anna Filipek

Nencki Institute of Experimental Biology, 3 Pasteur street, 02-093 Warsaw, Poland

We have purified to homogeneity a 10.5 kDa Ca^{2+}-binding protein from Ehrlich ascites tumour (EAT) cells (Kuźnicki & Filipek, 1987). This protein differs from S-100 protein, calbindin 9k, parvalbumin and on comodulin by several criteria such as electrophoretic mobility in SDS- or urea-polyacrylamide gels, amino acid composition, and lack of cross-reactivity with the antibodies specific to these Ca^{2+}-binding proteins. Recently, we found that the partial amino acid sequence of the 10.5 kDa Ca^{2+}- binding protein from EAT cells is homologous to that of human calcyclin, and therefore we call the mouse protein a calcyclin-like protein (Kuźnicki et al., 1989a). Calcyclin is the name given to the growth factor-inducible gene (Calabretta et al., 1986a; 1986b). It has been suggested that calcyclin protein is involved in the control of cell proliferation and may bind Ca^{2+}, as deduced from nucleotide sequence of the gene. To our best knowledge nobody has so far studied the protein itself.

The Ca^{2+}- binding parameters of the protein purified from EAT cells were analysed by direct and indirect methods (Kuźnicki and Filipek, 1987). Using gel filtration of the protein in the presence of $3x10^{-5}$M ^{45}CaCl$_2$ we were able to show that it binds one Ca^{2+}. The Ca^{2+}-binding of the calcyclin-like protein was directly visualised when the protein was subjected to SDS-PAGE, blotted onto nitrocellulose and incubated with ^{45}Ca^{2+} (Fig. 1). For comparison this figure shows also the electrophoretic mobility of rat parvalbumin on SDS-polyacrylamide gel and its ability to bind Ca^{2+} on nitrocellulose.

The binding of Ca^{2+} ions to the calcyclin-like protein was studied by several indirect methods. For example, tyrosine fluorescence intensity of the protein reversibly increased upon Ca^{2+} binding by 18% (Fig. 2). The titration curve of fluorescence intensity, plotted against the molar ratio of added Ca^{2+} to the calcyclin-like protein, suggested the existence of two Ca^{2+} binding sites per molecule. Since gel filtration indicated only one Ca^{2+}-binding site, we concluded that two binding sites exist which differ in their affinities for Ca^{2+} (Kuźnicki & Filipek, 1987).

The calcyclin-like protein from EAT cells changes its conformation upon Ca^{2+} binding as indicated by Ca^{2+}-dependent changes of: tyrosine fluorescence intensity, UV absorbance spectrum, mobility in urea PAGE and hydrophobicity. The changes in exposure of hydrophobic domain(s) have been shown by Ca^{2+}-dependent binding to phenyl-Sepharose (used for purification of the protein) and to fluphenazine-Sepharose (Kuźnicki & Filipek, unpublished). Binding of the calcyclin-like protein to fluphenazine in the presence of Ca^{2+} suggests that the protein may also bind to other calmodulin inhibitors such as trifluoperazine or W-7. If so, these drugs could be used to study the calcyclin-dependent cellular functions, in analogy to studies related to calmodulin-dependent processes. Calcyclin-like protein also binds Zn^{2+} (Filipek et al., 1989).

----
1. This paper is dedicated to the memory of Professor Witold Drabikowski who organized the First International Symposium on Calcium Binding Proteins held in Jablonna, Poland, in 1973.

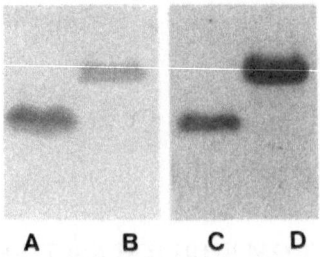

A B C D

Fig. 1 . Electrophoretic mobility and Ca^{2+}-binding of the calcyclin-like protein from EAT cells (A,C) and rat parvalbumin (B,D). (A,B): nitrocellulose blot stained with Ponceau Red; (B, D) autoradiogram of the same blot after incubation with $^{45}Ca^{2+}$.

Since the calcyclin-like protein was purified from tumour cells it was interesting to know whether it is a protein specific for tumour cells or also present in normal cells and tissues. This problem was studied using three methods: (1) Immunoblotting with a polyclonal, affinity purified antibody against calcyclin-like protein, (2) Purification of a protein from normal mouse tissues with similar properties, (3) Northern blotting with a full length calcylin cDNA as a probe.

The immunoblot of EAT cell extracts revealed two bands at the position of the monomer (10.5 kDa) and the dimer (21 kDa) of the calcyclin-like protein (Fig. 3). The dimer can be dissociated into the monomeric form by 2-mercaptoethanol (Filipek et al., 1989). As shown in Fig. 3 a 10.5 kDa immunoreactive protein is also present in low ionic strength extracts from mouse spleen, heart, skeletal muscle, stomach and cultured rat fibroblasts. The protein has also been found in extracts from mouse lung and kidney (not shown), but no positive reaction was detected in extracts from mouse brain (Fig. 3), intestine and liver (Kuźnicki et al., 1989b). No immunoreactivity was found in extracts from PTK_2 cells, mouse lymphocytes, X-63-Ag8 hybridoma cells and V_2 rabbit carcinoma cells.

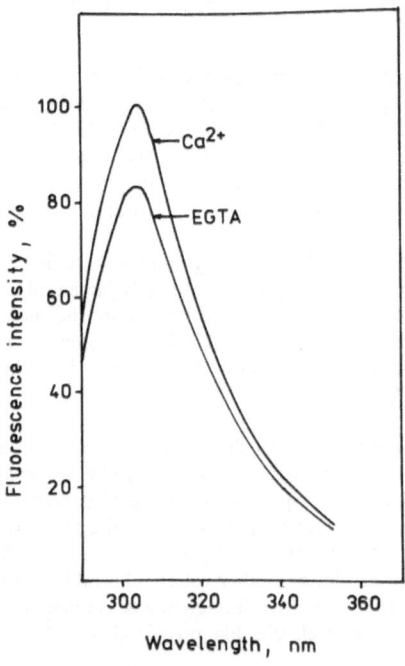

Fig. 2 . Ca^{2+}-dependent changes of tyrosine fluorescence intensity of the calcyclin-like protein (excitation at 280 nm).

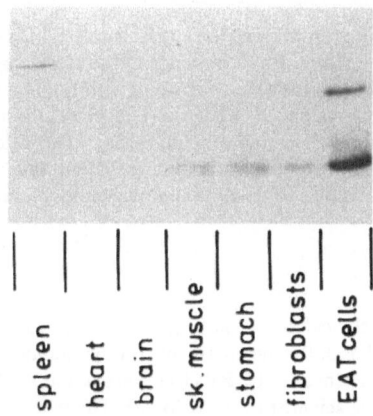

Fig. 3. Immunoblot with the antibody against the calcyclin-like protein from EAT cells and low ionic strength extracts of different mouse tissues and rat fibroblasts.

A calcyclin-like protein was also purified from mouse stomach (using the method developed for the purification of this protein from EAT cells) and was found to be identical to the EAT protein with respect to molecular weight, isoelectric point and Ca^{2+}-dependent conformational changes (Kuźnicki et al., 1989b). These results indicate that the calcyclin-like protein is present not only in tumour cells, but also in normal cells and tissues. This conclusion was confirmed by Northern blot analysis with the use of a full length calcyclin cDNA (Calabretta et al., 1986b). The hybridization data revealed a high level of calcyclin mRNA in EAT cells, much lower amounts in stomach and very little in other mouse organs studied.

The presence of the calcyclin-like protein in muscle raised the question whether the protein is present in myogenic cells. Immunocytochemistry of sections derived from mouse skeletal muscle revealed, at the light and electron microscopic level, that the calcyclin-like protein is not detectable in muscle cells, but is present in non-muscle cells, such as putative fibroblasts (Kuźnicki et al., 1989b).

We have also been looking for a target protein for the calcyclin-like protein using affinity chromatography (Fig. 4). A protein fraction from EAT cells was applied onto the calcyclin-Sepharose column in the presence of Ca^{2+}. The unbound proteins were eluted in buffers containing Ca^{2+} and high concentrations of NaCl. The proteins which bound to calcyclin in the presence of Ca^{2+} were subsequently eluted with the buffer containing EGTA (Fig. 4). The major protein band enriched by this method had an apparent molecular weight of about 36 kDa and was found to react with an antibody against mammalian calpactin (p36-p11 complex) in immunoblots. These results indicate that in EAT cells one of the possible target proteins of the calcyclin-like protein is immunologically similar to the members of the calpactin-lipocortin family (cf. Glenney et al., 1986).

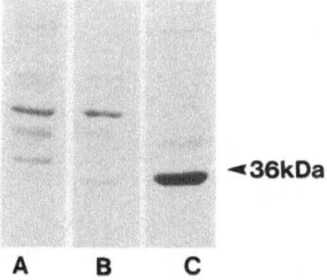

Fig. 4 . Purification of a calcyclin-target protein from EAT cells by affinity column chromatography. SDS-PAGE: A, fraction applied in the presence of Ca^{2+}; B, fraction of unbound proteins; C, fraction of proteins eluted in the presence of EGTA.

In summary, the calcyclin-like protein from EAT cells is a 10.5 kDa heat stable protein, which binds two Ca^{2+} each with different affinity. Upon Ca^{2+} binding, the protein changes its conformation exposing hydrophobic regions. In this conformation it is able to interact with fluphenazine and with a 36 kDa protein immunologically similar to mammalian calpactin. Calcyclin-like protein binds Zn^{2+} and forms dimers like other members of S-100 protein family (Kligman & Hilt, 1988). The calcyclin-like protein is present in several mouse tissues such as stomach, skeletal muscle, heart, spleen, lung and kidney, but seems to be absent from brain, intestine and liver as well as from some tumourigenic cell lines.

ACKNOWLEDGEMENTS

We thank Dr. V. Gerke for the calpactin antibody, Dr. L. Kaczmarek and B. Kamińska for Northern blot analysis and Prof. B. M. Jockusch for the comments on the manuscript. This work was supported by a grant from the Polish Academy of Sciences (CPBP 04.01). Part of the work was performed at the University of Bielefeld where J.K. was a guest scientist of the SFB 223. The work on homology of the Ca^{2+}-binding protein to human calcyclin was performed in Dr. C. W. Heizmann's laboratory in Zürich (see Kuźnicki et al., 1989a).

REFERENCES

Calabretta, B., Venturelli, D., Kaczmarek, L., Narni, F., Talpaz, M., Anderson, B., Beran M., and Baserga R., 1986, Altered expression of G_1-specific genes in human malignant myeloid cells, Proc. Natl. Acad. Sci. U.S.A., 83: 1495.

Calabretta, B., Battini, R., Kaczmarek, L., de Riel, J. K., and Baserga, R., 1986, Molecular cloning of the cDNA for a growth factor-inducible gene with strong homology to S-100, a calcium-binding protein, J. Biol. Chem, 261: 12628.

Filipek, A., Heizmann, C. W., and Kuźnicki, J., Zinc binding and dimer formation by calcyclin-like calcium binding protein from Ehrlich ascites tumour cells, in: "Proceedings of First European Symposium on Calcium Binding Proteins in Normal and Transformed Cells", R. Pochet, D.E.M. Lawson and C.W. Heizmann, eds., Plenum Press (1989)

Glenney J. R. Jr., Boudreau M., Galyean R., Hunter T., and Tack B., 1986, Association of the S-100-related Calpactin I light chain with NH_2-terminal tail of the 36-kDa heavy chain, J. Biol. Chem., 261: 10485.

Kligman D., and Hilt, D.C., 1988, The S-100 protein family, Trends in Biochem. Sci., 13: 437.

Kuźnicki, J., and Filipek, A., 1987, Purification and properties of a novel Ca^{2+}-binding protein (10.5 kDa) from Ehrlich-ascites-tumour cells, Biochem. J., 247: 663.

Kuźnicki, J., Filipek, A., Hunziker, P. E., Huber, S., and Heizmann, C. W., 1989a, Calcium binding protein from mouse Ehrlich ascites tumour cells is homologous to human calcyclin, Biochem. J., in press.

Kuźnicki, J., Filipek, A., Heimann, P., Kaczmarek, L., Kaminska, B., 1989b, Tissue specific distribution of the calcyclin-like 10.5 kDa calcium binding protein, FEBS Lett., in press.

INTERACTIONS OF S100 PROTEINS WITH PROTEIN KINASE

SUBSTRATES. BIOLOGICAL IMPLICATION

Jean Christophe Deloulme, Monique Sensenbrenner and Jacques Baudier

Centre de Neurochimie du CNRS, INSERM U. 44 , 5 Rue Blaise Pascal, 67084
Strasbourg, France

INTRODUCTION

S-100 proteins are a group of low molecular weight (10 kDa) acidic proteins highly concentrated in brain tissues (for a recent review see Kligman and Hild, 1988). S100 proteins purified from bovine brain, are a mixture of hetero- and homodimer of two types of subunit, α and ß with different amino acid composition (Isobe et al., 1977). The amino acid sequence of the α and ß subunits revealed the structural relationship of S100 with the calcium binding proteins of the EF-hand type (Isobe and Okuyama, 1978). Both subunits have one 30-residue putative EF- hand calcium binding domain (site I) in the N- terminal part and one typical 28-residue domain (site II) in the C- terminal part. Calcium- binding studies on bovine brain S100$\alpha\alpha$ and S100ßß (S100b) proteins confirmed the presence of two specific calcium-binding sites per subunits (Baudier et al., 1986a). The calcium binding sequence on the α and ß subunit have been studied by means of intrinsic fluorescence and absorption spectroscopy, binding of the Ca^{2+} analoge Tb^{3+} and H-NMR (in preparation), and showed that in both cases saturation of the typical Ca^{2+}-binding sites (site II) occured first followed by the binding of Ca^{2+} to the putative Ca^{2+}- binding site (site I). In the absence of monovalent cation the affinities of the typical sites IIα and IIß range between 10-20 μM and those of the putative sites Iα and Iß range between 100-400 μM. In the presence of physiological intracellular KCl concentrations the S100 protein affinities for calcium drastically decrease to 500 μM - 1 mM, which become probably not compatible with intracellular calcium concentrations. However, it has been shown that S100 protein affinities for calcium may also depend greatly on the quaternary and tertiary protein structures. Conformational effectors such as Zn^{2+} ions in the peculiar case of S100b, alkylation of Cys 85α or Cys 84ß, or interaction with target protein were proved to greatly increase the calcium binding affinities of sites IIα and IIß in the range of .5 - 10 μM and to decrease the antagonistic effect of KCl on calcium binding (Baudier et al., 1986a, 1986b, 1987a). It could then be postulated that S100 proteins, in vivo, may interact with other cellular components (ions, proteins, membranes) which affect their conformation and modulate their calcium binding properties (for a review see Baudier, 1988). By analogy with other Ca^{2+}- binding proteins, such as calmodulin, one might suppose that the biological functions of S100 proteins are related to Ca^{2+}- dependent processes within the cells. Several biochemical activities were previously suggested for S100 proteins. Best documented are promotion of calcium- dependent microtubule dissociation and inhibition of microtubule assembly (Baudier et al., 1982; Deinum et al., 1983; Endo and Hidaka, 1983; Donato et al., 1985), calcium- dependent activation of fructose- 1,6- bisphosphate aldolase (Zimmer and Van Eldik, 1987), stimulation of prolactin secretion from pituicytes (Ishikawa et al., 1983) and inhibition of protein kinase substrates phosphorylation (Qi and Kuo, 1984; Albert et al., 1984; Kligman and Patel, 1986; Baudier and Cole 1988a; Hagiwara et al., 1988). Authors have also suggested that S100 proteins might be involved in the maturation and differentiation of glial cells (Labourdette and Mandel, 1980) and others reported that a disulfide form of brain S100ß can stimulate neurite outgrowth in primary cultures of chicken cerebral cortex neurons (Kligman and Marshak, 1985). Kligman and Hild (1988) postulated that the various proposed and demonstrated effects of the S100 proteins in cell- type- differentiation and morphological differentiation may be mediated by the same basic process i.e. inhibition of protein phosphorylation.

The aim of this review is to summarize recent findings on the interaction of the S100 proteins with protein kinase substrates which emphasis a possible role of S100 proteins in the regulation of protein kinase substrate phosphorylation.

INTERACTION OF S100 PROTEINS WITH THE MICROTUBULE ASSOCIATED TAU PROTEINS

The diverse functions that microtubules perform in cells must often require the precise control of assembly equilibrium and polymer stability. These properties are affected by specialized proteins that selectively associate with microtubules, presumably for specific purposes. Brain microtubules polymerized in vitro contain a variety of low molecular weight associated proteins called tau. Tau proteins stimulate both nucleation and elongation of microtubules from purified tubulin and thus might play a major role in regulating tubulin polymerization. Freshly purified tau proteins exist in multiple states of phosphorylation (Lindwall and Cole, 1984b). The Ca^{2+}/CaM-dependent protein kinase II was recently identified as the enzyme that catalyses mode I phosphorylation of tau (Baudier and Cole, 1987). Mode I phosphorylation induces a conformational change in the tau protein structures that is responsible for a decrease in the mobility on SDS-PAGE and inhibits tau in its promotion of microtubule formation (Yamamoto et al., 1983; Lindwall and Cole, 1984a). Purified tau proteins can exist in a second mode of phosphorylation (mode II), which does not change the mobility of tau on SDS-PAGE. Protein kinase C appears to be responsible for mode II phosphorylation of tau in vitro (Baudier et al., 1987b). Recently, it was reported that phosphorylated tau proteins are the major antigenic component of the paired helical filaments that characterize degenerative human neurons in Alzheimer's disease (Grundke-Iqbal et al., 1986; Ihara et al., 1986). Therefore, there is considerable interest in understanding factors that might regulate tau phosphorylation and thus, probably their functions.

Few years ago we first reported that S100 protein, like calmodulin, induced total disassembly of microtubules at millimolar Ca^{2+} concentration and did so with even higher efficiency than did calmodulin (Baudier et al., 1982) and that Zn^{2+} ions at μM concentrations mimiked the effect of Ca^{2+} on S100b -induced inhibition of microtubule assembly (Deinum et al., 1983). Subsequently, Endo and Hidaka (1983) reported inhibition of microtubule assembly by S100 protein in a dose-dependent manner in the presence of 20 μM Ca^{2+}. These authors also demonstrated by affinity chromatography that phosphocellulose - purified tubulin failed to bind to S100-Sepharose columns whether Ca^{2+} was present or not, suggesting that S100 proteins might primarily bind to microtubule components rather than to tubulin. In support of this suggestion is the observation that purified tau proteins bind to S100b -Sepharose column in the presence of Ca^{2+} or Zn^{2+} (Baudier et al. 1987a, Baudier and Cole, 1988a). We confirmed the Ca^{2+}-dependent interaction of S100 proteins with tau proteins by cross-linking experiments through covalent binding between Cys 85α or Cys 84ß in the C- terminal region of the S100 subunits and Cys residue in tau proteins (Baudier and Cole, 1988b). We further demonstrated that such an interaction also inhibited mode I tau phosphorylation by the Ca^{2+}/CaM-dependent protein kinase as well as mode II phosphorylation by the protein kinase C. These results suggested that part of the S100 effect on microtubule assembly in vitro could be through the formation of S100-Ca^{2+} or S100-Zn^{2+}-tau complexes.

INTERACTION OF S100 PROTEINS WITH SPECIFIC PROTEIN KINASE C SUBSTRATES

Protein kinase C (PKC), a calcium-activated phospholipid-dependent kinase, is thought to play an important role in controlling several cellular processes. PKC is present ubiquitously in a variety of tissues and is most concentrated in the brain. Its role in the central nervous system is largely unknown, but several lines of evidence have implicated PKC in controlling the release of neurotransmitters and in the neuroplasticity. An understanding of the exact function of PKC in the brain requires the identification and characterization of their in vivo substrates.

In the brain only two PKC substrates have been formely identified and characterized : neuromodulin (also called GAP43, P57, B50 or F1) and p87 (Albert et al., 1984; Patel and Kligman, 1988; Alexander et al., 1988). Other in vitro substrates for PKC, such as the microtubule associated protein MAP-2 and tau, are also considered as putative in vivo substrates (Tsuyama et al., 1986). We recently observed, that although such solubility is generally uncommon among proteins, several brain proteins were found to be soluble in 2.5% perchloric acid (PCA), and many of them were in vitro substrates for PKC. Three of the PCA soluble brain PKC substrates were subsequently identified as neuromodulin, p87 and tau proteins (Baudier et al., 1989). Another low molecular weight, specific in vitro substrate for PKC, neurogranin (formely designated p17), soluble in PCA was further purified and characterized (submitted). A major point that links these PKC substrates is their common interactions with S100 proteins. Two different laboratories first reported that p87 interacted with S100 protein and that such an interaction inhibited the Ca^{2+}-dependent phosphorylation of p87 by PKC (Albert et al., 1984; Kligman and Patel, 1986). Interactions between neuromodulin and neurogranin with the S100-ß subunit have been studied in our laboratory.

Neuromodulin was first identified as an unusual calmodulin binding protein because it interacts with calmodulin in a calcium independent manner (Alexander et al., 1988). Interaction with calmodulin inhibits neuromodulin phosphorylation by PKC. We also observed that neuromodulin interacts with the S100b protein and that the S100-ß subunit can form a covalent complex with neuromodulin via disulfide bridges as was previously observed between S100-ß and tau proteins (Baudier et al., 1989). The formation of a covalent complex between S100-ß and neuromodulin was absolutly dependent on the presence of Ca^{2+}. However, on the contrary to calmodulin, S100b does not inhibit neuromodulin phosphorylation by PKC (unpublished results).

Neurogranin, a recently identified in vitro substrate for PKC purified in our laboratory, is a brain specific phosphoprotein specifically located in neurons of the cerebral cortex, hippocampus, striatum and few other discreet areas in the rat forebrain. Purified neurogranin exists essentialy as a mixture of disulfide-linked dimer and higher oligomers that can be reduced only in the presence of an excess of DTT (100mM) and EDTA (1 mM). Studies have shown that neurogranin phosphorylation in intact hippocampal slices, incubated with 32P, was increased when the slices were stimulated with phorbol ester suggesting that it is likely an in vivo substrate for PKC. Furthermore, the colocalization in rat brain of neurogranin and PKC I isozyme and the identical developmental expression of both proteins indicated that it is PKC I isozyme that may use neurogranin as substrate in vivo (submitted). Significantly, neurogranin also interacted with the Ca^{2+}/Zn^{2+}-S100b protein as it was observed for neuromodulin. Interaction between S100b and neurogranin was demonstrated by covalent cross-linking between Cys 84ß of the S100b and sulfhydryl groups on its targets (Fig. 1). The attack of S100b in rupturing disulfide bridges in neurogranin dimer ressembles that was previously reported for tau protein dimer and oligomers. We confirmed the interaction of S100 protein with neurogranin dimer by the observation that such an interaction also inhibited its phophorylation by PKC.

BIOLOGICAL IMPLICATIONS

Tau protein phosphorylation is thought to be a mechanism by which cells may regulate tubulin polymerization. Mode I phosphorylation of tau proteins by the Ca^{2+}/CaM-dependent protein kinase reduced its ability to promote microtubule assembly (Lindwall and Cole, 1984b; Yamamoto et al. 1983). It was also reported that mode I phosphorylated tau proteins are the major antigenic component of the paired helical filaments forming the neurofibrillary tangles that characterize degenerative human neurons in Alzheimer's disease. It is significant, then, that in Alzheimer's brain enrichment in phosphorylated tau proteins is accompanied by an impoverishment in microtubules. Our observation that S100 proteins inhibit tau protein phosphorylation by the $Ca^{2+}/calmodulin$-dependent kinase opens ʟ the possibility that S100, in vivo, is involved in the regulation or misregulation of microtubules through its binding to tau and its effect on tau phosphorylation. The gene for the S100-ß subunit was recently mapped on chromosome 21 (Allore

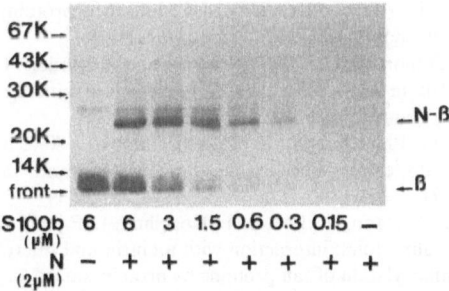

Fig. 1 . Cross-linking between neurogranin and the S100-ß protein. Neurogranin (2 μM) was mixed with increasing amounts of S100b as indicated and incubated in the presence of 0.5mM Ca^{2+} at 37°C for 30 min. Proteins were separated by 0.1% NaDodSO4-12.5% PAGE and analysed by Western blotting using antibodies directed against S100 protein. ß refers to the position of the S100-ß subunit, and N-ß to the covalent complex between neurogranin monomer and the S100-ß subunit.

et al., 1988). It was suggested that triplication of this chromosome in Down's syndrome may result in an increased gene dosage and overexpression of S100-ß. This is a proposed mechanism for the pathogenesis of nervous system abnormalities seen in this syndrome. Interestingly, there is indication that families with an inherited form of Alzheimer's disease share similar genetic defect on chromosome 21 than Down's patients and that both patients develop similar histopathological lesions, amyloid plaques and neurofibrillary tangles (Barnes, 1987). The possible involvement of S100 proteins in ethiology of Down's syndrome or Alzheimer's disease through its interaction with tau proteins should be investigated further.

S100 protein is also considered as a marker for several human tumors and strong sequence homology exists between S100-subunits and specific proteins whose synthesis is preferentially expressed in cells during growth and differentiation (for a review see Kligman and Hild, 1988). A striking sequence homology also exist between S100-subunit and the regulatory subunit of calpactin I, p10, a substrate for both tyrosine kinase and PKC (Gerke and Weber 1985). A recent report by Hagiwara et al.(1988) demonstrated that S100 interacts with calpactin I and inhibits the tyrosine phosphorylation of the substrate by tyrosine kinase. These data, combined with the previous finding that S100 protein inhibits specifically tau protein, neurogranin and P87 phosphorylation by PKC support the hypothesis that modulation of protein phosphorylation might be one of the S100 protein function.

Finally we have previously observed that the state of oxidation of cysteine residue (Cys 84ß or Cys 85α) has an influence on the monomer-dimer equilibrium that characterizes the quaternary structure of S100 proteins and on the protein affinitiy for calcium (Baudier et al. 1986b). Alkylation of these residues with thiol reagents resulted in a drastic increase of the S100 protein affinity for calcium which become, then, compatible with physiological intracellular calcium concentrations. The observed covalent cross-linking between cysteine on S100-subunit and sulfhydryl groups on protein kinase substrate (tau protein, neuromodulin and neurogranin) through disulfide bridges confirmed the presence of highly reactive sulfhydryl groups in S100 proteins and suggest that similar covalent interaction of S100 proteins with other cellular targets might well represent one of the biological active state of the S100 proteins.

REFERENCES

Albert, K.A., Wu, W.C., Nairn, A.C., and Greengard, P., 1984, Inhibition by calmodulin of calcium/phospholipid-dependent protein phosphorylation. Proc. Natl. Acad. Sci. USA, 81:3622.

Alexander, K.A., Wakim, B.T., Doyle, G.S., Walsh, K.A., and Storm, D.R., 1988, Identification and characterization of the calmodulin binding domain of neuromodulin, a neurospecific calmodulin-binding protein. J. Biol. Chem., 263:7544.

Allore, R., O'Hanlon, D., Price, R., Neilson, K., Willard, H.F., Cox, D.R., Marks, A., and Dunn, R.J., 1988, Gene encoding the ß subunit of S100 protein is on chromosome 21: Implication for Down Syndrome. Science, 239:1311.

Barnes, D., 1987, Defect in Alzheimer's is on chromosome 21. Science, 235: 846.

Baudier, J., Briving, C., Deinum, J., Haglid, K., Sorskog, L., and Wallin, M., 1982, Effect of S100 proteins and calmodulin on Ca^{2+}-induced disassembly of brain microtubule proteins in vitro. FEBS lett., 147: 165.

Baudier, J., Glasser, N., and Gerard, D., 1986a, Ions binding to S100 proteins: Calcium and zinc-binding properties of bovine brain S100α, S100a(αß), and S100b(ßß) protein; Zn^{2+} regulates Ca^{2+} binding on S100b protein. J.Biol.Chem., 261: 8192.

Baudier, J., Glasser, N., and Duportail, G., 1986b, Biman- and acrylodan- labeled S100 proteins. Role of cysteines-85 α and -84ß in the conformation and calcium binding properties of S100$\alpha\alpha$ and S100b(ßß) proteins. Biochemistry, 25: 6934.

Baudier, J., and Cole, R.D., 1987, Phosphorylation of tau proteins to a state like that in Alzheimer's brain is catalyzed by a calcium/calmodulin- dependent kinase and modulated by phospholipids. J.Biol.Chem., 262: 17577.

Baudier, J., Mochley-Rosen, D., Newton, A., Lee, S.H., Koshland, D.E., and Cole, R.D., 1987a, Comparison of S100b protein with calmodulin: interaction with melittin and microtubule-associated tau proteins and inhibition of phosphorylation of tau proteins by protein kinase C. Biochemistry, 26: 2886.

Baudier, J., Lee, S.H., and Cole, R.D., 1987b, Separation of the different microtubule-associated tau protein species from bovine brain and their mode II phosphorylation by Ca^{2+}/phospholipid- dependent protein kinase C. J. Biol. Chem., 262: 17584.

Baudier, J., and Cole, R.D., 1988a, Interactions between the microtubule-associated tau proteins and S100b regulate tau phosphorylation by the Ca^{2+}/calmodulin-dependent protein kinase II. J. Biol. Chem. 263: 5876.

Baudier, J., and Cole, R.D., 1988b, Reinvestigation of the sulfhydryl reactivity in bovine brain S100b(ßß) protein and the microtubule- associated tau proteins. Ca^{2+} stimulates disulfide cross-linking between the S100b ß subunit and the microtubule-associated tau(2) protein. Biochemistry, 27: 2728.

Baudier, J., 1988, S100 proteins: Structure and calcium binding properties, in "Calcium and calcium binding proteins," C. Gerday, R. Gilles, and L. Bolis, eds., Springer-Verlag, Berlin Heidelberg.

Baudier, J., Bronner, C., Kligman, D., and Cole, R.D., 1989, Protein kinase C substrates from bovine brain. Purification and characterization of neuromodulin, a neuron-specific calmodulin- binding protein. J. Biol. Chem. 264: 1824.

Deinum, J., Baudier, J., Briving, K., Rosengreen, L., Wallin, M., Gerard, D., and Haglid, K., 1983, The effect of S100a and S100b proteins and Zn^{2+} on the assembly of brain microtubule proteins in vitro. FEBS lett., 163: 287.

Donato, R., Isobe, T., and Okuyama, T., 1985, S100 proteins and microtubules: analysis of the effects of rat brain S100 (S100b) and ox brain S100a0, S100a and S100b on microtubule assembly - disassembly. FEBS lett., 186: 65.

Endo, T., and Hidaka, H., 1983, Effect of S100 protein on microtubule assembly - disassembly. FEBS lett., 161: 235.

Gerke, V., and Weber, K., 1985, The regulatory chain in the p36 kD substrate complex of viral tyrosine-specific protein kinases is related in sequence to the S100 protein of glial cells. EMBO J., 4: 2917.

Grundke-Iqbal, I., Iqbal, K., Tung, Y.C., Quinlan, M., Wisniewski, H.M., and Binder, L.I., 1986, Abnormal phosphorylation of the microtubule- associated protein tau in Alzheimer cytoskeletal pathology. Proc. Natl. Acad. Sci. USA, 83: 4913.

Hagiwara, M., Ochiai, M., Owada, K., Tanaka, T., and Hidaka, H., 1988, Modulation of tyrosine phosphorylation of p36 and other substrates by the S100 protein. J. Biol. Chem. 263: 6438.

Ihara, Y., Nukina, N., Miura, R., and Ogawara, M., 1986, Phosphorylated tau protein is integrated into paired helical filaments in Alzheimer's disease. J. Biochem., 99: 1807.

Ishikawa, H., Nagami, H., and Shirasawa, N., 1983, Novel clonal strains from adult rat anterior pituitary producing S100 protein. Nature, 303: 711.

Isobe, T., Nakajima, T., and Okuyama, T., 1977, Reinvestigation of extremely acidic proteins in bovine brain. Biochim. Biophys. Acta, 494: 222.

Isobe, T., and Okuyama, T., 1978, The amino acid sequence of S100 protein (Pap I-b protein) and its relation to the calcium-binding proteins. Eur. J. Biochem., 116: 79.

Kligman, D., and Marshak, D.R., 1985, Purification and characterization of a neurite extension factor from bovine brain. Proc. Natl. Acad. Sci. USA, 82: 7136.

Kligman, D., and Patel, J., 1986, A protein modulator stimulates C kinase- dependent phosphorylation of a 90K substrate in synaptic membranes. J. Neurochem., 47: 298.

Kligman, D., and Hild, D.C., 1988, The S100 protein family. TIBS, 13: 437.

Labourdette, G., and Mandel, P., 1980, Effect of norepinephrine and dibutyryl cyclic AMP on S100 protein level in C6 glioma cells. Biochem. Biophys. Res. Commun., 96: 1702.

Lindwall, G., and Cole, R.D., 1984a, Phosphorylation affects the ability of tau protein to promote microtubule assembly. J. Biol. Chem., 259: 5301.

Lindwall, G., and Cole, R.D., 1984b, The purification of tau protein and the occurrence of two phosphorylation states of tau in brain. J. Biol. Chem., 259: 12241.

Patel, J., and Kligman, D., 1988, Purification and characterization of an Mr 87,000 protein kinase C substrate from rat brain. J. Biol. Chem., 262: 16686.

Qi, D.F., and Kuo, J.F., 1984, S100 modulates Ca^{2+}-independent phosphorylation of an endogenous protein in brain. J. Neurochem., 43: 256.

Tsuyama, S., Bramblett, G.T., Huang, K.P., and Flavin,M., 1986, Calcium/ phospholipid-dependent kinase recognizes sites in microtubule - associated protein 2 which are phosphorylated in living brain and are not accessible to other kinases. J. Biol. Chem., 261: 4110.

Yamamoto, H., Fukunaga, K., Tanaka, E., and Miyamoto, E., 1983, Ca^{2+}- and calmodulin-dependent phosphorylation of microtubule-associated protein 2 and tau factor, and inhibition of microtubule assembly. J. Neurochem., 41: 1119.

Zimmer, D.B., and Van Eldik, L.J., 1987, Identification of a molecular target for the calcium-modulated protein S100: Fructose- 1,6-biphosphate aldolase. J. Biol. Chem., 261: 11424.

NOVEL AND SELECTIVE INHIBITORS OF CaM-KINASE II AND OTHER CALMODULIN-DEPENDENT ENZYMES

Hiroyoshi Hidaka, Masatoshi Hagiwara and Hiroshi Tokumitsu

Department of Pharmacology, Nagoya University, School of Medicine, Showa-ku, Nagoya 466, Japan

DEVELOPMENT OF CALMODULIN ANTAGONISTS

Twenty years has passed since two laboratories working independently on the purification of cyclic nucleotide phosphodiesterase reported the existence of a protein factor that activated enzymatic activity (1, 2). These early observations are credited the discovery of calmodulin. Biopharmacological studies using calmodulin antagonists such as naphthalenesulfonamides and phenothiazines have facilitated an insight into calmodulin function, not only in vitro but also in vivo. Phenothiazines were the first antagonists of calmodulin to be reported (3), although they have multiple macromolecular targets inside cells such as membranes (4), albumin (5), dehydrogenases (6), and dopamine receptors (7).

W-7 is the most potent calmodulin antagonist in naphthalenesulfonamides and the W-7-binding sites of calmodulin are also responsible for binding of phenothiazines (8). In addition to these, we have reported several compounds which exhibited calmodulin antagonistic effects such as HT-74 (9), bepridil (10), and CV-159 (11). However, since calmodulin exerted pleiotropic effects on various cellular function by activating multiple enzymes, "calmodulin antagonists" were not adequate to elucidate the specific role of Ca^{2+}/calmodulin-dependent enzymes including Ca^{2+}-dependent phosphodiesterase (Ca^{2+}-PDE), myosin light chain kinase (MLC-kinase), and calmodulin-dependent protein kinase II (CaM-kinase II) , in vivo. Therefore we tried to develop a potent and direct inhibitor of each calmodulin-dependent enzyme (Fig. 1).

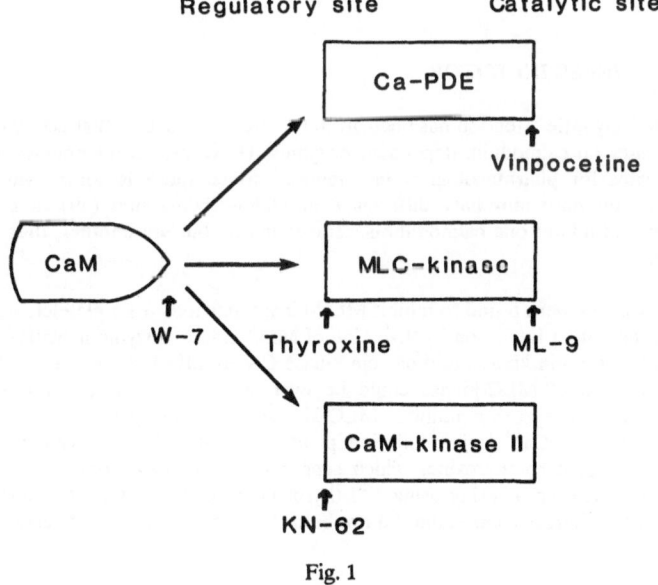

Fig. 1

DIRECT MODULATORS OF Ca^{2+}-PDE

Calmodulin was found to be a Ca^{2+}-dependent stimulator of cyclic nucleotide phosphodiesterase, but it was difficult to evaluate the physiological significance of the enzyme by using calmodulin antagonists such as naphthalenesulfonamides and phenothiazines. Therefore selective inhibitors and activators of Ca^{2+}-PDE should be potential tools for characterizing the properties and physiological functions of this enzyme. Vinpocetine, 14-ethoxycarbonyl-(3α, 16α ethyl)-14,15-eburnamenine, inhibited Ca^{2+}-PDE not by interacting with calmodulin but rather by interacting directly with the enzyme (12). The concentration of vinpocetine which produced 50% inhibition of the Ca^{2+}-PDE activity was approximately 21 μM, both in the presence and absence of Ca^{2+}/calmodulin complex. This compound was only a weak inhibitor of cGMP phosphodiesterase and cAMP phosphodiesterase, with IC_{50} values above 500 μM. On the other hand, we found that HA-542, 1-(2,4-dipiperidino-6-quinazoline-sulfonyl)-4-(2-ethoxy-2-phenylethyl) piperazine, and HA-543, 1-(2,4-dipiperidino-6-quinazolinesulfonyl)-4-cinnamylpiperazine are selective Ca^{2+}-independent activators of Ca^{2+}-PDE (13). This activation was also inhibited by phenothiazines or naphthalene-sulfonamides. Moreover, these synthetic activators of Ca^{2+}-PDE are potent inhibitors of cAMP and cGMP phosphodiesterases.

DIRECT INHIBITORS OF MLC-KINASE

Myosin light chain kinase (MLC-kinase) is a Ca^{2+}/calmodulin-dependent protein kinase which specifically phosphorylates myosin light chain. This reaction is obligatory for stimulation of actin-activated, myosin ATPase in nonmuscle cells and smooth muscle and is a prerequisite for contractile response (14). Recently we reported that 75% of 12 anti-gizzard smooth muscle MLC-kinase monoclonal antibodies cross-reacted with nonmuscle (platelet) MLC-kinase, but none of those reacted with skeletal muscle enzyme (15). Thus, we investigated the inhibitory effects of natural and synthesized substances on smooth muscle and nonmuscle MLC-kinase.

A derivative of W-7, named ML-9, 1-(5-chloronaphthalene-1-sulfonyl)-1H-hexahydro-1,4-diazepine hydrochloride, exhibited the selective inhibition toward smooth muscle and nonmuscle MLC-kinase (Ki = 3.8 μM), whereas higher concentration of the compound was needed to inhibit skeletal muscle MLC-kinase (Ki = 49 μM) and Ca^{2+}-PDE (Ki = 50 μM) (16). The Ki values for Ca^{2+}/calmodulin-dependent, and trypsin-treated Ca^{2+}/calmodulin-independent smooth muscle MLC-kinase activity were 3.8 μM and 4.0 μM, respectively. The mode of interaction of ML-9 with MLC-kinase was investigated by double reciprocal plots, and it revealed that ML-9 was a competitive inhibitor with ATP and noncompetitve inhibitor with the phosphate acceptor (myosin light chain) of the catalytic reaction. These data suggest that ML-9 has excellent selectivity toward nonmuscle and smooth muscle MLC-kinase and that the inhibitory effects of ML-9 is due to direct action on the active site of the enzyme and not on calmodulin. More recently we found that an iso-quinolinesulfonamide, HA-140 (17), and a flavonoid, quercetine (18), exhibited potent inhibitory effects on MLC-kinase. These drugs are structurally different from ML-9, but molecular mechanisms of the inhibition are the same.

NEW TYPE OF MLC-KINASE INHIBITOR

Thus far, relatively little attention has been given to developing agents that act by binding to the calmodulin recognition sites on calmodulin-dependent enzymes. However, recent findings suggest that these may provide useful sites for pharmacological intervention. Since there is some evidence that various calmodulin-dependent enzymes may have different calmodulin-binding sites (19), it may be possible to develop agents that would inhibit one calmodulin-sensitive enzyme but not another, thus producing a relatively selective action.

L- and D-thyroxines were found to inhibit MLC-kinase activities from platelet, smooth and skeletal muscles. Contrary to the potent inhibition by thyroxine of MLC-kinase, enzyme activities of casein kinase I and II, cAMP-dependent protein kinase, and protein kinase C were affected only weakly by thyroxine. This thyroxine-induced inhibition of MLC-kinase could be overcome by high concentrations of calmodulin. Kinetic analysis indicated that thyroxine inhibited MLC-kinase competitively with respect to calmodulin and the Ki value was 2.5 μM(20). Ca^{2+}/calmodulin-independent activity of the enzyme produced by trypsin treatment was not suppressed by thyroxine, which suggests that thyroxine does not affect the catalytic domain of MLC-kinase. The gel overlay using ^{125}I-thyroxine revealed that MLC-kinase evidenced ^{125}I-thyroxine binding activity, whereas calmodulin did not bind thyroxine. Moreover, digestion fragments which contain calmodulin-binding domain of MLC-kinase bound ^{125}I-thyroxine. These results indicated that thyroxine binds at or near calmodulin-binding domain of MLC-kinase and inhibits calmodulin-induced activation of the enzyme. Recently we found that thyroxine-coupled Sepharose was available to purify MLC-kinase instead of calmodulin-coupled Sepharose (21).

SELECTIVE INHIBITOR OF CaM-KINASE II

In 1980, we first reported the existence of Ca^{2+}/calmodulin-dependent protein kinase which phosphorylated myelin basic protein in rabbit brain (22). In 1981, Yamauchi and Fujisawa reported Ca^{2+}/calmodulin-dependent protein kinase II (CaM-kinase II) prepared from rat brain cytosol (23). CaM-kinase II was found to have relatively broad substrate specificities in nervous system (24) and to be a large multimeric enzyme composed of two related subunits with -α(50kDa) and β(60kDa)(25). Because of the complexity of activation mechanism and broad specificities of substrates, the physiological role of this kinase was not fully elucidated. Thus, an experimental approach using a specific inhibitor of CaM-kinase II would be helpful to clarify the physiological significance of the enzyme in living cells, especially in the highly complex phenomena of the central nervous system. Newly synthesized compound KN-62, 1-(NO-bis-1,5-isoquinolinesulfonyl)-N-methyl-L-tyrosil-4-phenylpiperazine, inhibited CaM-kinase II activity with Ki value of 0.9 μM, but did not give significant effects on the activities of protein kinase C, and cAMP-dependent protein kinase (26). An autoradiogram revealed that KN-62 inhibited the Ca^{2+}/calmodulin-dependent autophosphorylation of both α and β subunits of CaM-kinase II in the presence or absence of exogenous substrate. Kinetic analysis indicated that this inhibitory effect of KN-62 was competitive with respect to calmodulin. However, KN-62 did not significantly inhibit other Ca^{2+}/calmodulin-dependent enzymes such as Ca^{2+}-PDE and MLC-kinase, nor autophosphorylated Ca^{2+}/calmodulin-independent CaM-kinase II activity up to 100 μM. Moreover, KN-62-coupled Sepharose retained CaM-kinase II, but not calmodulin. These results indicate that KN-62 is a selective and direct inhibitor which recognizes the calmodulin-binding domain of CaM-kinase II. To clarify the physiological function of CaM-kinase II in living nerve cells, we examined the effect of KN-62 on PC-12 pheochromocytoma cells. KN-62 significantly suppressed 50 kDa protein phosphorylation induced by EGF or A-23187, which was immunoprecipitated with anti-CaM kinase II antibody suggesting that the 50 kDa protein is a subunit of CaM-kinase II in PC-12 cells and KN-62 blocked CaM kinase II activity in vivo. This compound would be useful as a molecular tool to elucidate the biological function of CaM-kinase II in vivo.

CONCLUSION

We have found various inhibitors of Ca^{2+}/calmodulin messenger system and these inhibitors can be classified into three types on the basis of sites of drug action as described in table 1. Direct pharmacological manipulation of living cells or tissues by these drugs may present a novel approach towards elucidation of the physiolobical role of Ca^{2+}/calmodulin cascade.

Table 1. Classification of novel inhibitors of Ca^{2+}/calmodulin cascade

	Drugs	Binding Site	Inhibition of Enzymes	Ref.
(I)	W-7	Calmodulin	All Calmodulin- Dependent Enzymes	8
	HT-74			9
	Bepridil			10
	CV-159			11
(II)	Vinpocetine	Regulatory Site of Enzyme	Ca^{2+}-PDE	12
	ML-9		MLC-kinase	16
	HA-140		MLC-kinase	17
	Quercetine		MLC-kinase	18
(III)	Thyroxine	Catalytic Site of Enzyme	MLC-kinase	19
	KN-62		CaM-kinase II	26

REFERENCES

1. S. Kakiuchi, and R. Yamazaki, Biochem. Biophys. Res. Commun. 41: 1104 (1970).
2. W.Y. Cheung, Biochem. Biophys. Res. Commun. 38: 533 (1970).
3. F. Honda, S. Katsuki, and N. Sakai, Jpn. J. Pharmacol. Suppl. 23: 27 (1973).
4. P. Seeman, Pharmacol. Rev. 24: 583 (1972).
5. S. Gabay, and P. C. Hung, in: "The Phenothiazines and Structurally Related Drugs", I. S. Forrest, D. J. Carr, and Usdin, eds., pp. 175-189, Raven Press, New York (1974).
6. F. M. Veronese, R. Bevilacqua, and I. M. Chaiken, Mol. Pharmacol. 15: 313 (1979).
7. J. Giesecke, and H. Hebert, Q. Rev. Biophys. 12: 263 (1979).
8. H. Hidaka, T. Yamaki, M. Naka, T. Tanaka, H. Hayashi, and R. Kobayashi, Mol. Pharmacol. 17: 66 (1980).
9. T. Tanaka, H. Umekawa, M. Saitoh, T. Ishikawa, T. Shin, M. Ito, Y. Kawamatsu, H. Sugihara, and H. Hidaka, Mol. Pharmacol. 29: 264 (1986).
10. H. Itoh, T. Ishikawa, and H. Hidaka, J. Pharmacol. Exp. Ther. 230: 737 (1984).
11. H. Umekawa, K. Yamakawa, K. Nunoki, N. Taira, T. Tanaka, and H. Hidaka, Biochemical. Pharmacol. 37: 3377 (1988).
12. M. Hagiwara, T. Endo, and H. Hidaka, Biochem. Pharmacol. 33: 453 (1984).
13. T. Tanaka, E. Yamada, T. Sone, and H. Hidaka, Biochemistry 22: 1030 (1983).
14. R. S. Adelstein, and E. Eisenberg, Annu. Rev. Biochem. 49: 921 (1980).
15. M. Hagiwara, H. Tokumitsu, K. Onoda, T. Tanaka, M. Ito, N. Kato, and H. Hidaka, J. Biochem. (in press) (1989).
16. M. Saitoh, T. Ishikawa, S. Matsushima, M. Naka, and H. Hidaka, J. Biol. Chem. 262: 7796 (1987).
17. M. Hagiwara, M. Inagaki, M. Watanabe, M. Ito, K. Onoda, T. Tanaka, and H. Hidaka, Mol. Pharmacol. 32: 7 (1987).
18. M. Hagiwara, S. Inoue, T. Tanaka, K. Nunoki, M. Ito, and H. Hidaka, Biochem. Pharmacol. 37: 2987 (1988).
19. C. B. Klee, and T. C. Vanaman, Adv. Protein Chem. 35, 213 (1982).
20. M. Hagiwara, S. Mamiya, and H. Hidaka, J. Biol. Chem. 264: 40 (1989).
21. S. Mamiya, M. Hagiwara, S. Inoue, and H. Hidaka, J. Biol. Chem. 264:8575 (1989).
22. T. Endo, and H. Hidaka, Biochem. Biophys. Res. Commun. 97: 553 (1980).
23. T. Yamauchi, H. Fujisawa, FEBS Lett. 129: 117 (1981).
24. M. B. Kennedy, Annu. Rev. Neurosci. 6: 493 (1983).
25. A. C. Narin, H. C. Hemmings, Jr., and P. Greengard, Annu. Rev. Biochem. 54: 931 (1985).
26. H. Tokumitsu, T. Chijiwa, M. Hagiwara, A. Mizutani, and Hidaka, H. (submitted).

STRUCTURAL DETAILS OF THE INTERACTION OF CALMODULIN WITH THE PLASMA MEMBRANE CA^{2+}-ATPase

Joachim Krebs, Thomas Vorherr, Peter James, Ernesto Carafoli, Theodore A. Craig* and D. Martin Watterson*

Laboratory of Biochemistry, Swiss Federal Institute of Technology (ETH), Zurich Switzerland

*Department of Pharmacology, Vanderbilt University and Laboratory of Cellular and Molecular Physiology, Howard Hughes Medical Institute, Nashville, Tennessee, 37232 USA

INTRODUCTION

The plasma membrane Ca^{2+}-ATPase is one of the enzymes which are regulated by calmodulin (for a recent review see Carafoli et al., 1988). The interaction between the two proteins has been studied in de tail. The CaM-binding domain of the ATPase has recently been identified (James et al., 1988, Verma et al., 1988), and it has been shown that the Ca^{2+}-ATPase can be fully stimulated by the C-terminal half of CaM (i.e. AA 78-148) but not by the N-terminal half (1-77; see Guerini et al., 1984). Furthermore, it could be demonstrated that the 3rd Ca^{2+}-binding loop of CaM (counted from the N-terminus) is essential for the activation of the ATPase (Guerini et al., 1984). Recently, chemically modified calmodulins provided evidence that arginine and methionine residues located in the C-terminal half of CaM are important for the interaction between CaM and the plasma membrane Ca^{2+}-ATPase (Guerini et al., 1987).

In this report we will describe the effects of amino acid replacements in the central helix or the 3rd Ca^{2+}-binding domain of calmodulin due to site directed mutagenesis on the stimulation of the erythrocyte Ca^{2+}-ATPase. Furthermore, NMR-spectroscopy experiments to characterize the interaction of calmodulin with synthetic peptides corresponding to the calmodulin-binding domain of the Ca^{2+}-ATPase will be described (James et al., 1988).

MATERIALS AND METHODS

CaM-deficient erythrocyte membranes (i.e. ghosts) have been prepared as described by Niggli et al. (1981) using recently outdated human blood from the local blood bank. The membranes were stored at -70°C until further usage. The Ca^{2+}-ATPase was isolated from erythrocyte membranes as described elsewhere (Niggli et al., 1981) and stored at -70°C in the elution buffer (20 mM Hepes pH = 7.2, 130 mM KCl, 1 mM MgCl$_2$, 2 mM EDTA, 0.05 % Triton X 100, 0.05 % Phosphatidylcholine, 2 mM DTT) in the presence of 10% glycerol and after increasing the final concentration of MgCl$_2$ up to 2 mM and of CaCl$_2$ up to 100 μM. The enzyme was assayed either in ghost membranes or in the isolated form essentially as described by Zurini et al. (1984). Protein concentration was determined as described (Zurini et al., 1984) using bovine serum albumin as standard.

The principles of the construction of a synthetic calmodulin gene, its expression in E. coli and the use of the cassette mutagenesis approach for site directed mutagenesis studies were outlined by Roberts et al., (1985) and by Craig et al. (1987), respectively. Details will be described elsewhere.

The nuclear magnetic resonance measurements were carried out at 360 MHz using a Bruker 360

AM machine at T=30°C. 500 transients of 8K sized spectra were usually signal-averaged in 1D-NMR experiments using a 90° observation pulse. The residual signal of the solvent was suppressed by a presaturation pulse. Further details are given elsewhere (Vorherr et al., 1989).

RESULTS AND DISCUSSION

As indicated above, the C-terminal half of CaM is essential for the activation of the plasma membrane Ca^{2+}-ATPase. Of further interest is that the ATPase responds differentially towards the chemical modification of certain amino acid residues (Guerini et al., 1987). So it was shown that modification of Arg-residues (i.e. 90/106) reduced the activation potential of CaM for the ATPase by 50% whereas Lys-modification did not. In addition, modification of Met-residues had a more pronounced effect if amino acids of the C-terminal half (e.g. met 109, 124) rather than those of the central helix or the N-terminal half were modified (Guerini et al., 1987). In line with these observations is the finding that the activation of the Ca^{2+}-ATPase (i.e. in the erythrocyte membrane as well as in the isolated form) by VU-1 CaM is practically identical with the activation of bovine brain CaM (Fig. 1) indicating that the amino acid replacements of VU-1, especially $M_{71} \rightarrow L$, $R_{86} \rightarrow K$, $M_{144} \rightarrow V$, and the absence of the N-acetyl group and trimethyllysine, are not critical for the ATPase. The resistance of the enzyme to certain changes in the central helix is further documented by a similar activation curve for VU-11 and VU-15 (data not shown) in which either K_{75} was replaced by P or the tripeptide KGK had been inserted between D_{80} and S_{81}, (typical for troponin C).

On the other hand, replacement of the three Glu-residues in positions 82-84 (=VU-8) by three Lys-residues reduced the efficiency of the activation of the ATPase about 5-fold (Fig. 1) indicating that the dipole-character of the central helix has a role in the cooperative properties of the 2 halves of CaM. This, in turn, would affect the activation potential for the ATPase.

A similar decrease in activation efficiency could be observed with VU-17, in which S101 is replaced

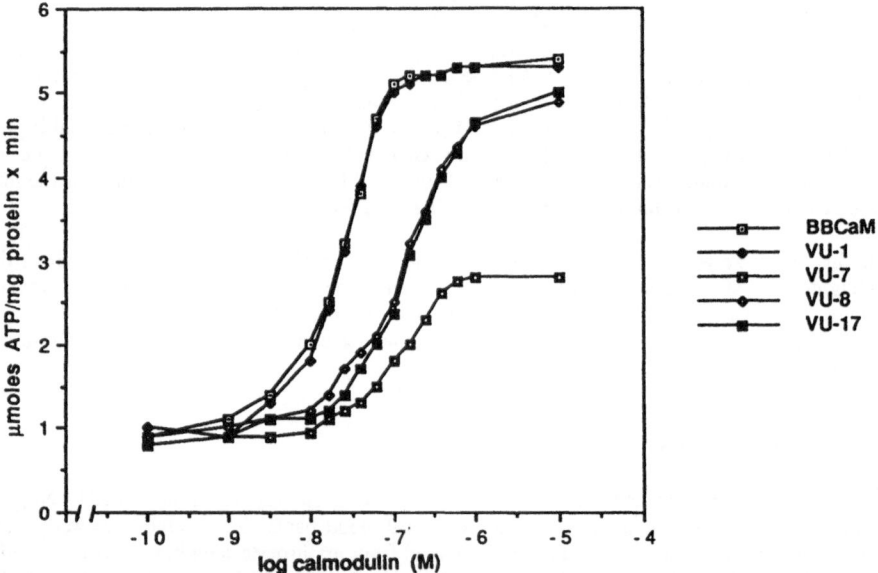

Fig. 1 . Stimulation of the purified Ca2+-ATPase by different calmodulins. The different calmodulins were obtained as described in the Materials and Methods section and tested for their ability to stimulate the ATPase by using the coupled enzyme assay (Zurini et al., 1984). BBCaM = calmodulin purified from bovine brain; VU-1 = genetically engineered calmodulin, used as standard, containing the following modifications as compared to BBCaM (Watterson et al., 1980; see also Roberts et al., 1985): $E_6 \rightarrow D$, $T_{70} \rightarrow N$, $M_{71} \rightarrow L$, $I_{85} \rightarrow L$, $R_{86} \rightarrow K$, $Y_{99} \rightarrow F$, $I_{130} \rightarrow V$, $M_{144} \rightarrow V$, $T_{146} \rightarrow M$. In addition, VU-1 is not acetylated at the N-terminus and does not contain trimethyllysine at position 115. The other engineered VU-calmodulins contain in addition the following modifications: VU-7 = $A_{88} \rightarrow P$, VU-8 = $EEE_{82-84} \rightarrow KKK$, VU-17 = $S_{101} \rightarrow F$.

Table 1. Synthetic peptides of the CAM-binding region of the Ca^{2+}-ATPase

C15W	L-R-R-G-Q-I-L-**W**-**F**-R-G-L-N-R-I
C20W	L-R-R-G-Q-I-L-**W**-**F**-R-G-L-N-R-I-Q-T-Q-I-K
C28W	L-R-R-G-Q-I-L-**W**-**F**-R-G-L-N-R-I-Q-T-Q-I-K-V-V-N-A-F-S-S-S

The synthesis procedures and the purification protocols are described by
Vorherr et al. (1989)

by an F (Fig. 1). This observation is of special interest since in a mutant of Paramecium tetraurelia which lacks calcium dependent potassium outflux the outward K^+-current could be restored by injecting wild-type CaM (Hinrichsen et al., 1986). These findings indicated that a mutation in CaM was responsible for this significant difference in the activation properties. Sequence studies indicated that the only difference between wild type CaM (Schaefer et al., 1987) and CaM from the panthophobiac mutant was the replacement of S101 by an F. Amino acids at this position (i.e. at the -X position according to Kretsinger's scheme) usually are ligated to Ca^{2+} using a water molecule as a bridge (Szebenyi and Moffat, 1986). Therefore, it may be possible that the exchange S101 → F somewhat influences the geometry for the 3rd Ca^{2+}-binding domain. This, in turn, could be responsible for the reduced efficiency in activating the Ca^{2+}-dependent plasma membrane ATPase.

The activation of the ATPase by CaM using VU-7, in which A_{88} is replaced by P (Fig. 1), is changed even more drastically. Since the structural properties of the 3rd loop of CaM should be intact in order to efficiently stimulate the ATPase it is of interest that this replacement in the middle of the N-terminal helix of the 3rd Ca^{2+}-binding domain which most probably changes the orientation of the former is much more effective than the replacement K_{75} → P or the activation of other enzymes by VU-7 (T.A. Craig and D.M. Watterson, unpublished observations). These results further support the conclusion that the plasma membrane Ca^{2+}-ATPase is especially sensitive to changes in the 3rd Ca^{2+}-binding domain of CaM.

In order to investigate the interaction of calmodulin with the ATPase in more detail we synthesized peptides corresponding to the calmodulin binding domain of the ATPase (James et al., 1988, Vorherr et al., 1989; see Table 1) and studied their interaction with calmodulin using circular dichroism, infrared spectroscopy, nuclear magnetic resonance and fluorescence techniques (Vorherr et al., 1989). As documented elsewhere most of these peptides interacted with calmodulin with binding constants similar to the native enzyme, i.e. in the nM range. Here we summarize some of these data.

Figure 2 compares the aromatic region of the ^{1}H-NMR spectra of Ca_4-calmodulin (2a) with those of the calmodulin-peptide complex (2b) and of the isolated peptide C15W (2c). This peptide interacts with calmodulin in a manner similar to that of the longer peptides C20W or C28W, albeit with lower affinity (Vorherr et al., 1989). Figure 2 shows that the ^{1}H-resonances of the free tryptophan of peptide C15W located between 7.4 and 7.6 ppm (2c) disappeared completely when calmodulin and the peptide interacted: the spectrum of the isolated peptide showed resonances at chemical shift positions of tryptophan or phenylalanine typical of the random coil conformation (Fig. 2c). As calmodulin was added in the presence of Ca^{2+} all of the tryptophan resonances shifted upfield disappearing under the broad envelope between 7.0 and 7.4 ppm (Fig. 2b). The resonances of calmodulin were apparently also influenced. This is especially evident for the resonances of Tyr_{138} which were significantly upfield shifted in the presence of the peptide. Similar results have been obtained with the peptides C20W and C28W, even if the poor solubility properties of the latter made NMR-measurements rather difficult.

As documented elsewhere (Vorherr et al., 1989) peptides C15W and C20W which display essentially a random coil conformation as free peptides (e.g. see Fig. 2c) adopt secondary structure, mainly α-helical, upon complex formation with calmodulin as indicated by CD spectroscopy. With respect to the identification of the site of interaction it was of interest to note that 2D-NMR NOESY experiments

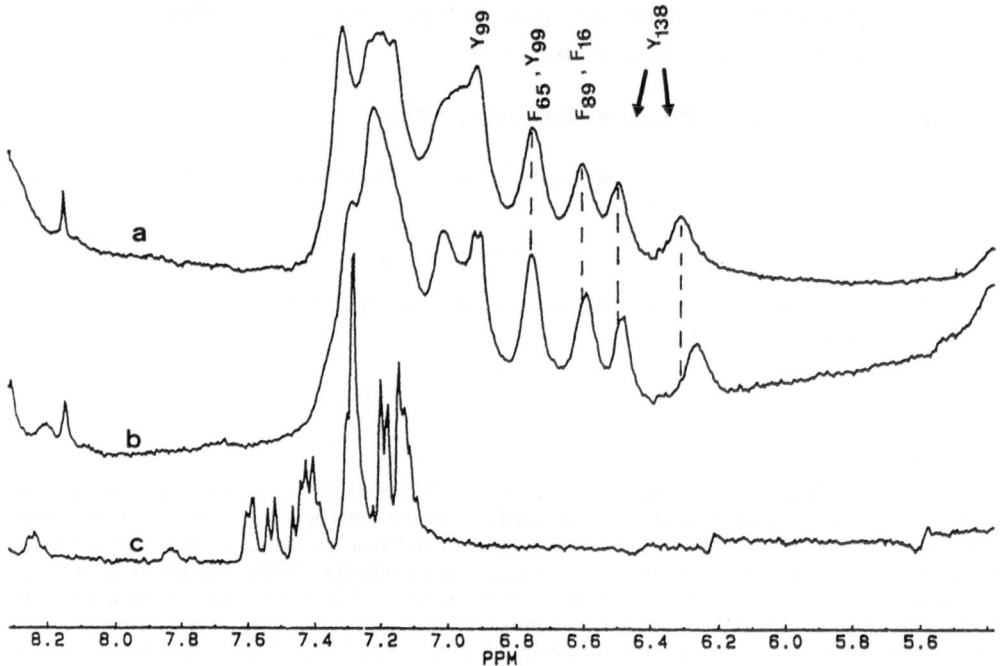

Fig. 2 . Aromatic portion of the ^{1}H-NMR spectra of calmodulin and the synthetic peptide C15W. The Figure shows the spectra of Ca$_4$-calmodulin (a), of the peptide-calmodulin complex (1:1; b) and of the peptide C15W (c). The spectra were recorded as described in the Materials and Methods section, further details are given by Vorherr et al. (1989).

(Vorherr et al., 1989) provided evidence that the C-terminal half of calmodulin interacted preferentially with the peptides. This is in excellent agreement with the conclusion (see above) that the C-terminal half of calmodulin is of special importance in the activation of the plasma membrane Ca^{2+}-pump.

This report summarizes our findings leading to the conclusion that the C-terminal half of calmodulin plays a dominant role in the activation of the plasma membrane Ca^{2+}-ATPase. The results presented show that several amino acids of the 3rd Ca^{2+}-binding domain of calmodulin are of special importance for the efficient interaction of the ATPase with calmodulin.

REFERENCES

Carafoli, E., Krebs, J., and Chiesi, M., 1988, Calmodulin in the transport of calcium across biomembranes, in: "Molecular aspects of cellular regulation", P. Cohen and C.B. Klee, eds., Vol. V, Elsevier, Amsterdam.

Craig, T.A., Watterson, D.M., Prendergast, F.G., Haiech, J., and Roberts, D.M., 1987, Site-specific mutagenesis of the -helices of calmodulin. Effects of altering a charge cluster in the helix that links the two halves of calmodulin, J. Biol. Chem., 262: 3278.

Guerini, D., Krebs, J., and Carafoli, E., 1984, Stimulation of the purified erythrocyte Ca^{2+}-ATPase by tryptic fragments of calmodulin, J. Biol. Chem., 259: 15172.

Guerini, D., Krebs, J., and Carafoli, E., 1987, Stimulation of the erythrocyte Ca^{2+}-ATPase and of bovine brain cyclic nucleotide phosphodiesterase by chemically modified calmodulin, Eur. J. Biochem., 170: 35.

Hinrichsen, R.D., Burgess-Cassler, A., Soltvedt, B.C., Hennessey, T., and Kung, C., 1986, Restoration by calmodulin of a Ca^{2+}-dependent K$^+$-current missing in a mutant of paramecium, Science, 232: 503.

James, P., Maeda, M., Fischer, R., Verma, A.K., Krebs, J., Penniston, J.T., and Carafoli, E., 1988, Identification and primary structure of a calmodulin binding domain of the Ca^{2+}-pump of human erythrocytes, J. Biol. Chem., 263: 2905.

Niggli, V., Adunyah, E.S., Penniston, J.T., and Carafoli, E., 1981, Purified (Ca^{2+}-Mg^{2+})-ATPase of the erythrocyte membrane. Reconstitution and effect of calmodulin and phospholipids, J. Biol. Chem., 256: 395.

Roberts, D.M., Crea, R., Malecha, M., Alvarado-Urbina, G., Chiarello, R.H., and Watterson, D.M., 1985, Chemical synthesis and expression of a calmodulin gene designed for site-specific mutagenesis, Biochemistry, 24: 5090.

Schaefer, W.H., Lukas, T.J., Blair, I.A., Schultz, J.E., Watterson, D.M., 1987, Amino acid sequence of a novel calmodulin from paramecium tetraurelia that contains dimethyllysine in the first domain, J. Biol. Chem., 262: 1025.

Szebenyi, D.M.E., and Moffat, K., 1986, The refined structure of vitamin D-dependent calcium-binding protein from bovine intestine. Molecular details, ion binding, and implications for the structure of the other calcium-binding proteins, J. Biol. Chem., 261: 8761.

Verma, A.K., Filoteo, A.G., Standord, D.R., Wieben, E.D., Penniston, J.T., Strehler, E.E., Fischer, R., Heim, R., Vogel, G., Mathews, S., Strehler-Page, M.-A., James, P., Vorherr, T., Krebs, J., and Carafoli, E., 1988, Complete primary structure of a human plasma membrane Ca^{2+} pump, J. Biol. Chem., 263: 14152.

Vorherr, T., James, P., Krebs, J., Enyedi, A., McCormick, D.J., Penniston, J.T., and Carafoli, E., 1989, The interaction of calmodulin with the calmodulin binding domain of the plasma membrane Ca^{2+}-pump, Biochemistry, in press.

Watterson, D.M., Sharief, F., and Vanaman, T.C., 1980, The complete amino acid sequence of the Ca^{2+}-dependent modulator protein (calmodulin) of bovine brain, J. Biol. Chem., 255: 962.

Zurini, M., Krebs, J., Penniston, J.T., and Carafoli, E., 1984, Controlled proteolysis of the purified Ca^{2+}-ATPase of the erythrocyte membrane. A correlation between the structure and the function of the enzyme, J. Biol. Chem., 259: 618.

ERYTHROCYTE Ca^{2+}-ATPase: ACTIVATION BY ENZYME

OLIGOMERIZATION VERSUS BY CALMODULIN

D. Kosk-Kosicka,[1] T. Bzdega,[1] A. Wawrzynow,[1] S. Scaillet,[1] K. Nemcek,[2] and J.D. Johnson[2]

[1] Department of Biological Chemistry, University of Maryland, School of Medicine
660 West Redwood Street, Baltimore, Maryland 21201, USA

[2] Department of Physiological Chemistry, The Ohio State University, Medical Center
Columbus, Ohio 43210, USA

SUMMARY

The subject of our studies is the mechanism of activation of the erythrocyte Ca^{2+}-ATPase. Using purified, detergent solubilized enzyme it was found that equivalent maximal Ca^{2+}-ATPase activity is obtained either upon addition of calmodulin or upon increase of enzyme concentration. Three independent methods, including Ca^{2+}-ATPase activity, polarization of the enzyme modified with an external fluorescent probe, and efficiency of fluorescence resonance energy transfer between enzyme molecules have established that the concentration dependent activation is due to enzyme oligomerization. The oligomers bind calmodulin with a lower stoichiometry (0.5 mol calmodulin/mol Ca^{2+}-ATPase), higher Ca^{2+} affinity (K_{Ca}=pCa 7.4), and higher cooperativity for Ca^{2+} (n_H=2.6) than the monomeric form (stoichiometry=1 mol calmodulin/mol Ca^{2+}-ATPase, K_{Ca}=pCa 7.0, n_H=1.1). The Ca^{2+} dependence of calmodulin binding and activation of monomers indicates that calmodulin binds before the Ca^{2+}-ATPase activity is exhibited, demonstrating that the activation of this enzyme form is totally dependent on calmodulin. In contrast, oligomers reveal very similar Ca^{2+} dependence for calmodulin binding and for Ca^{2+}-ATPase activity as well as for Ca^{2+} binding (assessed by tryptophan fluorescence), and for the oligomerization process (assessed by fluorescence energy transfer). The calmodulin antagonist drug 48/80 inhibits the calmodulin dependent activity of the monomers (I_{50}=1.4 $\mu g/ml$) but has no effect on the activity of oligomers, confirming that calmodulin plays no role in the activation of the oligomeric enzyme. Our studies indicate that the erythrocyte Ca^{2+}-ATPase can be activated by its high affinity, Ca^{2+} dependent binding of calmodulin or by a Ca^{2+} dependent oligomerization process which may involve calmodulin binding site.

The red cell Ca^{2+}-ATPase has been studied as the most experimentally accessible of the calmodulin regulated plasma membrane Ca^{2+}-ATPases (Schatzmann, 1982; Carafoli and Zurini, 1982). In the red cell the enzyme serves as the only Ca^{2+} extrusion system to maintain intracellular Ca^{2+} levels near 0.1 μM. Maintenance of this low Ca^{2+} level is important for normal physiological functions. Further, perturbation of intracellular Ca^{2+} has been linked to the etiology of certain disease states, such sickle cell or hypertension (AlJaboree et al., 1984; Penniston, 1984; Vincenzi et al., 1987).

The importance of regulation of intracellular Ca^{2+} is reflected in the complexity of regulation of the Ca^{2+}-ATPase. In addition to the well established stimulation of the Ca^{2+}-ATPase activity by calmodulin (see Schatzmann, 1982) we have recently discovered that the Ca^{2+}-ATPase is regulated by enzyme oligomerization (Kosk-Kosicka and Bzdega, 1988). Further, there is evidence for regulation by cAMP dependent phosphorylation (Neyses et al., 1985), proteolysis (Taverna et al., 1980; Enyedi et al., 1987), and acidic phospholipids (Niggli et al., 1981; Choquette et al., 1984). It is the goal of this work to elucidate the mechanism of Ca^{2+}-ATPase regulation by calmodulin and by oligomerization, in order to eventually understand the in vivo regulation of the enzyme in health and disease.

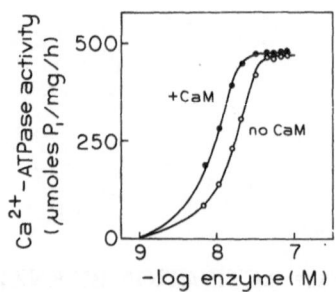

Fig. 1. Dependence of Ca^{2+}-ATPase activity on enzyme concentration in the presence (●) and absence (○) of calmodulin. The reaction mixture for the Ca^{2+}-ATPase activity contained: 50 mM Tris-maleate, pH 7.5, 130 mM KCl, 8mM $MgCl_2$, 1 mM EGTA and sufficient $CaCl_2$ to yield 17.75 μM free Ca^{2+}. The concentration of $C_{12}E_8$ was kept constant at 150 μM and the molar ratio of phosphatidyl-choline to the enzyme was constant at 300:1. Calmodulin, when present, was at a saturating concentration of 2 mol/mol enzyme. The reaction was started with 3 mM ATP, and after 30 min at 37°C various aliquots were withdrawn for subsequent colorimetric P_i measurement with malachite green (Kosk-Kosicka and Bzdega, 1988).

The mechanisms of Ca^{2+}-ATPase regulation by calmodulin and oligomerization are best studied with the purified, detergent soluble Ca^{2+}-ATPase in an in vitro system of mixed phospholipid/detergent micelles, which has been established to mimic the in vivo condition (Kosk-Kosicka et al., 1986a & b). We have a multifaceted approach to the problem, utilizing not only enzymatic activity measurements but also fluorescence methods to directly measure calmodulin-enzyme and enzyme-enzyme interactions.

RESULTS AND DISCUSSION

Enzyme concentration dependent oligomerization leads to the activation of the Ca^{2+}-ATPase activity

The Ca^{2+}-ATPase was purified from the erythrocyte ghosts and maintained in a soluble form with the nonionic detergent octaethyleneglycol mono-n-dodecyl ether ($C_{12}E_8$) in the presence of phosphatidyl-

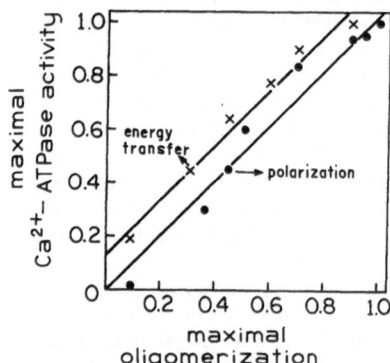

Fig. 2. Relationship between Ca^{2+}-ATPase activity and enzyme oligomerization as measured by fluorescence polarization (○) or fluorescence energy transfer efficiency (x). The Ca^{2+}-ATPase activity was measured in the absence of calmodulin at various enzyme concentrations as shown in Fig. 1. Fluorescence polarization of the FITC derivatized enzyme, and energy transfer between FITC-labeled (donor) and eosin-5-maleimide-labeled (acceptor)enzyme molecules were measured over the same range of enzyme concentrations in separate experiments. Values are expressed relative to the maximal values obtained at high enzyme concentration:, 0.08 and 6.6%, respectively (Kosk-Kosicka and Bzdega, 1988; Kosk-Kosicka et al.,1989b). The least-squares fits to the data for activity versus polarization and activity versus energy transfer have a slope of 1.032 and 1.036, an intercept of 0.00 and 0.13, and a correlation coefficient of 0.987 and 0.990, respectively.

choline. Initial studies of specific enzyme activity (expressed per mg protein) versus enzyme concentration were conducted with the expectation that a horizontal line would be obtained. However, such measurements in the absence of calmodulin yielded the sigmoid plot shown in Fig. 1. Various control experiments excluded a possibility that the low activity at low enzyme concentration could be explained by enzyme denaturation (Kosk-Kosicka and Bzdega, 1988). Inclusion of calmodulin in the assays (Fig. 1) produced a parallel sigmoid plot which was shifted towards lower enzyme concentrations. Thus, at low enzyme concentration the Ca^{2+}-ATPase activity is stimulated 2-3 fold by calmodulin, in agreement with observations by other authors (as reviewed by Schatzmann, 1982 or Carafoli and Zurini, 1982). On the other hand at concentrations above 30-40 nM the enzyme is fully active in the absence of calmodulin, and calmodulin addition has no apparent effect on the Ca^{2+}-ATPase activity. Our data suggested that the revealed gradual transformation from a calmodulin-dependent tot a calmodulin-independent form, with a K1/2 of 15 ± 5 nM enzyme, was due to protein-protein interactions of the Ca^{2+}-ATPase molecules.

The hypothesis of enzyme oligomerization was tested through physical methods to detect protein-protein interactions. The same dependence on enzyme concentration was observed for measurements of both polarization of the fluorescein-5'-isothiocyanate (FITC) derivatized enzyme and for efficiency of intermolecular fluorescence resonance energy transfer from the FITC-Ca^{2+}-ATPase to the eosin-5-maleimide-labeled Ca^{2+}-ATPase (Kosk-Kosicka and Bzdega, 1988; Kosk-Kosicka et al., 1989b). Derivatization of the Ca^{2+}-ATPase with the FITC, a lysine-specific probe, has been extensively characterized with the respect to the stoichiometry, pH dependence, and competition with ATP (Kosk-Kosicka, unpublished data); it has been further shown to bind specifically to the functionally important lysine 601 in the nucleotide binding domain (Filoteo et al., 1987). The observed increase of FITC polarization with increasing enzyme concentration was consistent with a decrease in probe mobility resulting from a conformational change induced by enzyme self-association and transduced to the FITC-binding site (Kosk-Kosicka and Bzdega, 1988). The increase in energy transfer as a function of increasing enzyme concentration reflected oligomerization of donor-labeled enzyme molecules (FITC) with acceptor-labeled molecules (eosin-5-maleimide). The secondary plot in Fig. 2 shows that the Ca^{2+}-ATPase activity increases linearly with the increase in oligomerization as measured by either polarization or energy transfer efficiency. Thus it is shown that the enzyme is activated by oligomerization.

Differences in calmodulin binding to monomers and oligomers

Both enzyme forms, called here monomers and oligomers (the oligomeric size has not been determined as yet), bind calmodulin as was found by calmodulin affinity chromatography (Kosk-Kosicka and Bzdega, 1988), and recently with fluorescent spinach calmodulin derivative, 2-(4-maleimidoanilino)-naphthalene-6-sulfonic acid-calmodulin (MIANS-CaM) (Kosk-Kosicka et al., 1989a). Fluorescence titrations

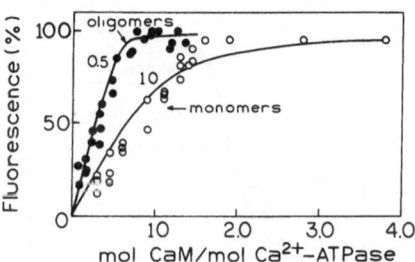

Fig. 3 . Stoichiometry of calmodulin binding to Ca^{2+}-ATPase monomers(o) and oligomers (●). The enzyme in either predominantly monomeric (12 nM) or oligomeric (40nM) form was titrated with MIANS-CaM. The assay medium was as described in Fig. 1. The fluorescence changes of MIANS-CaM upon binding to the enzyme were fitted by using an iterative nonlinear regression method, as described before (Kosk-Kosicka et al., 1989a). The fluorescence intensity of MIANS-CaM with the buffer was subtracted from fluorescence intensity of MIANS-CaM with Ca^{2+}-ATPase for each calmodulin concentration. 100% maximal fluorescence was equivalent to 50% increase of MIANS-CaM fluorescence intensity observed upon its binding to either monomers or oligomers, as measured at saturating MIANS-CaM concentrations. The best fits shown in Fig. 3 were obtained assuming a 1:1 and 0.5:1 stoichiometry of calmodulin binding to monomers and oligomers, respectively.

of monomers or oligomers with MIANS-CaM allowed us to reveal significant differences in calmodulin binding to the two enzyme forms. As shown in Fig. 3 by best fit of the experimental data the monomers appear to bind 1 mol calmodulin/mol Ca^{2+}-ATPase, in agreement with some earlier predictions (Graf and Penniston, 1981; Kosk-Kosicka and Bzdega, 1988). The oligomers are saturated near 0.5 mol calmodulin/mol Ca^{2+}-ATPase suggesting the possibility that oligomerization of the Ca^{2+}-ATPase monomers results in the elimination of one calmodulin binding site per every two associated enzyme molecules. The loss of CaM binding site upon Ca^{2+}-ATPase oligomerization may be due to its being buried upon association of enzyme molecules, either because it is on the interface and required for oligomerization or as an indirect result of conformational change that occurs during oligomerization and leads to full activation of the Ca^{2+}-ATPase.

We observed a further difference in the Ca^{2+} dependence of calmodulin binding and Ca^{2+} dependence of Ca^{2+}-ATPase activity of monomers as compared to oligomers. Fig. 4 A demonstrates that monomers must bind calmodulin (half-maximal at pCa 7.0, with little cooperativity $n_H = 1.1$) before they can exhibit Ca^{2+} dependent activation (half-maximal at pCa 6.6, with high cooperativity $n_H = 2.8$). In the case of enzyme oligomers (Fig. 4B), the Ca^{2+} concentration dependence of calmodulin binding is very similar to the Ca^{2+} dependence of Ca^{2+}-ATPase activity in the absence of calmodulin (both half-maximal near pCa 7.4, with $n_H = 3$), and calmodulin produces no increase in oligomer's activity. Further, activation of oligomers as a function of Ca^{2+} is very similar to the Ca^{2+} dependence of oligomerization (as measured by fluorescence energy transfer, Kosk-Kosicka et al., 1989b) and the binding of Ca^{2+} to the enzyme (as measured by tryptophan fluorescence, Kosk-Kosicka and Inesi, 1985)(see Fig. 4B). These data indicate that the Ca^{2+} dependent oligomerization process can, in the absence of calmodulin, produce a highly cooperative Ca^{2+} dependent activation of the enzyme. Activation by oligomerization actually occurs at lower Ca^{2+} ($K_{Ca} = $ pCa 7.4

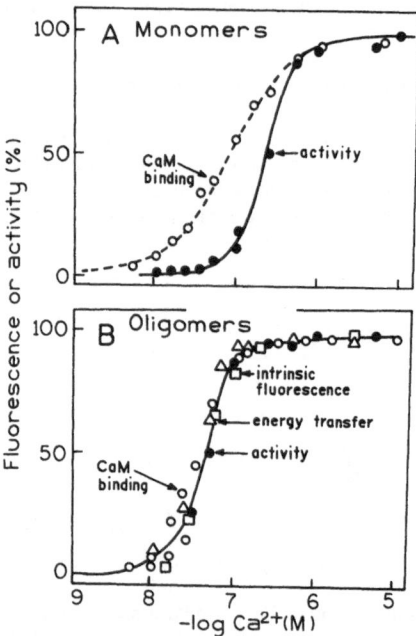

Fig. 4 . Comparison of Ca^{2+}-dependence of (A) calmodulin binding and activation of monomeric Ca^{2+}-ATPase and (B) calmodulin binding, Ca^{2+}-binding, oligomerization, and Ca^{2+}-ATPase activity of oligomeric enzyme.
Calmodulin binding (open circles) and Ca^{2+}-ATPase activity (filled circles) of the predominantly monomeric (12 nM) and oligomeric (40 nM) enzyme were studied as described in Figs. 1 and 3, respectively. Additionally the Ca^{2+} concentration dependence of tryptophan intrinsic fluorescence (□) and energy transfer efficiency (△) were measured for oligomers, as described in Kosk-Kosicka and Inesi, 1985; and Kosk-Kosicka et al., 1989b, respectively. The free Ca^{2+} concentration was calculated from Ca^{2+}/EGTA buffers as described before (Kosk-Kosicka and Bzdega, 1988).

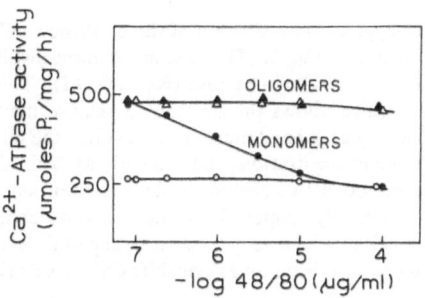

Fig. 5 . Inhibition of the Ca^{2+}-ATPase activity of monomers (o●) but not oligomers (▲△) by compound 48/80. The Ca^{2+}-ATPase assay was performed as in Fig. 1 with the inclusion of the specified concentrations of 48/80. The enzyme concentration was either 12 nM (predominantly monomeric) or 40 nM (oligomeric). The free Ca^{2+} was 17 μM, and calmodulin when present was 30 nM (filled symbols).

7.4) than activation by calmodulin (K_{Ca}=pCa 6.6). This might suggest that the former pathway, activation by oligomerization, is switched on at lower Ca^{2+} concentration than is activation by calmodulin. This difference might be of physiological importance if such oligomerization reactions occur in vivo. (For more detailed discussion see Kosk-Kosicka et al., 1989 b).

Is Ca^{2+}-ATPase activity of oligomers modulated by calmodulin in any way ?

Addition of calmodulin to the oligomerized enzyme caused no significant difference in any of the following measurements: (1) Ca^{2+}-ATPase activity (Fig. 1, data points for enzyme concentrations over 30-40 nM); (2) fluorescence polarization of the FITC modified enzyme (Kosk-Kosicka and Bzdega, 1988); (3) Ca^{2+}-ATPase tryptophan fluorescence intensity (Kosk-Kosicka and Inesi, 1985). Additional evidence for lack of effect of calmodulin on activation by oligomerization is shown in Fig. 5. The effect of the calmodulin antagonist, compound 48/80, on the Ca^{2+}-ATPase activity of monomers and oligomers was studied. Compound 48/80 is capable of inhibiting calmodulin stimulation of monomeric enzyme with I_{50}=1.4 μg/ml, which is comparable to I_{50}=0.85 μg/ml and 3.4 μg/ml reported for inhibition of the calmodulin stimulated activity of the membranous Ca^{2+}-ATPase (as measured in either inside out vesicles or red cell membranes preparations by Gietzen et al., 1983 and Rega et al., 1985, respectively). In contrast, compound 48/80 does not inhibit the Ca^{2+}-ATPase activity of oligomers. These results are consistent with our previous findings using gel electrophoresis and immunoassay, that enzyme activation at high enzyme concentrations (oligomer conditions) is not the result of calmodulin contamination of the Ca^{2+}-ATPase (Kosk-Kosicka et al., 1986a). Our data strongly indicates that while calmodulin does bind to the enzyme oligomers with high affinity and in a Ca^{2+} dependent manner (Figs. 3 and 4B), this binding has no effect on the Ca^{2+} dependent activation of the oligomerized enzyme.

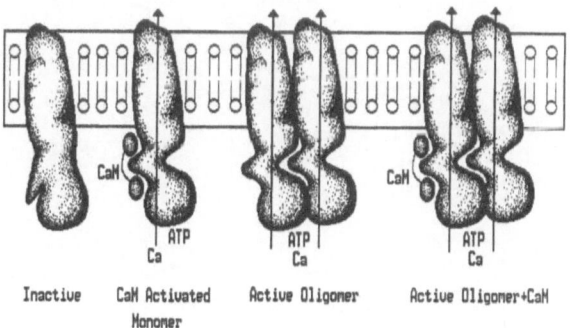

Fig. 6 . A schematic representation of the calmodulin and oligomerization induced activation of the Ca^{2+}-ATPase.

In conclusion our studies suggest that activation of the erythrocyte Ca^{2+}-ATPase can occur by two distinct processes. These are illustrated in Fig. 6. The first mechanism involves the Ca^{2+} dependent interaction of 1 mol of calmodulin per 1 mol of Ca^{2+}-ATPase ($K_d = 1.6$ nM). This interaction introduces a conformational change of the enzyme which allows the enzyme to achieve maximal Ca^{2+}-ATPase activity at pCa 6 in a highly cooperative fashion. The second mechanism involves the oligomerization of Ca^{2+}-ATPase molecules which is maximal at protein concentrations above 30-40 nM. This oligomerization process may involve one calmodulin binding site per every two enzyme molecules to produce an enzyme conformer which binds Ca^{2+} with high affinity and in a highly cooperative fashion, and which is capable of full Ca^{2+}-ATPase activity at pCa 7 in the absence of calmodulin. At present it is uncertain which process operates in vivo to ensure a most efficient Ca^{2+} extrusion by this pump as cytosolic Ca^{2+} levels rise within the cell.

REFERENCES

Al-Jabore, A., Minocherhomjee, A.M., Villalobo, A., and Roufogalis, B.D., 1984, in: "Erythrocyte Membranes 3: Recent Clinical and Experimental Advances", A.R. Liss, ed., New York.

Carafoli, E., and Zurini, M., 1982, Biochim. Biophys. Acta 683:279-301.

Choquette, D., Hakim, G., Filoteo, A.G., Plishker, G.A., Bostwick, J.R., and Penniston, J., 1984, Biochem. Biophys. Res. Commun. 125:908-915.

Enyedi, A., Flura, M., Sarkadi, B., Gardos, G., and Carafoli, E., 1987, J. Biol. Chem. 262:6425-6430.

Filoteo, A.G., Gorski, J.P., and Penniston, J.T., 1987, J. Biol. Chem. 262:6526-6530.

Gietzen, K., Adamczyk-Engelmann, P., Wutrich, A., Konstantinova, A., and Bader, H., 1983, Biochim. Biophys. Acta 736:109-118.

Graf, E., and Penniston, J.T., 1981, Arch. Biochem. Biophys. 210:257-262.

Kosk-Kosicka, D., and Inesi, G., 1985, FEBS Lett. 189:67-71.

Kosk-Kosicka, D., Scaillet, S., and Inesi, G., 1986a, J. Biol. Chem. 261:3333-3338.

Kosk-Kosicka, D., Scaillet, S., and Inesi, G., 1986b, Biophys. J. p547a.

Kosk-Kosicka, D., and Bzdega, T., 1988, J. Biol. Chem. 263:18184-18189.

Kosk-Kosicka, D., Bzdega, T., and Johnson, J.D., 1989a, Biochemistry., in press.

Kosk-Kosicka, D., Bzdega, T., Wawrzynow, A., 1989b, J. Biol. Chem., in press.

Neyes, L., Reinlib, L., and Carafoli, E., 1985, J. Biol. Chem. 260:10283-10287.

Niggli, V., Adunyah, E.S., Penniston, J.T., and Carafoli, E., 1981, J. Biol. Chem. 256:395-401.

Rossi, J. P. F. C., Rega, A. F., and Garrahan, P. J., 1985, Biochim. Biophys. Acta 816:379-386.

Penniston, J.T., 1984, in: "Calcium and Cell Function", W. Y. Cheung, ed., Academic Press, New York.

Schatzmann, H.J., 1982, in: "Membrane transport of calcium" E. Carafoli, ed., Academic Press, London.

Taverna R.D., and Hanahan, D.J., 1980, Biochem. Biophys. Res. Commun. 94:652-659.

Vincenzi, F.F., Morris, C.D., Kinsel, L.B., Kenny, M., and Mc Carron, D.A., 1986, Hypertension 8:1058-1066.

CALPAIN I ACTIVATES Ca^{2+} TRANSPORT BY THE HUMAN ERYTHROCYTE PLASMA MEMBRANE CALCIUM PUMP

Kevin K.W. Wang, Basil D. Roufogalis and Antonio Villalobo[*]

Laboratory of Molecular Pharmacology, Faculty of Pharmaceutical Sciences, University of British Columbia, Vancouver, B.C. V6T 1W5 Canada

[*]Established Investigator on leave from C.S.I.C., Instituto de Investigaciones Biomedicas Madrid, Spain

Many extracellular signals, including hormones, exert their effects on cells by elevating free Ca^{2+} concentration. For this reason, it is important for a resting cell to maintain a submicromolar concentration of free Ca^{2+} in the cytosol. In red cells, this function is largely provided by the plasma membrane-bound Ca^{2+}- translocating ATPase. Both Ca^{2+} translocating and ATP-hydrolytic activities of this enzyme are stimulated by calmodulin (CaM) via reversible binding (see Al-Jobore et al., 1981). We recently reported an irreversible means of activating the ATP-hydrolytic activity of the enzyme, involving the cytosolic Ca^{2+}-dependent protease (calpain I) (Wang et al., 1988a; 1988b). However, it is not yet established whether the calcium translocating activity of the enzyme is also activated by calpain I. In this study, we further examine the effect of calpain I on the liposome-reconstituted calcium pump.

EXPERIMENTAL PROCEDURES

Purification of the human erythrocyte Ca^{2+}-ATPase and calpain I were carried out as described earlier (Wang et al., 1988b). The purified Ca^{2+}-ATPase was reconstituted into phosphatidylcholine vesicles using a cholate-dialysis method described previously (Villalobo & Roufogalis, 1986). Treatment of the proteoliposomes (20-35 μg protein.ml^{-1}) with calpain was performed in 67 mM KCl, 50 mM potassium-Hepes (pH 7.4), 0.67 mM MgCl$_2$, 10 mM dithiothreitol at 25°C for 120 min (unless stated otherwise). Proteolysis was arrested with 200 μM leupeptin. Determination of the Ca^{2+}-ATPase activity was essentially as described previously (Wang et al., 1988b). Measurement of the initial rates of Ca^{2+} uptake and ATP hydrolysis was carried out at 37°C in 1.5 ml of 73 mM KCl, 17 mM potassium-Hepes (pH 7.4), 5 mM dithiothreitol, 730 μM MgCl$_2$, 130 μM NADH, 5 mM phosphoenolpyruvate, 1 μM calmodulin (when added), 20 units pyruvate kinase (Sigma), 90 units lactate dehydrogenase (Sigma), 27 μM ATP, 0.4 μM free Ca^{2+} and 10 μM Arsenazo III (Ca^{2+} uptake only). Monitoring of the rate of NADH oxidation (340 nm - 400 nm) or the disappearance of the Arsenazo-calcium complex (650 nm - 720 nm) was done on a dual wavelength spectrophotometer. In the experiment described in Table 1, ATP hydrolysis was monitored by a colorimetric method (see Wang et. al., 1988a) using the following conditions: proteoliposome (0.6 - 1.2 μg protein and 0.6-1.2 mg phospholipid were incubated at 37°C for 30 min in 0.4 ml of 55 mM Tris-maleate (pH 7.2), 66 mM KCl, 6.5 mM MgCl$_2$, 120 nM calmodulin (when added), 2 mM ATP and various concentrations of CaCl$_2$ and EGTA to reach desired free Ca^{2+} concentration.

RESULTS

Effect of calpain on the reconstituted calcium pump

The majority of previous work on the reconstituted calcium pump from human erythrocyte was carried out on asolectin vesicles (Haaker & Racker, 1979; Niggli et al., 1981; Villalobo & Roufogalis, 1986). However, since the acidic phospholipid components in asolectin mimick the effect of calmodulin, the stimulatory effect of calmodulin on Ca^{2+} transport has not been demonstrated. This is the first report demonstrating that, at low free Ca^{2+} concentration (0.4 μM), calmodulin stimulated the initial rates of both Ca^{2+} uptake and ATP-hydrolytic activity in untreated proteoliposomes 5 fold (500 compared to 90 nmol.mg prot.$^{-1}$.min^{-1}) (Fig. 1) and 6 fold (559 compared to 84 nmol.mg prot.$^{-1}$.min^{-1}), respectively (Fig. 2). The basal initial rate of Ca^{2+} uptake (in the absence of calmodulin) was found to be increased by calpain pretreatment in a concentration-dependent manner (Fig. 1). At the highest calpain concentration used (4.5 unit.ml^{-1}), the basal initial rate of Ca^{2+} uptake was increased about 3 fold (350 compared to 90 nmol.mg prot.$^{-1}$.min^{-1}) and became insensitive to calmodulin (Fig. 1). In parallel, the initial rate of ATP hydrolysis by the calcium pump was increased 4 fold after calpain (2.0 unit.ml^{-1}) treatment (from 84 to 435 nmol.mg prot.$^{-1}$.min^{-1}) and became largely independent of calmodulin (Fig. 2).

Fragmentation of the reconstituted calcium pump by calpain

To determine the molecular changes which accompany the proteolytic activation, fragmentation of the reconstituted calcium pump by calpain was examined (Fig. 3). The intact 136 kDa calcium pump was transformed mainly into 124 kDa and 127 kDa fragments, in the absence or the presence of calmodulin, respectively, similar to the pattern observed previously in the purified non-reconstituted enzyme (Wang et al. 1988b). After 120 min of proteolysis, a portion of unfragmented enzyme was still present, which was likely to represent calcium pump molecules that are either not properly orientated into the liposome bilayer or are outwardly-orientated (with respect to Ca^{2+}-translocation across the proteoliposomes).

Calcium dependence of the intact and calpain-treated calcium pump

The intact reconstituted calcium pump was studied with respect to its Ca^{2+} dependence. Table 1 illustrates that the affinity for Ca^{2+} of the initial rate of calcium uptake by the calcium pump was increased by

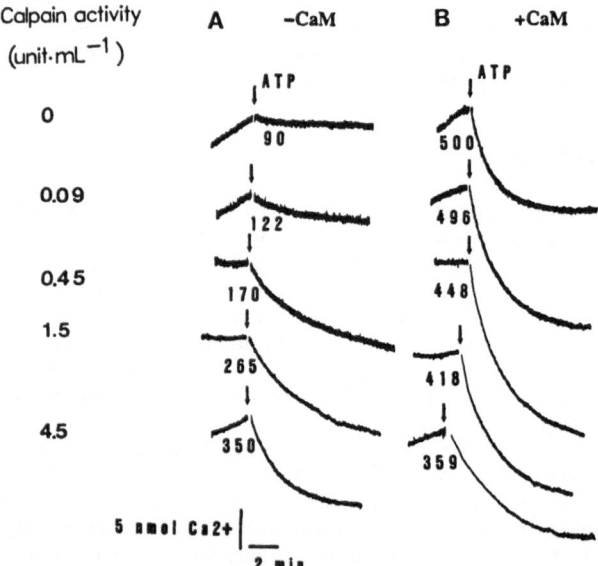

Fig. 1. Effect of calpain I on the initial rate of calcium uptake of the reconstituted calcium pump. Proteoliposomes (55 μg protein and 49 mg phospholipid) were incubated at 25°C for 120 min in 67 mM KCl, 50 mM potassium-Hepes (pH 7.4), 0.67 mM MgCl$_2$, 10 mM dithiothreitol, 400 μM free Ca^{2+} in the absence or the presence of various concentrations of calpain (as indicated). 200 μM leupeptin was added to arrest the proteolysis before the proteoliposome suspensions were subjected to calcium uptake measurement in the absence (A) or the presence of 1 μM calmodulin (B). 27 μM ATP and 60 nM A23187 were added as indicated on the traces. Numbers beside the time course curves represent initial rates of Ca^{2+} uptake in nmol.mg prot.$^{-1}$.min^{-1}. Results are representative of two separate experiments.

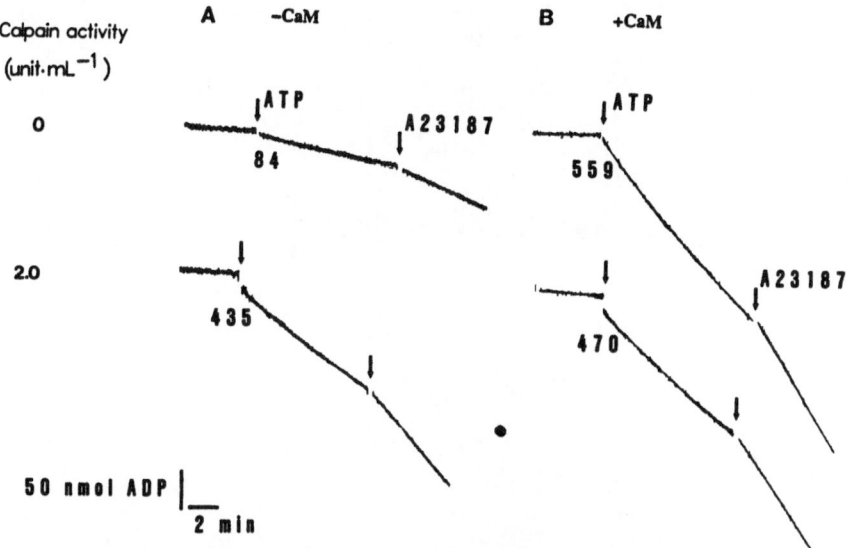

Fig. 2. Effect of calpain I on the initial rate of ATP-hydrolysis by the reconstituted calcium pump. Proteoliposomes (55 μg protein and 49 mg phospholipid) were pretreated with either 0 or 2.0 unit.ml^{-1} calpain (as indicated), as described in Fig. 1 before being subjected to ATP-hydrolysis measurement in the absence (A) or the presence of 1 μM calmodulin (B). 27 μM ATP and 60 nM A23187 were added as indicated on the traces. Numbers beside the time course curves represent initial rates of ATP hydrolysis in nmol.mg prot.$^{-1}$.min^{-1}. Results are representative of three separate experiments.

Table 1. Calcium dependence of the intact and calpain-treated calcium pump

Treatment	CaM in assay	Ca^{2+} uptake		ATP-hydrolysis	
		V_{max}	$K_{0.5}(Ca)$	V_{max}	$K_{0.5}(Ca)$
		(nmol.mg prot.$^{-1}$.min^{-1})	(μM)	(nmol.mg prot.$^{-1}$.min^{-1})	(μM)
- calpain	-	230	4	450	2.5
	+	697	0.3	875	0.2
+ calpain	-	479	0.4	593	0.3
	+	503	0.4	680	0.3

calmodulin. The $K_{0.5}$ for Ca^{2+} was decreased from 4 μM in the absence of calmodulin to 0.3 μM in its presence. Calmodulin also increased V_{max} from 230 to 697 nmol.mg prot.$^{-1}$.min^{-1} (Table 1). As expected, Proteoliposomes (50-100 μg protein plus 50-100 mg phospholipid) were pre-incubated in the absence (-) and the presence (+) of 2.0 unit.ml^{-1} of calpain, as described in Fig. 1. The treated proteoliposomes were then subjected to measurement of initial rate of Ca^2 uptake or ATP hydrolysis, at various concentrations of free Ca^{2+} in the absence (-) or presence (+) of calmodulin. V_{max} values and apparent $K_{0.5}(Ca)$ values were then calculated. The values shown are the averages of two separate experiments.

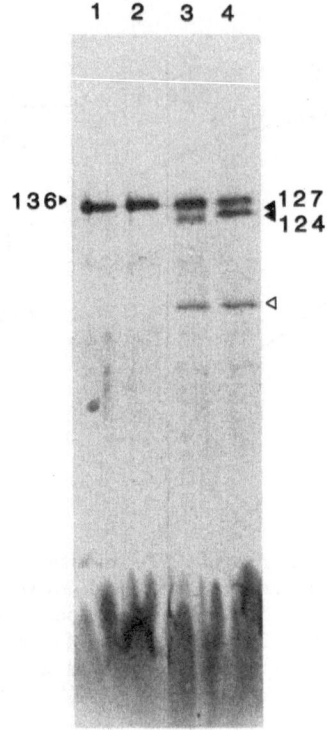

Fig. 3 . Calpain-mediated proteolytic fragmentation of the reconstituted calcium pump. Proteoliposomes (156 μg protein plus 112 mg phospholipid) were treated with either none (lane 1 & 2) or 2.0 units.ml^{-1} of calpain (lane 3 & 4) as in Fig. 1 in the absence (lane 1 & 3) or the presence (lane 2 & 4) of 130 nM calmodulin. The treated proteoliposomes were subjected to SDS-gel electrophoresis followed by Coomassie Blue staining. The numbers beside the lanes represent relative molecular masses in kDa and the open triangle indicates the bands of calpain and its fragment. Results are representative of five separate experiments.

after calpain treatment (in the absence of calmodulin), the calcium pump showed high affinity for Ca^{2+} (K$_{0.5}$ around 0.4 μM) even in the absence of calmodulin (Table 1). The addition of calmodulin only mildly increased the V$_{max}$ value from 479 to 503 nmol.mg prot.$^{-1}$.min^{-1}. The Ca^{2+}-dependence of the ATP-hydrolytic activity of the reconstituted Ca^{2+}-ATPase was also examined.

The intact calcium pump exhibited the low Ca^{2+} affinity (K$_{0.5}$ 2.5 μM), low V$_{max}$ (450 nmol.mg prot.$^{-1}$.min^{-1}) state in the absence of calmodulin and the high Ca^{2+} affinity (K$_{0.5}$ 0.2 μM), high V$_{max}$ (875 nmol.mg prot.$^{-1}$.min^{-1}) state when assayed in the presence of calmodulin (Table 1). Proteolysis by calpain appeared to transform the enzyme mostly into the permanently high Ca^{2+} affinity (0.3 μM), high V$_{max}$ (593-680 nmol.mg prot.$^{-1}$.min^{-1}) state irrespective of the calmodulin content in the assay medium (Table 1).

DISCUSSION

Recently, it was demonstrated by Au (1987) in pig erythrocytes and by us (Wang et al., 1988a) in human erythrocytes that the ATP hydrolytic activity of the erythrocyte plasma membrane-bound Ca^{2+}-ATPase was proteolytically activated by calpain I. This activation was accompanied by the loss of cal-modulin stimulation. We reported also that the presence of calmodulin during calpain treatment of the plasma membrane protected the Ca^{2+}-ATPase against proteolytic activation (Wang et al., 1988a).

In a later study (Wang et al., 1988b), we further characterized the molecular changes of the Ca^{2+}-ATPase during calpain treatment. A scheme of the fragmentation of the Ca^{2+}-ATPase summarizing these

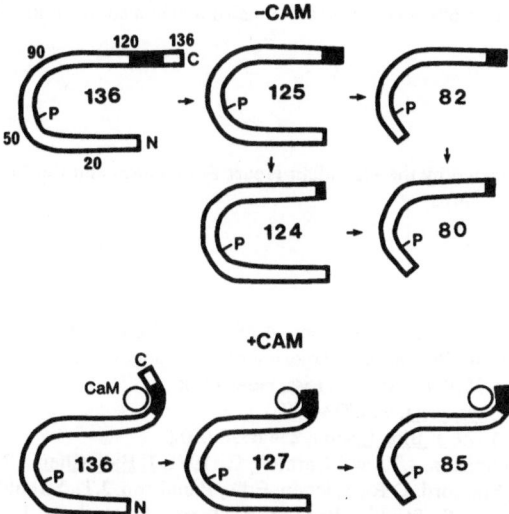

Fig. 4. Fragmentation scheme of the plasma membrane calcium pump produced by calpain I in the absence (-) and the presence (+) of calmodulin (CAM). The letters N and C represent the amino and carboxy-terminus of the amino acid sequence, respectively. The numbers beside the polypeptide represent the approximate amino acid residue numbers. The large number in the centre of each polypeptide represents its relative molecular mass. The letter "P" represents the acylphosphate site and the darkened region represents the calmodulin-binding domain (Verma et al., 1988). The calmodulin molecule is represented by a circle.

results is illustrated in Fig. 4. The native Ca^{2+}-ATPase (136 kDa) has its acylphosphate site located about 50 kDa from the N-terminal, and its calmodulin-binding domain 10 kDa from the C-terminal end (Verma et al., 1988). In the absence of calmodulin initial fragmentation appears to occur at the C-terminal, producing the 125 kDa and 124 kDa active fragments. Since its calmodulin-binding domain is mostly preserved, the 125 kDa fragment is sensitive to calmodulin. On the other hand, with most of the calmodulin-binding domain truncated, the 124 kDa fragment expresses calmodulin-independent activity. A slower cleavage seems to occur at the N-terminal side. The removal of a fragment of about 40 kDa from the N-terminal (leaving the acylphosphate-forming active site intact) produces two active fragments: a 82 kDa fragment originating from the 125 kDa fragment and an 80 kDa fragment arising from the 124 kDa fragment. In the presence of calmodulin, initial cleavage near the C-terminal end produces a higher molecular weight fragment (127 kDa). A subsequent cleavage about 40 kDa from the N-terminal end produces an 85 kDa active fragment. Both the 127 kDa and 85 kDa fragments can still bind and be stimulated by calmodulin (Fig. 4). In the present report, we have demonstrated that calpain treatment of the liposome-reconstituted Ca^{2+}-ATPase produced predominately the 124 kDa and the 127 kDa fragments, in the absence or the presence of calmodulin, respectively (Fig. 3), whilst the 80 kDa and 85 kDa forms were produced in only small amounts. The potent proteolytic activation of the Ca^{2+}-pumping ATPase and the loss of calmodulin-sensitivity (Fig. 1 and 2) again suggested that in the absence of calmodulin calpain produced mostly the 124 kDa calmodulin-independent fragment of the Ca^{2+}-ATPase, rather than the 125 kDa calmodulin-binding fragment.

This is also the first report showing clearly that calpain treatment of the plasma membrane calcium pump not only activates its ATP hydrolytic activity but also its primary physiological function: the calcium translocation activity (Fig. 1 and Table 1). Such irreversible activation was accompanied by the loss of calmodulin-dependence.

Since calpain I is found in human erythrocyte cytosol, proteolytic activation of the calcium pump by calpain can potentially occur in these cells. Under certain stress or pathological conditions, an uncontrolled and persistent elevation of the calcium level could activate calpain I, thereby subsequently activating the calcium pump irreversibly. Such an effect could be a defensive mechanism to reduce the cytosolic calcium concentration to resting levels.

ACKNOWLEDGEMENTS

This study was supported by the Canadian Heart Foundation and the Medical Research Council of Canada.

REFERENCES

Al-Jobore, A., Mauldin, D., Minocherhomjee, A.M., and Roufogalis, B.D., 1981, in: "Erythrocyte Membranes 3: Recent Clinical and Experimental Advances", W.C. Kruckeberg, J.W. Eaton, and G.J. Brewer, eds., pp 757-773, Alan R. Liss, New York.
Au, K.S., 1987, Biochim. Biophys. Acta, 905:273-278.
Haaker, H., and Racker, E., 1979, J. Biol. Chem., 254:6598-6602.
Niggli, V., Adunyah, E.S., Penniston, J.T., and Carafoli, E., 1981, J. Biol. Chem., 256:395-401.
Verma, A.K., Filoteo, A.G., Stanford, D.R., Wieben, E.D., Penniston, J.T., Strehler, E.E., Fischer, R., Heim, R., Vogel, G., Mathews, S., Strehler-Page M.-A., James, P., Vorherr, T., Krebs, J., and Carafoli, E., 1988, J. Biol. Chem., 263:14152-14159.
Villalobo, A., and Roufogalis, B.D., 1986, J. Membrane Biol., 93:249-258.
Wang, K.K.W., Villalobo, A., and Roufogalis, B.D., 1988a, Arch. Biochem. Biophys., 260:696-704.
Wang, K.K.W., Roufogalis, B.D., and Villalobo, A., 1988b, Arch. Biochem. Biophys., 267:317-327.

INHIBITORY REGULATION BY CALCIUM ION OF MYOSIN ATPASE

ACTIVITY : BINDING OF CALCIUM ION AND

PHOSPHORYLATION OF MYOSIN

Kazuhiro Kohama[1] and Tsuyoshi Okagaki[2]

[1]Department of Pharmacology, Gunma University School of Medicine, Maebashi 371

[1,2]Department of Pharmacology, Faculty of Medicine, University of Tokyo, Tokyo 113

[2]The Physical Laboratories, Nihon University at Narashino, Funabashi 274, Japan

INTRODUCTION

In 1980, we prepared native actomyosin from the lower eukaryote, *Physarum*, by referring to published methods (see Kohama and Ebashi, 1986, for a review). Our preparation of actomyosin showed Ca-sensitive ATPase activity with novel characteristics : the activity was highest in EGTA-containing solutions and was depressed by increasing concentrations of Ca^{2+} (Kohama et al., 1980). Thus far, regulation by Ca^{2+} has exclusively been shown to involve activation, and our result is the first example of an inhibitory mode of regulation by Ca^{2+} (Ebashi et al., 1982). We further purified myosin from the *Physarum* actomyosin by newly developed methods (Kohama and Kendrick-Jones, 1986; Kohama et al., 1986), and we found that this myosin is composed of a pair of heavy chains with molecular weight of 230 Kd each in SDS PAGE and two pairs of light chains having weights of 18 Kd and 14 Kd in SDS PAGE. Furthermore, the isolated myosin is phosphorylated, having about 4 mol Pi per mol of myosin (Kohama and Kendrick-Jones, 1986).

RESULTS

Inhibitory effect of Ca ions on the ATPase activity of myosin from Physarum

The actin-activated ATPase activity of *Physarum* myosin was highest in Mg-ATP solutions that contained EGTA and was reduced with increases in the concentration of Ca^{2+} (Fig.1). The half-maximal inhibition occurred in the μM range of Ca^{2+}, suggesting the physiological involvement of this Ca^{2+}-mediated inhibition in the motility of *Physarum*. The inhibitory effect of Ca^{2+} on the ATPase activity was confirmed by observing the movement of *Physarum* myosin along actin filaments; the movement was fast in Mg-ATP solutions that contained EGTA and slow in solutions that contained Ca^{2+} (Table 1).

Ca^{2+} enhances the actin-activated ATPase activity of the myosin prepared from scallop striated muscle (Kendrick-Jones et al., 1970). Accordingly, scallop myosin did move on actin filaments in an Mg-ATP solution that contained Ca^{2+} while it did not move in such a solution that contained EGTA (Table 1). It is well established that scallop myosin binds Ca^{2+} with high affinity to generate its active form (see Szent-Görgyi, 1980, for a review). Measurement of ^{45}Ca bound to *Physarum* myosin in the presence of Mg^{2+} indicated that the Ca-binding properties were similar to those of scallop myosin (Kohama and Kendrick-Jones, 1986; Kohama et al., 1987; see also Table 2). Therefore, we speculate that binding of Ca^{2+} to and its release from Physarum myosin inhibits and restores, respectively, the ATPase activity of the Physarum myosin.

Table 1 . Velocity of movement of myosin along actin filaments

	Myosin (μm/sec mean ± standard error)	
	Physarum	Scallop
EGTA	1.40 ± 0.08 (n=11 cells)	0.00 (n=3 cells)
Ca^{2+}	0.41 ± 0.08 (n=11 cells)	1.22 ± 0.39 (n=3 cells)

Actin cables in a *Chara* internodal cell were exposed by internal perfusion. Latex beads coated with either *Physarum* or scallop myosin were injected into the cell with EGTA solution (3 mM $MgCl_2$, 1 mM ATP, 1 mM DTT, 30 mM PIPES at pH 7.0, 170 mM sorbitol, and 5 mM EGTA, adjusted to pH 7.0 with 66 mM KOH) or with a Ca^{2+} solution (3 mM $MgCl_2$, 1 mM ATP, 1 mM DTT, 30 mM PIPES at PH 7.0, 170 mM sorbitol, 5 mM EGTA and 5 mM $CaCl_2$, adjusted to pH 7.0 with 76 mM KOH). Velocities (μm/sec) of the movement of the beads were calculated from observations under Nomarski optics (Kohama and Shimmen, 1985). Binding of Ca^{2+} to control (Phosphorylated) and phosphatase-treated (Dephosphorylated, see the legend to Figure 1) myosins was determined by the equilibrium dialysis method in 0.5 M KCl, 1 mM $MgCl_2$, 40 mM Tris-HCl at pH 7.5 and 30 μM $CaCl_2$, which contained 0.3 μCi/ml of $^{45}CaCl_2$ as a tracer. The total concentrations of calcium in this medium was determined by atomic absorption spectrophotometry. Protein concentrations of the myosins were determined by the Coomassie brilliant blue (Bio-Rad protein assay) using bovine serum albumin as a standard.

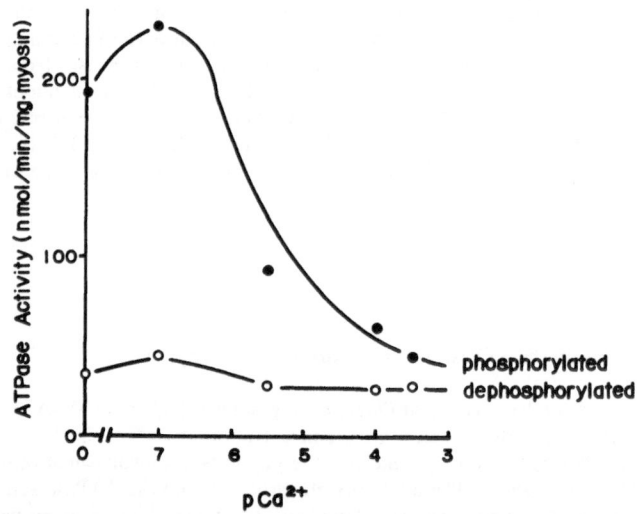

Fig. 1 . *Physarum* myosin (12.0 mg/ml) was incubated in 0.5 M KCl, 2 mM $MgCl_2$, 1 mM EGTA, 8 mM phosphate buffer (pH 6.3), 1 mM DTT and protease inhibitors (PMSF and E64) with (dephosphorylated) or without (phosphorylated) 0.47 mg/ml potato acid phosphatase (Sigma, 1150 U/mg protein) for 2 hr at 25°C and then subjected to gel-filtration column chromatography in 0.5 M NaCl, 0.1 mM EGTA, 14 mM ß-mercaptoethanol and 20 mM Tris-HCl buffer at pH 7.5 in a HPLC. The phosphorylated and dephosphorylated myosins were subjected to the assay of actin-activated ATPase activity measurement of libelated Pi (Youngberg, 1930). Reaction mixture contained 80 μg/ml *Physarum* myosin, 80 μg/ml rabbit skeletal muscle actin, 13 mM KCl, 1mM ATP, 2 mM $MgCl_2$, 20 mM Tris-HCl at pH 7.5, 0.1 mM DTT and 0.1 mM EGTA-Ca buffer for 10 min at 25°C.

Table 2 . ^{45}Ca^{2+}-binding to _Physarum_ myosin (mol Ca^{2+}/mol myosin)

Phosphorylated	Dephosphorylated
1.05	1.11
1.44	1.29
1.32	1.23
m = 1.27 ± 0.16	m = 1.21 ± 0.07 (Mean ± s.d.)

Structure and function of Ca-binding light chains (CaLc)

As suggested by ^{45}Ca-overlay on Western blots (Maruyama et al., 1984) after SDS PAGE of _Physarum_ myosin, the 14-Kd light chain appears to be the sole Ca-binding subunit of the myosin (Kohama et al., 1985).

This CaLc was isolated and purified from the myosin (Kohama et al., 1988b). Figure 2 shows the complete amino acid sequence of CaLc, which was determined by the analysis of protein (Kobayashi et al., 1988). CaLc is composed of 147 amino acid residues and its molecular weight is 16,131 dalton. An EF-hand consensus sequence (Krestinger and Nockolds, 1973) for the Ca-binding domain can be assigned to two positions, indicating that CaLc is a member of the family of Ca-binding proteins. Since CaLc can bind to the myosin heavy chains of rabbit skeletal muscle in place of the alkali light chain of the myosin, CaLc can also be classified as a member of the family of alkali-light chains (see Kohama, 1987, and Kohama, 1988, for reviews). It is noteworthy that CaLc has 40 % homology with vertebrate calmodulin, which might be expected from the activating effect of CaLc on phosphodiesterase activity in the presence of Ca^{2+} (Kohama, 1988). This activating mode of regulation by Ca^{2+} is quite the opposite of the inhibitory mode of regulation by Ca^{2+} of the ATPase activity of _Physarum_ myosin (Fig.1) even though CaLc is involved in both modes of regulation. Thus, we speculate that CaLc works as a mere Ca-receptive protein in _Physarum_ myosin and that the heavy chains and/or the 18-Kd light chains are responsible for the inhibitory mode of regulation by Ca^{2+} of the ATPase activity.

Effects of dephosphorylation of Physarum myosin

Physarum myosin was incubated with potato acid phosphatase, and then the myosin was separated from the phosphatase by gel-filtration chromatography. This treatment removed about 1-2 mol of Pi per mol of myosin. The actin-activated ATPase activity of the dephosphorylated myosin was low and independent

```
                5                 10            X     Y     Z   20  -X       -Y      25
Ac-Thr Ala Ser Ala Asp Gln Ile Gln Glu Cys Phe Gln Ile Phe Asp Lys Asp Asn Asp Gly Lys Val Ser Ile Glu

   -Z              30             35             40              45             50
Glu Leu Gly Ser Ala Leu Arg Ser Leu Gly Lys Asn Pro Thr Asn Ala Glu Leu Asn Thr Ile Lys Gly Gln Leu

            55             60             65             70             75
Asn Ala Lys Glu Phe Asp Leu Ala Thr Phe Lys Thr Val Tyr Arg Lys Pro Ile Lys Thr Pro Thr Glu Gln Ser

         80             85  X     Y     Z    -Y    -X  95      -Z      100
Lys Glu Met Leu Asp Ala Phe Arg Ala Leu Asp Lys Glu Gly Asn Gly Thr Ile Gln Glu Ala Glu Leu Arg Gln

         105            170            115            120            125
Leu Leu Leu Asn Leu Gly Asp Ala Leu Thr Ser Ser Glu Val Glu Glu Leu Met Lys Glu Val Ser Val Ser Gly

         130            135            140            145
Asp Gly Ala Ile Asn Tyr Glu Ser Phe Val Asp Met Leu Val Thr Gly Tyr Pro Leu Ala Ser Ala
```

Fig. 2 . Complete amino acid sequence of CaLc (Kobayashi et al, 1988). The loop sequence (X, Y, Z, -X, -Y, -Z) of an EF-hand sequence for the Ca-binding proteins (Krestinger and Nockolds, 1973) is identified at two positions. The N-terminal was blocked by acetylation (Ac-).

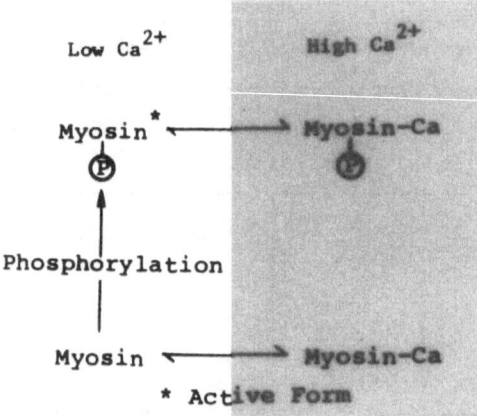

Fig. 3. A speculative model of the role of Ca^{2+} and phosphorylation. 1. *Physarum* myosin is active only in the phosphorylated form at low concentrations. 2. Ca-binding activity is independent of the degree of phosphorylation. 3. Myosin can be phosphorylated only at low concentrations of Ca^{2+}.

of the concentrations of Ca^{2+} (Fig.1)[1]. However, the Ca-binding property was not affected by such treatment. The catalytic site for ATPase activity is in the heavy chains (Kohama et al., 1988a). Thus, we speculate that the heavy chains of the dephosphorylated myosin can not respond to the Ca^{2+} sequestered by CaLc.

Efforts to purify myosin kinases

Myosin was phosphorylated both on the heavy chains and on the 18-Kd light chains. The phophorylation of the latter was only partial, but that of the former occurred at multiple sites (see Kohama, 1987, for a review). We have made significant progress with the purification of the kinases for the heavy chains and 18-Kd light chains from *Physarum* (Kohama and Okagaki, 1989). Column chromatography on hydroxylapatite of a fraction containing kinase activities for both the heavy and the 18-Kd light chains is able to separate the two activities (not shown). The partially purified kinases for the heavy and light chains are both Ca-sensitive. Like the ATPase activity of *Physarum* myosin, their activities are reduced by increases in the concentration of Ca^{2+}.

DISCUSSION

Binding of Ca^{2+} to myosins and phosphorylation of myosins are the physiological mechanisms involved in the regulation of the ATPase activity of myosins. Myosins from various sources have been classified according to the ways in which their ATPase activities are regulated by these mechanisms (Kendrick-Jones and Scholey, 1981). Skeletal myosin is a typical example of a myosin whose activity is modified neither by Ca^{2+} nor by phosphorylation. The activity of scallop myosin is regulated by binding of Ca^{2+} to myosin and not by phosphorylation of myosin, but the opposite is true of smooth muscle myosin. As shown in Figure 1, *Physarum* myosin is a first example of a myosin that is subject to both regulatory mechanisms.

----

1. The ATP activity of *Physarum* myosin in the text means actin-activated Mg-ATPase activity. Unlike the activity, ATPase activity of myosin in the absence of actin, i.e., K^+-EDTA ATPase, Ca-ATPase, and Mg-ATPase, were not affected by the treatment with phosphatase (not shown), indicating that the results shown in Figure 1 were not due to the non-specific denaturation of the myosin. The dephosphorylation is also known to modify the assembly of the myosin; the dephosphorylated myosin was more soluble than the control, i.e., the phosphorylated myosin (Takahashi et al., 1983; Kohama, 1987).

Figure 3 is a speculative model of the way in which the two mechanisms may be related to each other. The only active form of *Physarum* myosin is the phosphorylated form at low concentrations of Ca^{2+}. The myosin at high concentrations of Ca^{2+} or in the dephosphorylated forms is no longer active. Physarum myosin is phosphorylated only at low concentrations of Ca^{2+}. At present, however, we do not know how the dephosphorylation of the myosin by endogenous phosphatase(s) is affected by Ca^{2+}.

REFERENCES

Ebashi, S., Nakamura, S., Nakasone, H., Kohama, K., and Nonomura, Y., 1982, in "Calcium Modulators", T. Godfraind, et al., eds. Elsevier, Amsterdam pp. 39-49.

Kendrick-Jones, J., Lehman, W., and Szent-Görgyi, A.G., 1970, J. Mol. Biol., 54:313.

Kendrick-Jones, J., and Scholey, J.M., 1981, J. Muscle Res. Cell Motil., 2:347.

Kobayashi, T., Takagi, T., Konishi, K., Hamada, Y., Kawaguchi, M., and Kohama, K., 1988, J. Biol. Chem., 263:305.

Kohama, K., 1987, Adv. Biophys., 23:149.

Kohama, K., 1988, In: "Calcium Signal and Cell Responses", K. Yagi, and T. Miyazaki, eds., Japan Sci. Soc. Press, Tokyo/Springer-Verlag, Berlin, pp. 95-105.

Kohama, K., and Ebashi, S., 1986, In: "Molecular Biology of Physarm polycephalum", W.F. Dove et al., eds., Plenum Press, N.Y. pp. 175-190.

Kohama, K., and Kendrick-Jones, J., 1986, J. Biochem., 99:1433.

Kohama, K., Kobayashi, K., and Mitani, S., 1980, Proc. Japan Acad., 56B:591.

Kohama, K., Kohama, T., and Kendrick-Jones, J., 1987, J. Biochem., 102:17.

Kohama, K., Oosawa, F., Ito, T., and Maruyama, K., 1988b, J. Biochem., 104:730.

Kohama, K., and Okagaki, T., 1989, In: "Sarcomeric and Non-Sarcomeric Muscles : Basic and Applied Research Prospects for the 90s". U. Carraro, ed., Unipress Padova, Padova pp. 733-738.

Kohama, K., Shimmen, T., 1985, Protoplasma, 129:88.

Kohama, K., Sohda, M., Murayama, K., and Okamoto, Y., 1988a, Protoplasma [Suppl.2] 37-47.

Kohama, K., Takano-Ohmuro, H., Tanaka, T., Yamaguchi, Y., and Kohama, T., 1986, J. Biol. Chem., 261:8022.

Kohama, K., Uyeda, T.Q.P., Takano-Ohmuro, H., Tanaka, T., Yamaguchi, T., and Maruyama, K., 1985, Proc. Japan Acad., 61B:501.

Krestinger, R.H., and Nockolds, C.E., 1973, J. Biol. Chem., 248:3313.

Maruyama, K., Mikawa, T., and Ebashi, S., 1984, J. Biochem., 95:511.

Szent-Görgyi, A.G., 1980, In: "Muscle Contraction : Its Regulatory Mechanism", Ebashi et al., eds., Japan Sci. Soc. Press, Tokyo/Springer-Verlag, Berlin pp. 375-389.

Takahashi, K., Ogihara, S., Ikebe, M., and Tonomura, Y., 1983, J. Biochem., 93:1175.

Youngberg, G.E., 1930, J. Lab. Clin. Med., 16:158.

PARVALBUMIN EXPRESSION IN NORMAL AND MUTANT

XENOPUS EMBRYOS

Brian K. Kay

Department of Biology, University of North Carolina at Chapel Hill, Chapel Hill
NC 27599-3280, USA

INTRODUCTION

Our laboratory is interested in understanding the developmental regulation of the Ca^{2+}-binding protein, parvalbumin, in embryos. We wish to learn how the various members of the parvalbumin gene family are differentially expressed, and what the individual isoforms contribute to muscle cell structure and function. Parvalbumin is a member of the troponin C superfamily and is evolutionarily related to a number of Ca^{2+}-binding proteins, including calmodulin, troponin C, regulatory myosin light chains, oncomodulin, intestinal vitamin D-dependent calcium-binding protein, and Spec 1 protein (Goodman et al., 1979; Kretsinger, 1980; MacManus et al., 1983; Hardin et al., 1985).

Among adult tissues, parvalbumin (PV) is most abundant in fast-twitch muscles, although the role of this protein is unknown. In fast-twitch muscles of amphibians and fish, concentrations of 1 mM of this soluble protein have been measured (Wnuk et al., 1982), making it even more abundant than troponin C. These observations have lead to the hypothesis that PV may enable poikilotherms to contract and relax their muscles at cold temperatures rapidly, even though the ATPase and pumping activity of the sarcoplasmic reticulum are considerably reduced in the cold. In a model of muscle contraction-relaxation, PV is thought to remove Ca^{2+} from the troponin C after contraction, and thereby accelerate muscle relaxation in fast-twitch muscles (Haiech et al., 1979; Gillis et al., 1982; Heizmann et al. 1982). PV has also been detected in several other tissues as well; it has been isolated from developing bones, some endocrine glands, and GABA-nergic neurons of the brain (for review, see Heizmann and Berchtold, 1987). The role of PV in these tissues is unknown, although it is assumed that this protein is modulating the levels of free Ca^{2+} in the cytoplasm.

We have chosen the African clawed frog, Xenopus laevis, as model system for study of parvalbumin's role in embryonic and adult muscles. This organism is very popular in the study of the biochemistry of vertebrate development because eggs are plentiful and undergo rapid, stereotyped development in the laboratory. Its embryology is well described and has been the preparation of choice for several important studies in developmental biology. Its tadpoles have a large array of skeletal muscles for biochemical and cytological analysis, and the large size of its embryos permits microinjection of proteins and nucleic acids. Finally, there is a large body of information regarding gene expression in its oocytes and embryos (Dawid et al., 1983).

Over the past 2 years, we have examined the developmental regulation of PV during embryonic development of Xenopus. We have answered two questions: a) since Ca^{2+} has been implicated in numerous embryonic processes, does PV play a role in early embryonic cell division and migration?, and b) are different isoforms of PV expressed during development and what is the pattern of expression in embryonic muscles ? In summary, our work has shown that PV protein and mRNA are absent in early frog embryos, and first appear at stage 24 (two days old) when myotomal muscles are at the end of their cellular differentiation (Kay et al., 1987). The protein is expressed only in certain skeletal muscle cells, which we have shown by two different histochemical stains to correspond to fast-twitch fibers (Schwartz and Kay, 1988). In addition, we have biochemical and molecular genetic evidence that at least five different isoforms of PV exist in Xenopus, each encoded by a different gene. Furthermore, these genes appear to be developmentally regulated, with one isoform expressed in embryonic skeletal muscles and combinations of the other four expressed differently in various adult tissues.

In this chapter, we report our current effort toward understanding the role of PV in the differentiation of embryonic muscles. Our strategy has been to utilize cellular and molecular biological methods in defining the structure of the various PV isoforms expressed in Xenopus, and describing their developmental regulation in both normal and mutant tadpoles.

MATERIAL AND METHODS

1. Antibodies and Recombinant DNAs

Rabbit anti-Xenopus PV antiserum was the gift of Drs. I. Dawid and Y.-H. Chien and its specificity has been documented (Kay et al., 1987). Mouse monoclonal antibodies, which recognized epitopes unique to either the slow-twitch and fast-twitch isoforms of chicken myosin, were provided by Dr. F. Stockdale.

2. Immunoblots and Immunofluorescence

Extracts were prepared from lysates of whole tadpoles in 2% SDS - 50 mM Tris, pH 6.8 - 5 mM EGTA, resolved by polyacrylamide gel electrophoresis and transferred to nitrocellulose (Towbin et al., 1979). Antigens were detected with anti-Xenopus PV antiserum, ^{125}I-protein A, and film autoradiography. The polyclonal and monoclonal antibodies were used to stain Xenopus embryonic tissues by immunofluorescence in six mm thick paraffin sections or whole mounts (fixed in Bouin's fixative or 2% paraformaldehyde) according to published procedures (Schwartz and Kay, 1988). Rhodamine labelled α-bungarotoxin was obtained from Molecular Probes (Eugene, OR) and used according to published procedures (Kay et al., 1988).

3. In situ hybridizations

Coverslips with attached myotomal cells were prepared for in situ hybridization experiments according to published methods (Lawrence and Singer, 1986), after 30 minutes fixation in fresh 2% paraformaldehyde. The coverslips were incubated for six hours with a ^{3}H-labelled RNA prepared in vitro (Melton et al., 1984) with bacteriophage T7 RNA polymerase and a pGEM-7zf recombinant containing a 318 base pair Eco RI fragment, which carries most of the Xenopus PV coding region (Kay et al., 1987). Hybrids were detected with autoradiographic film emulsion after one to two week's exposure at 4°C.

EXPERIMENTAL RESULTS AND DISCUSSION

1. PV is expressed in the majority of tadpole myotomal cells

During development of the skeletal muscle system in animals, two primary muscle fiber types form (i.e., slow-twitch and fast-twitch). The two fiber types differ both physiologically and biochemically, with the overall consequence that a large muscle, composed of mixtures of the two types of muscle fibers, can be programmed to have geometrically precise contracture. The major physiological differences between the two muscle fiber types are in the speed of contraction and their resistance to fatigue. The two fiber types also differ in their oxidative capacities, myoglobin content, energy metabolism, mitochondria number, and lipid content. The two fiber types can histochemically be distinguished by certain mitochondrial stains, and pH optima for myosin ATPase. These histochemical changes are based on biochemical differences as well; for example, the myosin heavy chain exists as two isoforms, and each isoform is exclusively expressed in either the fast- or slow-twitch muscle cells (Gauthier et al., 1978).

We have examined the geometry of PV expressing cells in developing embryos. Figure 1A shows a stage 37 tadpole, approximately three days following fertilization. At the caudal end of the tail of such an embryo, there are two different skeletal muscle cell morphologies; both the short, fat cells and the long, slender cells are arrayed in a highly stereotyped arrangement (Fig. 1B). When this region of the tail is

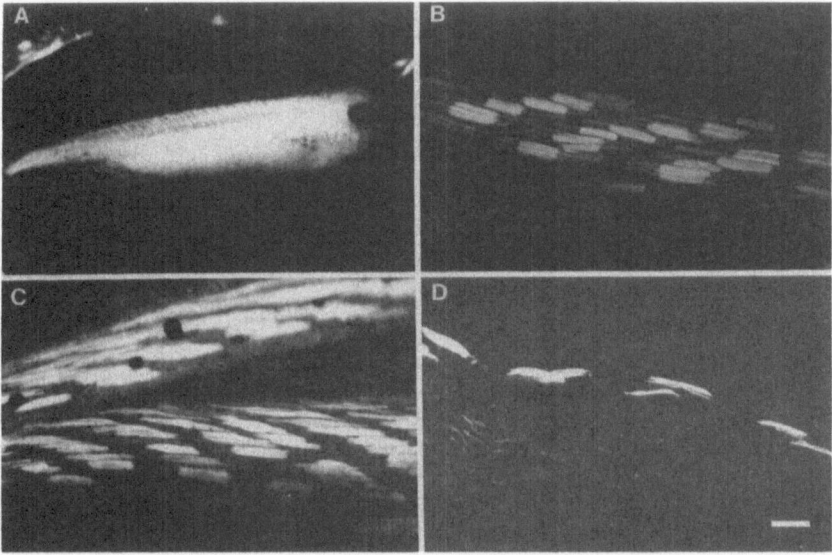

Fig. 1 . Immunofluorescence staining of PV and myosin antigens in embryonic tail muscle cells. The fast- and slow-twitch myosin isoforms were detected with separate monoclonal antibodies (F59 and S46; Crow and Stockdale, 1986). Panels A-D correspond to a stage 37 tadpole, phalloidin stained tail strip (paraformaldehyde fixed), PV antibody stained tail strip (fixed in Bouin's fixative) and anti-slow myosin stained tail strip (Bouin's fixed), respectively. Bar equals 600 μm in (A), 25 μm in (B-C), and 50 μm in (D).

immunostained for PV, only the short, fat cells react (Fig. 1C). It is well known that PV expression in other vertebrates is restricted to fast-twitch muscle cells (Celio & Heizmann, 1982). We have confirmed this by immunostaining tail strips for either the fast- or slow-twitch myosins with specific monoclonal antibodies (Crow and Stockdale, 1986). In Figure 1D, only the long, slender fibers stain for the slow-twitch myosin isoform. As will be seen below, the bulk of the myotomal cells stain positive for the fast-twitch myosin isoform. Since different fixatives are necessary for PV and myosin antigen detection in whole mounts, these staining experiments have been performed separately.

2. Both PV+ and PV- embryonic cells will differentiate in vitro

To evaluate the PV expression in embryonic muscle cells, we have cultured the cells from embryos and monitored PV protein and mRNA synthesis. Myotomal cells were removed from stage 20 embryos after extensive collagenase treatment and cultured in plastic petri dishes by standard procedures (Peng and Nakajima, 1978). Such cells flatten on the dish and as mononucleated cells they differentiate into appropriate looking skeletal muscle cells. Figure 2 shows the detection of PV protein by immunofluorescence staining in these cells. While the vast majority of cells are PV-plus, cultures often contained a small number of negative staining cells. Generally, the long, thin fibers were found to lack PV, just like they do in the tadpole tail. To determine whether or not the PV gene is inactive in these PV-minus cells, we performed in situ hybridization with ^{3}H-labelled RNA probes which corresponded to the PV anti-sense orientation. As seen in Figures 2c and 2D, most cells contained PV mRNA in their cytoplasm, except for the long, thin cells that had some immunofluorescence only at their ends which may be due to trapping.

Since the hybridization conditions are incompatible with antigen detection experiments, we have not double-localized both the PV protein and RNA species in individual myotomal cells. When in situ hybridizations were performed with ^{3}H-labelled RNA probes of the sense orientation, there were only background levels of silver grains over the field of myotomal cells (not shown). Thus, it appears that 1) PV expression in cultured myotomal cells is programmed to occur without continuous input, and 2) regulation is at the level of transcription or messenger RNA accumulation.

3. Analysis of the Unresponsive Mutant in Xenopus

Mutations that have reduced PV levels and apparent muscular defects have been previously identified in mice. In the mouse mutant, "arrested development of right response" (adr), animals are deficient in PV content, and display aberrant series of tetanic contracture following the termination of nerve stimulation, unlike normal muscles which cease contractures once motorneuron stimulation ends (Stuhfauth et al., 1984). Moreover, when adr muscles are repeatedly stimulated there are abnormal after-contractions. Defects in the proper expression or function of PV may be linked to, or influence the progress of, certain types of muscular dystrophy. In C57BL/6J dy^{2J}/dy^{2J} dystrophic mice, the levels of PV decrease significantly, and the resulting elevated levels of free sarcoplasmic Ca^{2+} may stimulate various Ca^{2+}-activated proteases (Klug et al., 1985).

Recently, we have investigated one available Xenopus behavioral mutant, "unresponsive" (unr), and found an interesting change in the cellular distribution of PV. This mutation was identified in Xenopus by homozygosing the genome of eggs activated with X-ray damaged sperm (Reinschmidt and Tompkins, 1984). Homozygous unr/unr embryos have the following phenotype: the animals, while morphologically normal, lack skeletal muscle activity, and are unresponsive to external stimulation. Only skeletal muscles appear affected, as the cardiac muscles function properly in the tadpole. Interestingly, this mutation is not lethal, because with time the tadpoles regain full motor activity and swim normally, and the adults are fully fertile. Based on electrophysiological studies, the defect in early embryos appears to be not in the nervous system, but somewhere distal to the neuromuscular junction (Dudek et al., 1987).

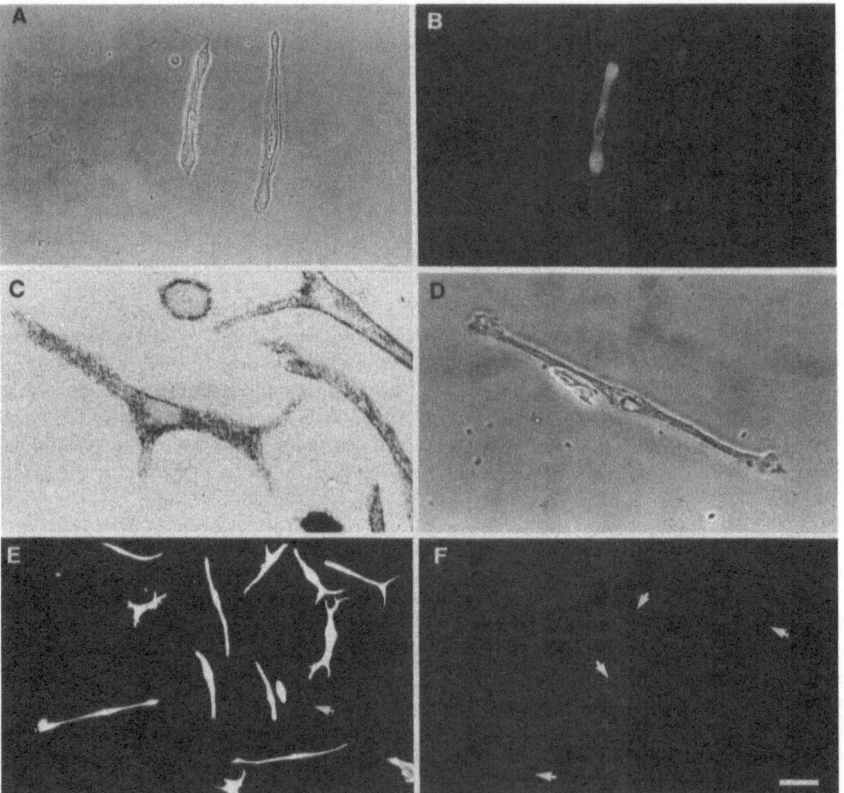

Fig. 2 . Immunofluorescence staining and *in situ* hybridizations of culture myotomal cells. Panel A is a phase contrast micrograph of two myotomal muscle cells after one week in culture. Panel B shows the same cells, after immunofluorescence staining for PV after fixed in Bouin's solution; only the cytoplasm stains of many cells. Panels C and D show *in situ* hybridizations of cells from the same slide obtained with an anti-sense ^{3}H-RNA probe. Note in these bright field views that the silver grains in the emulsion are over the myoplasm of the cells in panel C and are absent over the long, thin cell in panel D. Panels E and F demonstrates the staining of cultured cells fixed in paraformaldehyde and reacted with anti-fast and anti-slow myosin monoclonal antibodies, respectively; arrows mark muscle cells that failed to stain. The bar in (A-B), (C-D), (E), and (F) represents 25, 13, 50 and 25 μm, respectively.

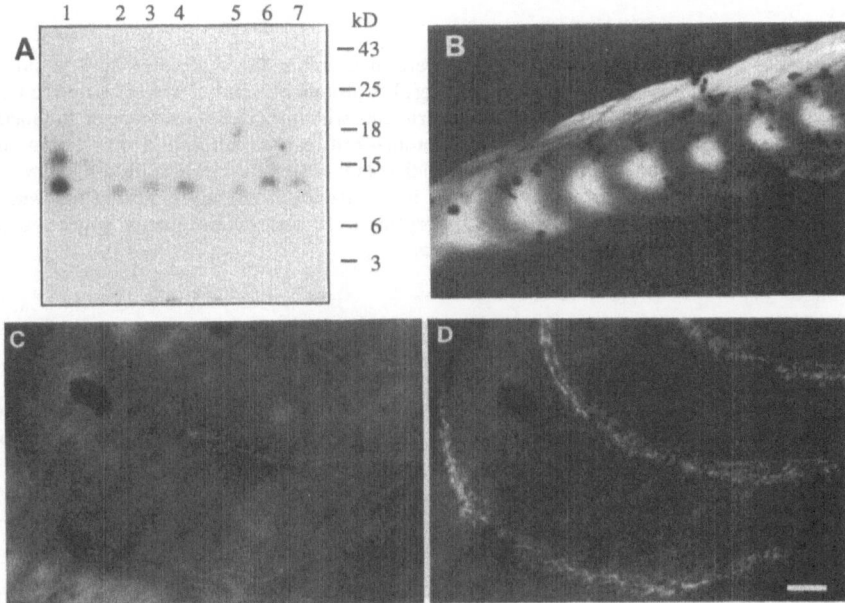

Fig. 3 . Biochemical and cytological evaluation of PV expression in paralyzed (unr/unr) and normal tadpoles. A) Autoradiogram of an immunoblot of cellular extracts after reaction with rabbit anti-*Xenopus* parvalbumin antibodies and ^{125}I-protein A. Lane 1 corresponds to protein lysates prepared from an adult gastrocnemius muscle, lanes 2-4 contain lysates of three different unr/unr tadpoles, and lanes 5-7 correspond to three identically staged tadpoles obtained from matings of an outbred adult pair. B) Immunofluorescence staining of PV in unr/unr myotomal muscles by whole mount staining. C) and D) correspond to whole mount staining for NCAM and acetylcholine receptors, respectively. The black dots evident in the tail strip are melanocytes, and the bar corresponds to 100 and 8 μm in (B) and (C-D), respectively.

A simple hypothesis regarding this recovery is that the embryonic isoform is encoded by the unr gene, and that a second isoform is expressed in tadpoles to restore motor activity. To test this possibility we have quantitated the amount of PV in mutant and wild-type tadpoles at the same stage of development (staged according to Nieuwkoop and Faber, 1967). Figure 3A shows that the amount of the embryonic PV detected by immunoblots of whole lysates of three individuals appears to be normal.

Next, we compared the cellular distribution of antigens in preparations of normal and mutant tadpole. By whole mount immunofluorescence staining methods (Schwartz and Kay, 1988), we discovered that PV is concentrated at the longitudinal ends of many myotomal muscle cells in paralyzed unr/unr tadpoles, rather than the homogeneous distribution of PV throughout the myoplasm of wild-type cells (Fig. 3B). Other proteins appear to be normally distributed in the mutants however, such as assembled myofibrils and acetylcholine receptors, as evidenced by proper phalloidin (not shown), NCAM (Fig. 3C) and x-bungarotoxin staining (Fig. 3D). In addition, the paralyzed myotomal muscles do not express NCAM protein widely on their cell surfaces, as do adult muscles after denervation (Kay et al., 1988).

We suspect that PV's unusual arrangement in the myotomal muscles provides some clues toward understanding the basis of the unresponsive mutation. At the moment, we do not believe that the UNR gene encodes embryonic PV for two reasons. First, our western blots and immuno-fluorescence staining experiments demonstrate that unr/unr embryos express the appropriate embryonic PV isoform. Second, the skeletal paralysis appears to involve slow-twitch fibers which normally lack PV. However, the abnormal cellular distribution of PV in myotomal cells of unr/unr embryos suggests that either the genetic defect occurs in a gene sequence involved in the correct cellular distribution of PV or that muscle activity is necessary for proper cellular distribution of PV. We favor the first hypothesis, as frog embryos reared in the presence of tubocurarine, at levels sufficient to cause skeletal muscle paralysis, still developed morphologically normal and expressed PV at proper levels (Schwartz and Kay, 1988).

SUMMARY

The levels and cellular distribution of PV were analyzed at the molecular level in normal and mutant Xenopus laevis embryos. In developing tadpoles, PV protein expression was restricted to fast-twitch myotomal cells, as shown by immunofluorescence experiments with myosin isoform-specific monoclonal antibodies. Both PV expressing and non-expressing myotomal cells can be cultured in vitro without neurons, and non-expressing cells have been shown to lack hybridizable levels of PV mRNA. In a paralyzed mutant, unresponsive, the level of PV protein was normal, but its cellular distribution was not. Thus, while neural input may not be necessary for PV expression in embryonic cells, one gene sequence appears needed for proper cellular distribution of PV in tadpole tail muscles.

ACKNOWLEDGMENTS

We appreciate the generous gifts of unresponsive tadpoles and anti-myosin antibodies from Drs. R. Tompkins and F. Stockdale, respectively. We thank Dr. H.B. Peng for preparing the cell cultures of myotomal cells. This work was funded in part by a March of Dimes Basil O'Connor Fellowship (5-576) and a grant from the American Cancer Society (CD-253)

REREFENCES

Celio, M.R. and Heizmann, C.W., 1982, Calcium binding protein parvalbumin is associated with fast contracting muscle fibers, Nature, 297:504.

Crow, M.T. and Stockdale, F.E., 1986, Myosin expression and specialization among the earliest muscle fibers of the developing avian limb, Dev. Biol., 113:238.

Dawid, I.B., Kay, B.K. and Sargent, T.S., 1983, Gene expression during Xenopus laevis development, in: "Gene structure and regulation in development", pp. 171-182. S. Subtelny and F.C. Kafatos, eds., Lis, Inc., New York.

Dudek, F.E., Ide, C.F. and Tompkins, R., 1987, Unresponsive, a behavioral mutant in Xenopus laevis. Electrophysiological studies of the neuromuscular system, J. Neurobiology, 18:237.

Gauthier, G.F., Lowrey, S. and Hobbs, A.W., 1978, Fast and slow myosin in developing muscle fibers, Nature, 275:25.

Gillis, J.M., Thomason, J., Lefevre, J. and Kretsinger, R.H. 1982, Parvalbumins and muscle relaxation: a computer simulation study, J. Muscle Res. Cell Motility, 3:377.

Goodman, M., Pechere, J.-F., Haiech, J. and Demaille, J.G. 1979, Evolutionary diversification of structure and function in the family of intracellular calcium-binding proteins, J. Mol. Evol., 13:331.

Haiech, J., Derancourt, J., Pechere, J.-F and Demaille, J.G. 1979, Magnesium and calcium binding to parvalbumins: evidence for differences between parvalbumins and an explanation of their relaxing function, Biochemistry, 18:2752.

Hardin, S.H., Carpenter, C.D., Hardin, P.E., Bruskin, A.M. and Klein, W.H., 1985, Structure of the Spec1 gene encoding a major calcium-binding protein in the embryonic ectoderm of the sea urchin Strongylocentrotus purpuratus, J. Mol. Biol., 186:243.

Heizmann, C.W. and Berchtold, M.W., 1987, Expression of parvalbumin and other Ca^{2+}-binding proteins in normal and tumor cells: a topical review, Cell Calcium, 8:1.

Heizmann, C.W., Berchtold, M.W. and Rowlerson, A.M., 1982, Correlation of parvalbumin concentration with relaxation speed in mammalian muscles, Proc. Natl. Acad. Sci., 79:7243.

Kay, B.K., Shah, A.J. and Halstead, W.E., 1987, Expression of the Ca^{2+}-binding protein, parvalbumin, during embryo development of the frog, Xenopus laevis, J. Cell Biol., 104:841.

Kay, B.K., Schwartz, L.M., Rutishauser, U., Qui, T.H. and Peng, H.B., 1988, Patterns of N-CAM rexpression during myogenesis in Xenopus laevis. Development, 103:463.

Klug, G., Reichmann, H. and Pette, D., 1985, Decreased parvalbumin contents in skeletal muscles of C57BL/6J (dy^{2J}/dy^{2J}) dystrophic mice, Muscle Nerve, 8:576.

Kretsinger, R.H., 1980, Structure and evolution of calcium-mediated proteins, CRC Crit. Rev. Biochem., 8:119.

Lawrence, J.B. and Singer, R.H., 1986, Intracellular localization of messenger RNAs for cytoskeletal proteins, Cell, 45:407.

MacManus, J.P., Watson, D.C. and Yaguchi, M., 1983, The complete amino acid sequence of oncomodulin - a parvalbumin-like calcium-binding protein from Morris hepatoma 5123tc, Eur. J. Biochem., 136:9.

Melton, D.A., Krieg, P.A., Rebagliati, M.R., Maniatis, T., Zinn, K. and Green, M. R., 1984, Efficient in vitro synthesis of biologically active RNA and RNA hybridization probes from plasmids containing the bacteriophage Sp6 promoter, <u>Nuc. Acids Res</u>., 12:7057.

Nieuwkoop, P. and Faber, J., 1967, "Normal tables of Xenopus laevis (Daudin)", 2nd ed. North-Holland Publishing Co., Amsterdam.

Peng, H.B. and Nakajima, Y., 1978, Membrane particle aggregates in innervated and noninnervated cultures of Xenopus embryonic muscle cells, <u>Proc. Natl. Acad. Sci. USA</u>, 75:500.

Reinschmidt, D.C. and Tompkins, R., 1984, Unresponsive, a new behavioral mutant in Xenopus laevis, <u>Differentiation</u>, 26:189.

Schwartz, L.M. and Kay, B.K., 1988, Differential expression of the Ca^{2+}-binding protein parvalbumin during myogenesis in Xenopus laevis, <u>Dev. Biol</u>., 128:441.

Stuhfauth, I., Reininghaus, J., Jockusch, H. and Heizmann, C.W., 1984, Calcium-binding protein, parvalbumin, is reduced in mutant mammalian muscle with abnormal contractile properties, <u>Proc. Natl. Acad. Sci. USA</u>, 81:4814.

Towbin, H., Staehelin, T. and Gordon, J., 1979, Electrophoretic transfer of proteins from polyacrylamide gels to nitrocellulose sheets: procedures and some applications, <u>Proc. Natl. Acad. Sci. USA</u>, 76:4350.

Wnuk, W., Cox, J.A. and Stein, E.S., 1982, Parvalbumins and other soluble high-affinity calcium-binding proteins from muscle. in: "Calcium and Cell Function", vol. II. pp. 243-278. W.Y. Cheung, ed., Academic Press, New York.

CALRETININ AND OTHER CaBPs IN THE NERVOUS SYSTEM

John Rogers, Masood Khan and Jon Ellis

Physiological Laboratory, University of Cambridge, Downing street, Cambridge CB2 3EG
UK

SUMMARY

At least three CaBPs are abundant in various types of nerve cells : calbindin-D28, calretinin, and parvalbumin. The sequence of chick calretinin, from cDNA clones, is 60% homologous to that of chick calbindin. The genomic calretinin gene has also been partially sequenced. Calretinin is a protein of 29-30 kilodaltons. Antisera have been raised against ß-galactosidase-calretinin fusion proteins, and used to compare the distribution of calretinin with that of calbindin by two-colour immunofluorescence. Some sections have also been stained for parvalbumin. In chick brain and retina, the three proteins are largely in different neurons. Calbindin and calretinin are particularly abundant in some sensory nuclei, and co-expression is more common in peripheral sensory neurons. In rat brain, and in retinae of rat, cat, and salamander, some of the expression patterns are conserved, but some are not. In the chick embryonic retina, some cells show a transient phase of calbindin immunoreactivity during development.

RESULTS

Calretinin is homologous to calbindin-D28

Calretinin was identified by the serendipitous cloning of a cDNA from chick retina (Rogers, 1987). The cDNA clone encoded four "EF-hands" which were most homologous to the first four domains of calbindin-D28. This cDNA clone was incomplete, ending at an internal EcoRI site. We have now obtained genomic sequence from an exon covering this EcoRI site (Wilson et al., 1988), and also two cDNA clones which are almost full length, from a library of chick retina cDNA in λ gt10. Curiously, these cDNA clones are circularly permuted; the cDNA appears to have been circularized during synthesis then cleaved at the same internal EcoRI site. However, the validity of the sequence is established by the identity of the coding sequences in the two cDNA clones and by the overlap with the genomic exon sequence (Fig. 1a). The chick calretinin sequence is now complete except for the N-terminus, where preliminary genomic sequence suggests that only 4 or 5 codons are missing. The sequence is 60% homologous to that of chick calbindin (Fig. 1b). Calbindin is the only CaBP with which calretinin shows almost perfect alignment. The most conserved segments between calretinin and calbindin are also those which are most conserved between chick and rat calbindin, and they are not the calcium-binding sites. They are the segments following sites II, IV, and V, as well as the altered site VI which does not bind calcium in calbindin.

The conservation of these segments suggests that they perform some function other than simply binding calcium. The same conclusion has been reached from the corresponding human sequences (Parmentier, this volume). Because of the ⩾78% conservation of both proteins between chicks and mammals, it is not yet clear whether there are any sequences selectively conserved only in one protein or the other, which might indicate different functions.

The gene has introns in discordant positions

The 5' two-thirds of the calretinin gene was cloned from a chick genomic library and partially sequenced, thus identifying the positions of 8 introns (Wilson et al., 1988). In the regions sequenced, all the intron positions are in exactly the same places as in the calbindin gene. Although some fall in similar

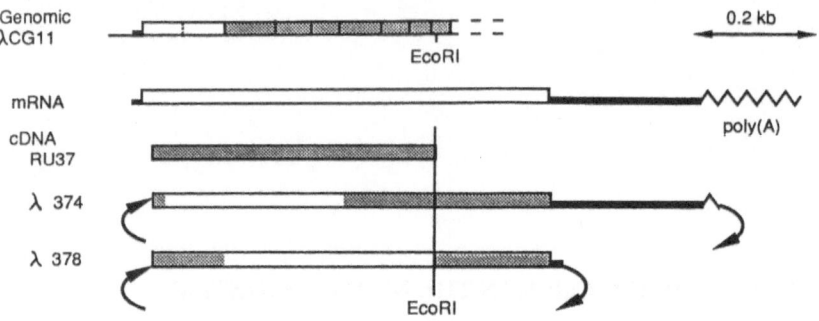

```
Genomic
λCG11
                                    EcoRI                          0.2 kb

mRNA
                                                                poly(A)
cDNA
RU37

λ  374

λ  378

                                    EcoRI
```

```
CALR:    ...APHLHLADVSASQFLDVWRH  FDADGNGYIEGKEL  ENFFQELESARKGTGVD
            **      ** **     * *   * *****  ****  ** ***  *** * *
CALB:    MTAETHLQGVEISAAQFFEIWHH  YDSDGNGYMDGKEL  QNFIQELQQARKKAGLD
          ** ***   * ******* *     * ** **  ****  ** ****  *******
CALB:    MAESHLQSSLITASQFFEIWLH   FDADGSGYLEGKEL  QNLIQELLQARKKAGLE

         SKRDSLGDKMKEFMHK  YDKNADGKIEMAEL  AQILPTEENFLLCFR-
          *  **  *          * **  ********  **
         -----LTPEMKAFVDQ  YGKATDGKIGIVEL  AQVLPTEENFLLFFRC
          * ****  ****       **   *********  *  ********** ***
         -----LSPEMKTFVDQ  YGQRDDGKIGIVEL  AHVLPTEENFLLLFRC

         QHVGSSSEFMEAWRR  YDTDRSGYIEANEL  KGFLSDLLKKANRPY
          *   **   **  **   ** *  ** *    **  * ** *** ***
         QQLKSSEDFMQTWRK  YDSDHSGFIDSEEL  KSFLKDLLQKANKQI
          *****   ** ****  ** *****  ***  * ****** ****
         QQLKSCQEFMKTWRK  YDTDHSGFIETEEL  KNFLKDLLEKANKTV

         DEAKLQEYTQTILRM  FDMNGDGKLGLSEM  SRLLPVQENFLLKFQ
           ** ***   ***    ** *  **** * *   *********  ***
         EDSKLTEYTEIMLRM  FDANNDGKLELTEL  ARLLPVQENFLIKFQ
          * ** ***   **    ** ****** ***  *********** ***
         DDTKLAEYTDLMLKL  FDSNNDGKLQLTEM  ARLLPVQENFLLKFQ

         GMKLSSEEFNAIFAF  YDKDGSGFIDEHEL  DALLKDLYEKNKK
          * *    ***   *   ** ** * *** **  ******* *****
         GVKMCAKEFNKAFEM  YDQDGNGYIDENEL  DALLKDLCEKNKK
          * *** ********   *************  ***********
         GIKMCGKEFNKAFEL  YDQDGNGYIDENEL  DALLKDLCEKNKQ

         EMSIQQLTNYRRSIM  NLSDG-GKLYRKEL  EVVLCSEPPL*
          *   *   * ***    **** ***** **   ** *
         ELDINNLATYKKSIM  ALSDG-GKLYRAEL  ALILCAEEN*
          ******   **** **  ***** *****   * **** *   *
         ELDINNISTYKKNIM  ALSDG-GKLYRTDL  ALILSAGDN*
```

Fig. 1. *(Upper)* Structures of genomic clone (λCG11) and three cDNA clones for chick calretinin. The box represents the coding region, with intron positions marked in the genomic clone. Shaded regions have been sequenced. *(Lower)* Sequence of calretinin from chick cDNA clones, aligned with sequence of calbindin from chick (middle rows; Wilson et al, 1985) and rat (bottom rows; Yamakuni et al, 1986). "EF-hand" sequences are aligned vertically.

positions between domains, others fall within the domains, at different positions in each domain, sometimes interrupting conserved coding sequences. A comparison with other available gene sequences from the CaBP superfamily shows few coincident intron positions, implying that the introns were inserted separately in different branches of the superfamily (Wilson et al., 1988).

Calretinin is a 30-kilodalton CaBP

Fusion proteins were made from plasmids in bacteria, containing ß-galactosidase coupled to either two or four domains from calretinin. On a protein gel blot, both fusion proteins bound [45]Ca (Rogers, 1987).

Natural calretinin was identified as a protein of 29-30 kd by three experiments. First, chick retina mRNA was hybrid-selected with a calretinin cDNA clone and translated in vitro, yielding a 29-kd product (Rogers, 1987). Second, antisera were raised against the ß-galactosidase-calretinin fusion proteins, and were used to immunoprecipitate an in vitro translation from chick retina; a 29-kd protein was specifically precipitated (unpublished results). Third, the same antisera were used on protein gel blots ("Western blots") of extracts of chick retina and brain, and labelled the same 29-kd protein (Rogers, 1989a).

The molecular weight deduced from the calretinin cDNA sequence (Fig. 1b), allowing for an extra 3 amino-acids at the N-terminus, would be about 30800, compared to 30018 for chick calbindin (Wilson et al., 1985).

In situ hybridization in chick brain : calretinin and calbindin are largely in different neurons

The chick brain was first surveyed by in situ hybridization with cDNA probes for calretinin and calbindin (the latter donated by Dr. Eric Lawson). The two mRNAs were found to be present in many types of neurons, with little overlap in their distributions (Rogers, 1987).

Antisera against Calretinin

Antisera were raised in rats and in a rabbit, against the ß-galactosidase-calretinin fusion proteins described above (Rogers, 1989a). On protein gel blots of extracts from chick retina or brain, they label only calretinin (apart from two bands which are displayed by all sera including non-immune sera). On immunohistochemistry to chick retina or brain, all the antisera give the same distinctive staining patterns, which are generally consistent with the pattern of calretinin mRNA distribution as deduced from in situ hybridization. The antisera do not detect calbindin.

In contrast, antisera against calbindin which have been used for immunohistochemical surveys of the chick brain (Jande et al., 1981a; Roth et al., 1981) appear to detect calretinin as well. On protein gel blots or on immunoprecipitation, they detect not only 28-kd calbindin but also a 29-kd band which comigrates with calretinin (Pochet et al., 1985; Parmentier et al., 1987). On immunohistochemistry, they stain some of the brain nuclei in which in situ hybridization detects mRNA for calretinin but not calbindin (Rogers, 1987). For the experiments described here, calbindin antiserum was donated by Dr. Eric Lawson (Jande et al., 1981b).

The antisera against the chick proteins react well with the homologous proteins in mammals, and the specificities appear to be the same as in chick : anti-calretinin does not detect calbindin, whereas anti-calbindin does detect calretinin, according to protein gel blots (Parmentier et al., 1987) and immunohistochemistry (see below, and Résibois et al., this volume).

This pattern of cross-reaction is consistent with the protein sequences. Some of the most similar parts of the two proteins are in domains V and VI, which were present in the natural calbindin used to raise the calbindin antisera, but not in the fusion proteins used to raise calretinin antisera.

Fortunately, the calbindin antisera stain calretinin sufficiently weakly in our immunohistochemistry conditions that the two can usually be distinguished by two-colour immunofluorescence. It was therefore possible to survey the chick and rat nervous systems by this method to compare the distribution of the two proteins.

A few regions were also examined with antisera against parvalbumin (gift of Dr. Claus Heizmann; Heizmann and Celio, 1987), CBP-18 (gift of Dr. Claude Klee; Manalan and Klee, 1984), GABA (mAb 3A12, gift of Dr. Peter Streit; Matute and Streit, 1986), and various neuropeptides (commercial antisera).

Immunochemistry was performed as described (Rogers, 1989a) on free-floating sections of brain or retina fixed with 4% paraformaldehyde (plus 0.5% glutaraldehyde for GABA).

Immunohistochemistry in the chick nervous system

In the chick central nervous system, the calretinin antisera gave striking staining of many types of neurons (Fig. 2-4). Two-colour immunofluorescence revealed that few neurons were positive for both calretinin and calbindin, except in a few regions.

197

Most of the positive neurons occur in the following areas.

- *Sensory nuclei* : These are among the most conspicuous locations for both proteins (Rogers, 1989a). All the sensory modalities are associated with many neurons positive for one or other protein, either in secondary sensory neurons (e.g. calretinin in the auditory pathways) or in interneurons(e.g. calbindin in the optic tectum). Fig. 2 shows an example from the spinal cord. Only in the nucleus solitarius are there many double-positive neurons.

- *Cerebellum* : It was surprising to see abundant calretinin immunoreactivity not only in ascending mossy and climbing fibres, but also in stellate cells, although there is no detectable calretinin mRNA in the cerebellar cortex (Rogers, 1989b). The origin of this immunoreactivity is still unresolved. Purkinje cells contain large amounts of calbindin and parvalbumin (Jande et al., 1981b; Roth et al., 1981; Braun et al., 1986).

- *Basal forebrain* : In the septal nuclei, separate regions are strikingly positive for calretinin or for calbindin (data not shown).

- *Reticular formation* : Up to 20% of neurons are positive for calretinin in some regions.

- *Hypothalamus* : Unlike other regions of the brain, the hypothalamus contains many cells positive for both calretinin and calbindin, as well as for each separately (Fig. 3).

- *Cerebral hemispheres* : Here, immunoreactivity appears to be confined to scattered cells, probably interneurons. Most of the cerebral hemispheres is crisscrossed by a network of calbindin-positive multipolar cells, except for the palaeostriatum where there are cells positive for calretinin instead (Fig. 4a); the latter cells also contain parvalbumin (data not shown; see also Braun et al., this volume).

In the peripheral sensory neurons, calretinin and calbindin are abundant. In contrast to the situation in the brain, they are often present together - in many of the ganglion cells of dorsal root ganglia, inner ear, and retina (Rogers, 1989a). Many other cell types in the retina are positive for calretinin (Fig. 5) or for calbindin, but not for both.

Parvalbumin is generally seen in different neurons from calretinin and calbindin in chick retina and brain, although there are some striking exceptions as noted above.

Immunohistochemistry in the rat nervous system

The rat brain has previously been surveyed for calbindin immunoreactivity (Jande et al., 1981a;

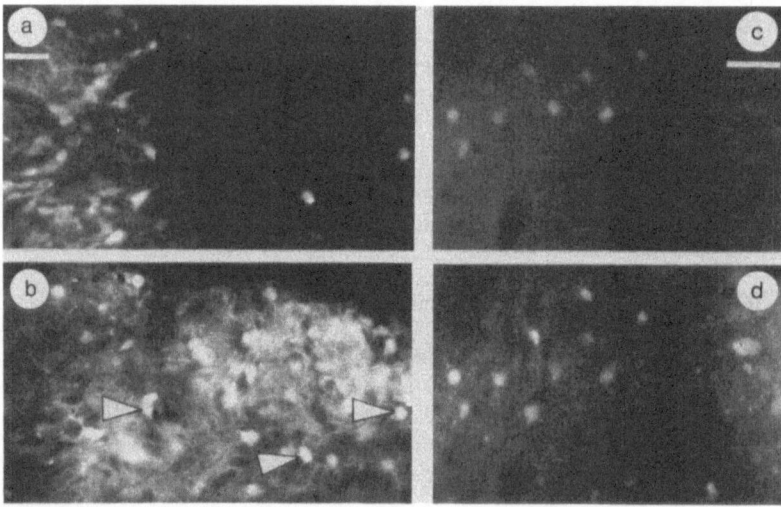

Fig. 2 . Substantia gelatinosa of spinal cord (transverse sections), from chick *(a,b)* and rat *(c,d)*, stained with antisera for calretinin *(a,c)* and calbindin *(b,d)*. Most cells are positive for one protein or the other in chick (a few double-positive cells are arrowed) but positive for both in rat. Dorsal is up, medial towards the middle of the page. Bars are 25 μm.

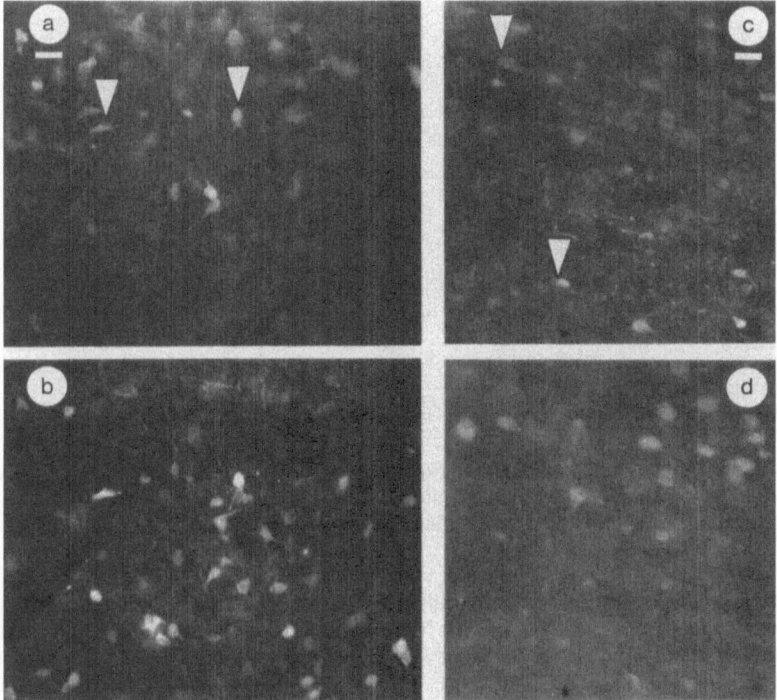

Fig. 3. Regions of lateral hypothalamic area (parasagittal sections), from chick *(a,b)* and rat *(c,d)*, stained with antisera for calretinin *(a,c)* and calbindin *(b,d)*. In each species some cells are positive for one protein or the other, and many other cells for both (examples arrowed). Bars are 25 μm.

Baimbridge and Miller, 1982; Feldman and Christakos, 1983; Garcia-Segura et al., 1984). In order to distinguish the distributions of calretinin and calbindin, a complete survey of the rat brain has now been done using peroxidase immunohistochemistry (Résibois et al., this volume and manuscript in preparation), supplemented with two-colour immunofluorescence in most areas. In comparison with the chick, many general features of the CaBP distribution are conserved in the rat, but there are also many differences. Cells expressing both proteins seem to be more common in the rat brain.

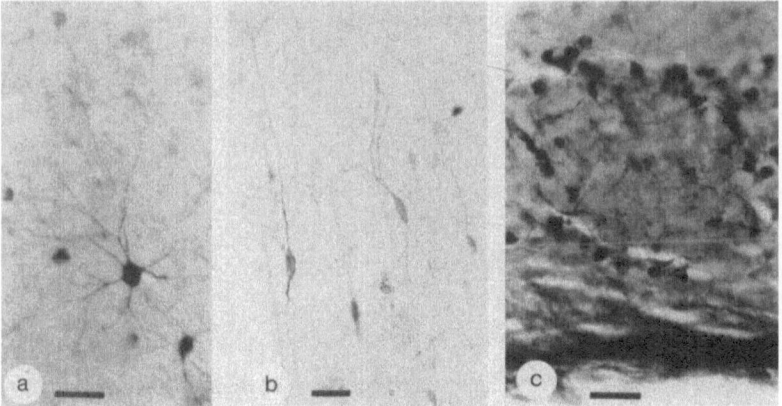

Fig. 4. Examples of neurons in the telencephalon, stained for calretinin by the "Vectastain" peroxidase method. *(a)* Chick palaeostriatum augmentum. *(b)* Rat cerebral cortex. *(c)* Rat olfactory bulb, showing one glomerulus; periglomerular cells and some olfactory nerve fibres are stained. Bars are 25 μm.

Fig. 5 . Chick retina, stained for calretinin with a fluorescein-labelled second antibody, and viewed with the confocal laser scanning microscope (White et al, 1987). Layers are labelled as follows : HC, horizontal cells; AC, amacrine cells; IPL, inner plexiform layer; GC, ganglion cells.

In comparison with the chick : - *Sensory nuclei* : These also show much immunoreactivity in the rat, but more neurons are double-positive, for example in the auditory nuclei (Résibois et al., this volume) and in the spinal cord (Fig. 2). In the olfactory bulb, antisera for calretinin, calbindin, and CBP-18 (Manalan and Klee, 1984) strongly stain many periglomerular cells (local circuit neurons) (Fig. 4c). The calretinin is in different cells from the calbindin and from the CBP-18, and cal retinin immunoreactivity is also seen in granule cells. Parvalbumin immunoreactivity is also strong in the olfactory bulb, but in the tufted cells (data not shown).

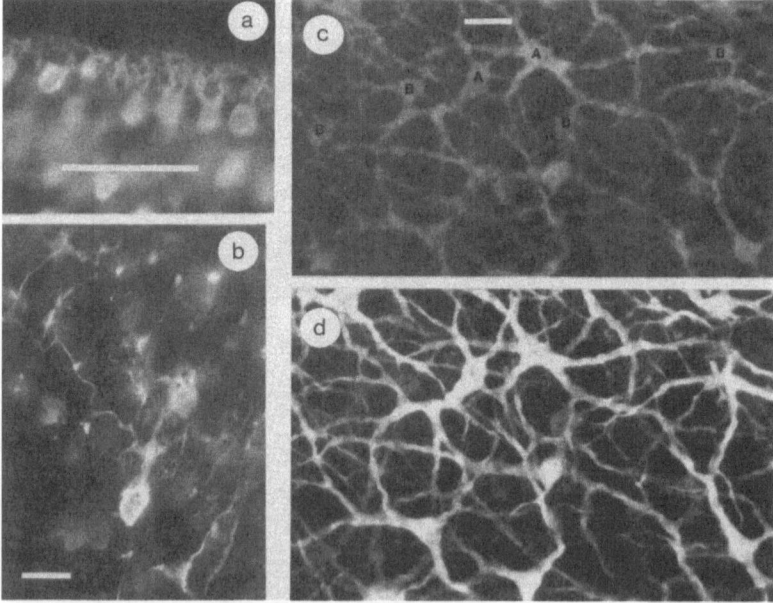

Fig. 6 . Retinal horizontal cells. *(a,b)* Oblique sections of retina of chick *(a)* and salamander *(b)*, stained for calretinin, showing horizontal cells with the photoreceptor layer unstained to top right. The bright spots in *b* are Landolt clubs. *(c,d)* Tangential section of cat retina, stained for calretinin *(c)* and calbindin *(d)*, showing horizontal cells of types A and B (examples labelled). Calbindin is very abundant in type A cells but weak in type B cells; calretinin is abundant in both, as is parvalbumin (not shown). Bars are 25 μm.

- *Cerebellum* : Whereas the distribution of calbindin and parvalbumin is conserved (Jande et al., 1981a,b; Baimbridge and Miller, 1982; Celio and Heizmann, 1981), that of calretinin is totally different; in the rat it is seen in the granule cells and Lugaro cells (Rogers, 1989b).

- *Basal forebrain and midbrain* : These regions contain many cells positive for calretinin and calbindin together, as well as for each separately.

- *Hypothalamus* : The preceding description applies to the rat hypothalamus, as it did for the chick (Fig. 3).

- *Cerebral hemispheres* : The cerebral cortex and hippocampus show scattered bipolar interneurons positive for calretinin or calbindin but not both (Fig. 4b). The calretinin-positive neurons have been compared with those positive for vasoactive intestinal peptide, somatostatin, or GABA, by two-colour immunofluorescence. In each case the distributions overlap slightly (and extensively with GABA in the ventral hippocampus), but not completely.

Retinae of chick, rat, cat, and salamander : some expression patterns are conserved

Retinae of these four species were examined by two-colour immunofluorescence. In chick, cat, and salamander (larval Ambystoma tigrinum), at least some cones are positive for calbindin, and none for calretinin. In each species, there are many amacrine cells and ganglion cells positive for one or other protein or for both. Most strikingly, in each species there is conspicuous labelling of the horizontal cells and their network of fibres, but there is no consistency as to which CaBP is present : it is calretinin in chick and salamander, calbindin in rat, and both proteins plus parvalbumin in all the horizontal cells in cat (Fig. 6). These generalizations have been extended to other species (Pasteels et al., this volume; Röhrenbeck et al., 1987).

Retina of chick embryo : an early phase of calbindin immunoreactivity

Retinae of chick embryos were examined to see when calretinin and calbindin appeared during development. In each cell type, the appropriate immunoreactivity appears several days after the cells differentiate, and the CaBP pattern appears virtually mature by 2-3 days before hatching. At comparatively early stages (13 to 16 days incubation), there are two to three times as many calbindin-immunoreactive ganglion cells as in postnatal retina, with more extensive dendrites; their decrease parallels the known ganglion cell death and dendritic "pruning" during development. At these stages there are also many more calbindin-immunoreactive amacrine cells than later. It is not known how much cell death occurs among the amacrine cells, but the decrease probably includes turnoff of calbindin expression in some cells, as at early stages calbindin immunoreactivity is coexpressed with calretinin in a conspicuous amacrine type that is later exclusively calretinin-positive.

DISCUSSION : FUNCTIONS OF CaBPs IN NEURONS

It seems likely that calretinin and calbindin were originally neuronal proteins. Calretinin has so far only been found in neurons (Rogers, 1987), and for calbindin, expression in the brain is conserved throughout the vertebrates while expression in other organs is more variable (Parmentier et al., 1987). Immunohistochemistry indicates that both proteins are generally present throughout the cytosol of the cells that contain them.

For calbindin, the in vitro calcium affinity and the in vivo abundance suggest that it should bind 99% of the free calcium entering the cytosol; thus it should be a very effective calcium buffer. This could have many effects on the activities of neurons. The most straightforward would be protection (Baimbridge et al., this volume). However, this alone does not explain why there should be different CaBPs in different sets of neurons, with some aspects of their distribution being conserved through a wide range of vertebrate species.

The surveys of CaBP-positive neuronal types do not reveal obvious common features which could explain the distributions. These proteins are present in many neurons of the sensory systems, in particular, but the positive cells include a wide range of electrophysiological types : fast-responding and slow-responding, excitatory and inhibitory, long-range and short-range. For example, on the one hand, in the avian auditory pathway calretinin is conspicuous in the sensory hair cells and in the secondary and tertiary projection neurons which transmit their signals with remarkable temporal precision (Takahashi et al., 1987;

Roger, 1989a). On the other hand, various CaBPs are conspicuous in the periglomerular and granule cells of the olfactory system, and the horizontal and amacrine cells of the retina, many of which are involved mainly in dendrodendritic communication for slow local inhibition of the adjacent sensory projections. It is not evident what property of horizontal cells requires the invariable presence of CaBP(s).

In view of the conserved diversity of expression in neurons, and the conserved amino-acid sequences in regions which are not primarily calcium-binding sites, it increasingly seems probable that calretinin and calbindin perform unknown functions beyond the simple buffering of calcium. There is little evidence for separate functions for the two proteins, but the conserved expression of calbindin in cones, for example, suggests that some functions may be specific to only one of these proteins.

ACKNOWLEDGEMENTS

We thank E. Lawson, P. Wilson, C. Heizmann, C. Klee, and P. Streit for clones and antisera; J. White, B. Amos, S. Hunt, J. Kilmartin, and J. Fawcett for the use of equipment; A. Résibois, B. Pasteels, and R. Pochet for collaboration on immunohistochemical surveys. This work was supported by the Medical Research Council, the Nuffield Foundation, and the Wellcome Trust.

REFERENCES

Baimbridge, K.G., and Miller, J.J., 1982, Immunohistochemical localization of calcium-binding protein in the cerebellum, hippocampal formation and olfactory bulb of the rat. Brain Res., 245: 223-229.

Braun, K., Schachner, M., Scheich, H., and Heizmann, C.W., 1986, Cellular localization of the CaBP parvalbumin in the developing avian cerebellum, Cell Tissue Res., 243:69-78.

Celio, M.R., and Heizmann, C.W., 1981, CaBP parvalbumin as a neuronal marker, Nature, 293:300-302.

Feldman, S.C., and Christakos, S., 1983, Vitamin D-dependent calcium-binding protein in rat brain: biochemical and immunocytochemical characterization. Endocrinology, 112: 290-302.

Garcia-Segura, L.M., Baetens, D., Roth, J., Norman, A.W. and Orci, L., 1984, Immunohistochemical mapping of calcium-binding protein immunoreactivity in the rat central nervous system, Brain Res., 296:75-86.

Heizmann, C.W., and Celio, M.R., 1987, Immunolocalization of parvalbumin, Methods in Enzymology, 139:552-570.

Jande, S.S., Maler, L., and Lawson, D.E.M., 1981, Immunohistochemical mapping of vitamin D-dependent calcium-binding protein in brain, Nature, 294:765-767.

Jande, S.S., Tolnai, S., and Lawson, D.E.M., 1981, Immunohistochemical localization of vitamin D-dependent CaBP in duodenum, kidney, uterus and cerebellum of chickens, Histochemistry, 71:99-116.

Manalan, A.S., and Klee, C.B., 1984, Purification and characterization of a novel CaBP (CBP-18) from bovine brain, J. Biol. Chem., 259:2047-2050.

Matute, C., and Streit, P., 1986, Monoclonal antibodies demonstrating GABA-like immunoreactivity. Histochemistry, 86:147-157.

Parmentier, M., Ghysens, M., Rypens, F., Lawson, D.E.M., Pasteels, J.L., and Pochet, R., 1987, Calbindin in vertebrate classes : immunohistochemical localization and Western blot analysis, General and Comparative Endocrinology, 65:399-407.

Pochet, R., Parmentier, M., Lawson, D.E.M. & Pasteels, J.L., 1985, Rat brain synthesizes two 'vitamin D-dependent' calcium-binding proteins, Brain Research, 345:251-256.

Rogers, J.H., 1987, Calretinin : a gene for a novel calcium-binding protein expressed principally in neurons, Journal of Cell Biology, 105:1343-1353.

Rogers, J.H., 1989a, Two CaBPs mark many chick sensory neurons, Neuroscience, in press.

Rogers, J.H., 1989b, Immunoreactivity for calretinin and other CaBPs in cerebellum, Neurosciences, in press.

Roth, J., Baetens, D., Norman, A.W., and Garcia-Segura, L.-M., 1981, Specific neurons in chick central nervous system stain with an antibody against chick intestinal vitamin D-dependent calcium-binding protein. Brain Res., 222: 452-457.

White, J.G., Amos, W.B., and Fordham, M., 1987, An evaluation of confocal versus conventional imaging of biological structures by fluorescence light microscopy, Journal of Cell Biology, 105:41-48.

Wilson, P.W., Harding, M., and Lawson, D.E.M., 1985, Putative amino acid sequence of chick calcium-binding protein deduced from a complementary DNA sequence. Nucl. Acids Res., 13: 8867-8881.

Wilson, P.W., Rogers, J., Harding, M., Pohl, V., Pattyn, G., and Lawson, D.E.M., 1988, Structure of chick chromosomal genes for calbindin and calretinin. <u>J. Mol. Biol.</u>, 200: 615-625.

Yamakuni, R., Kuwano, R., Odani, S., Kiki, N., Yamaguchi, Y., and Takahashi, Y., 1986, Nucleotide sequence of cDNA to mRNA for a cerebellar CaBP, spot 35 protein, <u>Nucl. Acids Res</u>., 14:6768.

IMMUNOHISTOCHEMICAL DETECTION OF 28KDa CALBINDIN IN HUMAN TISSUES

Roberto Buffa, Paolo Mare', Maurizio Salvadore and Ambrogio Gini

Dipartimento di Patologia Umana ed Ereditaria dell'Universita' di Pavia, Sezione di Anatomia Patologica 2, viale Borri 57, Varese, Italia

INTRODUCTION

We recently investigated (Buffa et al., 1989) tissues from some avians (quail, duck and chicken) and some mammals (rat, guinea pig, cat and dog) in order to localize the cellular source of 28KDa calbindin, using a polyclonal antiserum raised against a chicken antigen. Endocrine cells reacting selectively with this calbindin antiserum were found in the anterior lobe of the pituitary, in the adrenal medulla, and in the gastric mucosa of all mammalian species examined, reactivity was also found in the parafollicular cells from the mammalians studied and from the duck thyroid, in some insular cells from the pancreas of both mammals and avians and in the endocrine cells from the rectal mucosa of quails. Our results are in accordance with those reported in the literature (Roth et al., 1982; Kondo et al., 1985, Pochet et al., 1987).

Although some species variability was observed in the distribution of the immunoreactivity, we found rather unexpectedly, that most of the endocrine cells from human adult tissues failed to react with the calbindin antiserum from the chicken. These finding may be either related to technical problems (e.g. overfixation, selective influence of fixatives) or to subtle antigenic differences (Baimbridge et al., 1982; Sonnenberg et al., 1984) between chicken and human calbindins.

In this report, we describe the immunocytochemical localization of 28KDa calbindin in many human tissues from different organs.

MATERIALS AND METHODS

Antisera and Antibodies. A monospecific antiserum against chicken vitamin D dependent duodenal Calbindin was prepared according to Spencer et al. (1976); its main charactheristics are reported in the literature (for a review see Wilson et al., 1988). Monoclonal antibody LK2H10 directed against human chromogranin A was generously donated by Dr. R.V. Lloyd (Pathology Department, Michigan University, Ann Arbor, U.S.A.). The source and the main characteristics of hormone antisera used are summarized in table 1.

Tissues. Fragments from adrenal, pancreas, gut, kidneys, pituitary, thyroid, parathyroid were obtained from surgical specimens or taken during autopsies (pituitaries and all the tissues from foetuses and newborns). Some of the tumors were taken from the files of the Department of Morbid Anatomy of the Pavia University. The following fixatives were used: PAF (neutral picric acid-formaldehyde), AAF (70% alcohol - 5% acetic acid - 4% formaldehyde) or buffered formaldehyde (4% in 0,1 M pH 7.3 phosphate buffer) according to Rindi et al (1986).

Immunohistochemistry. Paraffin sections after dewaxing were hydrolyzed with an HCl solution (0.2 N for 1 - 10 hours, 37°C) and immunostained with an avidin-biotin technique (ABC reagents from Vector, Burlingame, CA, USA). The primary reagents were applied overnight in O.15M Tris buffer-saline pH 7.4 supplemented with human fresh serum (0.1%) from healthy donors.

Table 1. Antihormone antisera

code	antigen	dilution	Source
00476	Adrenocorticotropic hormone	1:1000	Ortho, Raritan, N.J., USA
00484	Growth hormone	1:1000	Ortho, Raritan, N.J., USA
00487	Prolactin	1:2000	Ortho, Raritan, N.J., USA
00486	Luteinizing hormone	1:1000	Ortho, Raritan, N.J., USA
00481	Follicle stimulating hormone	1:1000	Ortho, Raritan, N.J., USA
429	CLIP (cort.like inter. peptide)	1: 800	J.M.Polak, London, UK
00477	Calcitonin	1:1000	Ortho, Raritan, N.J., USA
00492	Thyroid stimulating hormone	1:1000	Ortho, Raritan, N.J., USA
00483	Glucagon	1:8000	Ortho, Raritan, N.J., USA
S1	Insulin	1: 800	Sorin, Saluggia, Italy
00488	Somatostatin	1:2000	Ortho, Raritan, N.J., USA
R6902	Gastrin releasing peptide	1:1000	N.Yanaihara, Shizuoka, Japan
12494	C-terminal Gastrin	1:2000	J.Walsh, Los Angeles, USA
43H2T	Serotonin	1:1000	Immunonuclear, Stillwater, MN., USA
2021	Secretin	1:1000	Mylab, Malmo, Sweden
221	Pancreatic polypeptide	1:2000	M.M.T.O'Hare, Belfast, UK
2KLH10	Chromogranin A	1:1000	R.V.Lloyd, Ann Arbor, MI, USA

3-3' diaminobenzidine (0.02% in 0.15M TBS pH 7.6) and 4 Cl-1 Naphtol (0.035% in 0.15M TBS pH 7.6) were used to detect peroxidase activity.

Pertinent specificity tests were performed according to Polak and Van Noorden (1983). In double immunostaining experiments the sections were treated as detailed elsewhere (Buffa et al., 1988).

RESULTS

Immunoreactivity for 28kDa calbindin in human tissues was affected by the different laboratory procedures and required, except for the kidney and the nervous tissues, a demasking treatment with hydrochloric acid. This step proved to be critical for all the tissues examined. In some cells calbindin immunoreactivity was cytoplasmatic and diffuse, in others it showed a preferential localization for the nuclei; sometimes both the nuclei and the cytoplasm reacted with the antiserum. (Table 2 and Fig. 1). Staining was prevented by the substitution of calbindin antiserum with non immune or an unrelated immune rabbit serum.

The identity of some calbindin immunoreactive cells was recognised owing to their particular morphology and their distribution (e.g. parafollicular cells of the thyroid, noradrenalin cells of the adrenals, the cells of the distal convoluted tubules of the nephrons) or with the use of the double immunostaining technique. In the pituitary a rather strong calbindin immunoreactivity was found in the nuclei of the cells reacting with the CLIP antiserum, occasionally in some gonadotroph cells, but never in prolactin, growth hormone and corticotroph cells. A slight immunoreactivity was occasionally detectable in thyrotroph cells. In the pancreas calbindin was localized in both nuclei and cytoplasm of pancreatic polypeptide cells from the uncinate process and from insular tumors. The cells of the oxyntic mucosa stained with calbindin antiserum showed consistent immunoreactivity for Chromogranin A and no reactivity for serotonin antibodies; these cells were tentatively identified as ECL cells. No calbindin immunostaining was detected in the pyloric mucosa. The duodenal calbindin was localised in the cytoplasm of secretin cells. In the thymus of newborns, calbindin immunoreactivity was associated with the S-100 protein and this was detectable throught the cytoplasm and in nucleus of epithelial reticular cells of the cortex and in Hassal's corpuscle, calbindin was also found in the reticular cells of the marginal zone of the newborn spleen. A few nuclei from the seminiferous tubules of the testis and from the bronchiolar epithelia of newborns reacted for calbindin.

The non neoplastic adult breast failed to react with calbindin antiserum which reacted with minor cellular population of breast tumors (both ductal and lobular type). Calbindin immunoreactivity was detectable in the majority of the medullary carcinomas from the thyroid of our collections although the percentage of the positive cells showed a great variability.

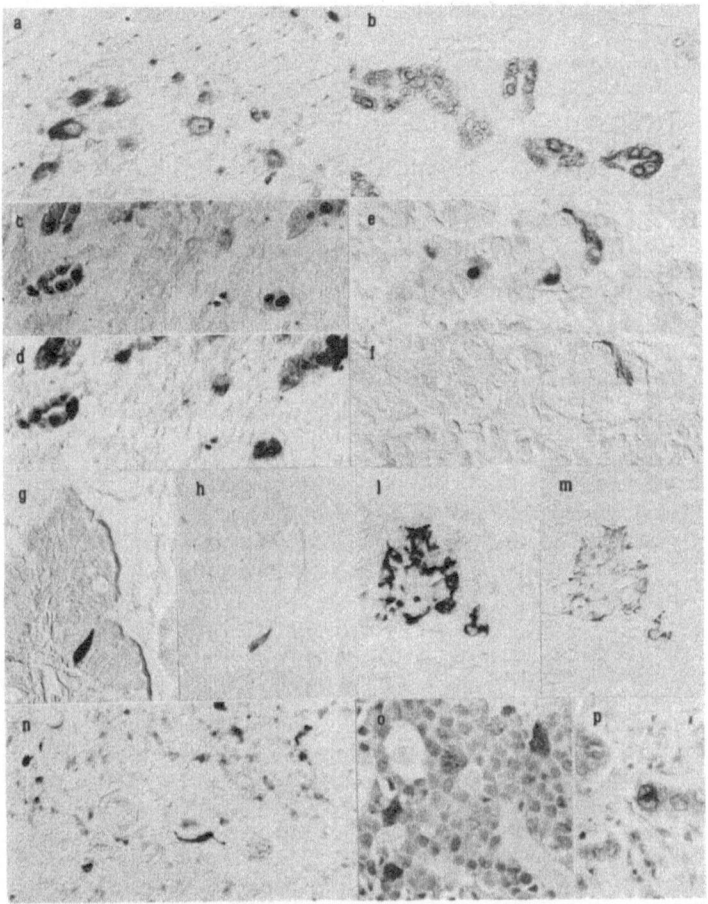

Fig. 1 . 28kDa calbindin immunoreactivity in Purkinje cells of the cerebellar cortex(a) and in the distal convoluted tubules of the kidney (b). Calbindin immunostain of both nuclei and cytoplasms (c) of the pituitary cells identified as CLIP cells after restaining of the same section with a CLIP antiserum; no calbindin reactivity in the cell (arrow) immunoreactive with a TSH antiserum (e,f). The Secretin cell of the duodenum (h) and the pancreatic polypeptide cells (m) of the juxtaduodenal portion of the pancreas show calbindin immunoreactivity (g and l). Calbindin immunoreactivity in the thymus, in a thyroid medullary carcinoma and in a breast ductal carcinoma. Immunoperoxidase o and p hematoxylin counterstain; a, b, c, d, e, f, g, h, o, p, x 400, l, m, n, x 250. Nomarski optic.

DISCUSSION

In this study, a demasking treatment with hydrocloric acid proved to be useful for the immunohistochemical detection of the 28KDa calbindin of the human tissues. 28kda calbindin was found in neurons as well as in endocrine and non endocrine cells of humans, its distribution has been found to be the same as what has been reported for other animal species (Jande et al.,1981, Roth et al.,1981; Schreiner et al.,1983; Schreiner et al.,1986; Pochet et al.,1987; Buffa et al.,1988). The pituitary cells identified as alpha-MSH cells by the CLIP (Corticotropin-like intermediate lobe peptide) antiserum show an heavier reaction than TSH secreting cells, although the latter are reported to have higher concentration of receptors for the 1-25(OH)$_2$ vitamin D3 (Stumpf et al.,1987). GH and prolactin cells are devoid of calbindin immunoreactivity like most

Table 2. 28kDa Calbindin immunoreactivity in human tissues

Central Nervous System
Neuronal bodies ++ C/N
Purkinje +++C/N
Ext.granular layer +++N

Pituitary
LH -
FSH -
TSH +/-
ACTH -
ACTH adenoma 0/2
CLIP +++N
CLIP adenoma 1/1

Thyroid*
Parafollicular* +
Medullary carcinoma 18+19
Follicular* -

Lymphonode -

Parathyroid -

Kidney
Glomerular tub. -
Proximal tubules -
Distal tubules +++N/C

Pancreas
Glucagon -
Insulin -
Somatostatin -
Pancreatic polyp. +++N/C
PPoma 2/2

Stomach
ECL-histamine +++N/C
Gastrin -
Somatostatin -

Intestines
Secretin +++C
Colecystokinin -
Gast. inhibit. pept. -
Motilin -

Breast
Ductules -
Ductal carcinoma 2/5
Lobules -
Lobular carcinoma 1/1

Thymus*
Cortical reticular ++
Hassal's Corpuscle ++

Spleen* -

Adrenals
Noradrenalin (?) ++
Adrenalin -

Lung* -

Testis*
Sertoli(?) ++N

+++ Strong ++ Moderate + Weak immunoreactivity
* Tissues from newborns or foetuses
N Nuclear staining, C Cytoplasmatic staining

of the gonadotrophs; occasionally a few of the latter have shown a very weak staining. The thyroid parafollicular cells are weakly stained by the anti-calbindin antiserum during the neonatal period and lack reactivity in the adults; however most of the medullary carcinomas of adults show a more or less consistent population of calbindin immunoreactive cells.

The pancreatic calbindin is in the pancreatic polypeptide secreting cells of both newborns and adults; moreover the gut calbindin is localized in the duodenal secretin cells and in the oxyntic ECL (Hakanson et al.,1986) cells (indirect evidence) but it is not present in the gastric antrum and in the jejunum, ileum and large bowel. Therefore in humans as well as in other mammalians species studied by Buffa et al. (1988), calbindin is in endocrine cells which have a direct or an undirect role in calcium homeostasis like the parafollicular cells which produce calcitonin (Copp 1964), the ECL cells which are a target of the gastrin cells, these cells induce hypocalcemia either releasing calcitonin (Care et al.,1971) or by a mechanism localized in the oxyntic mucosa (Schnlak and Kaplan, 1975). Secretin cells in turn are known to be able to release parathyroid hormone (Fischer et al.,1973; Windeck et al., 1978).

In newborns calbindin was co-localized with the protein S-100 in the reticular cells of the cortex and of the Hassal's corpuscle; this finding is in keeping with the report of Schreiner et al. (1986) on the chicken thymus. These authors demonstrated that this calbindin immunoreactivity was age and diet dependent and suggested for vitamin D a role in monitoring the microenvironnement of the maturing T cells.Our results on the detection of calbindin in the testis need further investigations as we lack a direct evidence for the precise localization of the cell type involved. At present this result is undirectly validated by evidence of the presence of vitamin D receptors in the components of the seminiferous tubules (Levy et al.,1985).

The use of an antiserum elicited against an heterologous antigen coupled to a demasking treatment of the tissues (which may solubilize the share of calbindin more loosely bound) may lower the sensitivity of the immunochemical technique; so we cannot exclude that the distributions of the human calbindin could be wider than we have reported.

ACKNOWLEDGEMENTS

This work was supported in parts by grants from the Italian Consiglio Nazionale delle Ricerche (Progetto Finalizzato TECNOLOGIE BIOMEDICHE). The authors wish to thank Dr D.E.M. Lawson (Institute of Animal Physiology, Babraham, Cambridge, U.K.) for his generous gift of calbindin antiserum. We are also indebted with Doctors J.M.Polak, R.V.Lloyd, N.Yanaihara and M.M.T.O'Hare for their generous supply of the antihormone antisera.   We thank   G.Goglione for printing photographs and S.Tibiletti for typing the manuscript.

REFERENCES

Baimbridge, R.G., Miller, J.T, and Parker, C.O., 1982, Calcium-binding protein distribution in the rat brain. Brain Res., 239:519.

Buffa, R., Mare', P., Salvadore, M., Solcia, E., Furness, J.B., and Lawson, D.E.M., 1989, Calbindin 28KDa in endocrine cells of known or putative calcium-regulating function. Histochemistry, 91:107.

Care, A.D., Bruce, J.B., Bodklins, J., Kennedy, A.D., Conaway, H., and Anast, C.S., 1971, Role of pancreozimin-cholecystokinin and structurally related compounds as calcitonin secretagogues. Endocrinology, 89:262.

Copp, D.H., 1964, Parathyroid, calcitonin and control of plasma calcium. Recent Progr. Horm. Res., 20:59.

Fischer, J.A., Blum, G.N., and Binswanger, V. 1973, Parathyroid hormone response to epinephrine: in vivo J. Clin. Invest., S2:2434.

Hakanson, R., Bottcher, G., Ekblad, E., Panula, P., Simonsson, M., and Dohlsten, M., 1986, Histamine in endocrine cells of the stomach. A survey of several species using a panel of Histamine antibodies. Histochemistry, 86:5.

Jande, S.S., Maler, L., and Lawson, D.E.M., 1981, Immunohistochemical mapping of vitamin D - dependent calcium-binding protein in brain. Nature, 294:765.

Jande, S.S., Tolnai, S., and Lawson, D.E.M.,1981, Immunohistochemical localization of vitamin D - dependent calcium-binding protein in duodenum, kidney, uterus and cerebellum of chickens. Histochemistry, 71:99.

Kondo, H., Kuramoto, H., Iwanaga, T., and Fujita, T., 1985, Cerebellar Purkinje cell-specific protein like immunoreactivity in nor-adrenalin-chromaffin cells and ganglion cells but not in adrenalin-chromaffin cells in the rat adrenal medulla. Arch. Histol. Jpn., 48:421.

Levy, F.O., Eikvar, L., Froysa, A., Cervenka, J., Yoganathan, T., and Hansson, V., 1985, Testicular Calcitriol receptors; appearance during development and presence in adult testicular cell cultures. Vitamin D. A Chemical,Biochemical and Clinical Update pp. 151, 1985 Walter de Gruyter & Co, Berlin.

Pochet, R., Pipeleers, D.G., and Malaisse, W.J., 1987, Calbindin 27KDa: preferential localization in non B-islet cells of the rat pancreas. Biol. Cell., 61:155.

Polak, J.M., and Van Noorden, S., 1983, Immunocytochemistry: applications in pathology and biology. In: Immunocytochemistry today. Techniques and practice. J.M. Polak, and S. Van Noorden, eds., J. Wright, London.

Rindi, G., Buffa, R., Sessa, F., Tortora, O., and Solcia, E., 1986, Chromogranin A,B, and C Immunoreactivities of mammalian endocrine cells.Distribution,distinction from costored hormones/prohormones and relationship with the argyrophil component of the secretory granules. Histochemistry, 85:19.

Roth, J., Bonner-Weir, S., Norman, A.V., and Orci, L., 1982, Immunocytochemistry of vitamin D-dependent calcium-binding protein in chick pamcreas: exclusive localization in B-cells. Endocrinology 110:2216.

Schreiner, D.S., Jande, S.S., and Lawson, D.E.M., 1986, Immunocytochemical localization of vitamin D dependent calcium-binding protein. <u>Brain Res</u>., 222:452.

Schreiner, D.S., Jande, S.S., Parkers, C.O., Lawson, D.E.M., and Thomasset, M., 1983, Immunocytochemical demonstration of two vitamin D-dependent calcium-binding proteins in mammalian kidney. <u>Acta Anat</u>., 117:1.

Schulak, J.A., and Kaplan, E.L., 1975, The importance of the stomach in gastrin-induced hypocalcemia in the rat. <u>Endocrinology</u>, 96:1217.

Sonnenberg, J., Pansini, A.R., and Christakos, S., 1984, Vitamin D dependent rat renal calcium-binding protein : development of a radioimmunoassay, tissue distribution and immunologic identification. <u>Endocrinology</u>, 115:640.

Spencer, R., Chapman, M., Emtage, J.S., and Lawson, D.E.M., 1978, Production and properties of vitamin D induced mRNA from chick calcium-binding protein. <u>Eur. J.Biochem</u>., 71:393.

Stumpf, W.E., and O'Brien, L.P., 1987, Autoradiographic studies with 3H,1-25 dihydroxy vitamin D3 in thyroid and associated tissues of the neck region. <u>Histochemistry</u>, 87:53.

Wilson, P.W., Rogers, J., Harding, M., Pohl, V., Pattyn, G., and Lawson, D.E.M., 1988, Structure of chick chromosomal genes for Calbindin and Calretinin. <u>J. Mol. Biol</u>. 200:615.

Windeck, R., Brown, E.M., Gardner, D.G., and Aurbach, G.D., 1978, Effect of gastrointestinal hormones on isolated bovine parathyroid cells. <u>Endocrinology</u>, 103:2020.

COMPARISON BETWEEN RAT BRAIN CALBINDIN- AND CALRETININ-IMMUNO-REACTIVITIES

A. Résibois,[1] F. Blachier,[2] J.H. Rogers,[3] D.E.M. Lawson,[4] and R. Pochet[1]

[1]Laboratoire d'Histologie, Faculté de Médecine, Université Libre de Bruxelles
2 rue Evers, 1000 Bruxelles, Belgique

[2]Laboratoire de Médecinè Expérimentale, Faculté de Médecine, Université Libre
de Bruxelles, 2 rue Evers, 1000 Bruxelles, Belgique

[3]Physiological Lab, University of Cambridge, Downing street, Cambridge CB2 4EG, UK

[4]Institute of Animal Physiology and Genetics Research, Babraham Hall
Cambridge CB2 4AT, UK

INTRODUCTION

Several calcium-binding proteins are present in the central nervous system including the closely related protein calbindin-D 28K (Taylor, 1974; Baimbridge et al., 1982), and calretinin (Rogers, 1987). The existence of calretinin became apparent with the demonstration by immunoblotting that rat cerebral extracts contained two proteins cross-reacting with calbindin antiserum (Pochet et al. 1985). The two proteins differed in size with one being about 2 kDa larger than the other. Subsequently, calretinin was cloned and sequenced (Rogers, 1987) and appeared to be identical to the larger protein recognized by anti-calbindin. Because of the high degree of homology between calbindin and calretinin (Rogers, 1987; Wilson et al., 1988 and Parmentier, 1989), antiserum against either protein may cross-react with the other and therefore the immunohistochemical mapping in brain must be re-assessed. The existence of a single immunoreactive protein band in gels made from some brain extracts does not rule out the possibility that calbindin antiserum used recognized calretinin. Indeed, calretinin-like immunoreactivity is negative in large rat brain areas such as cerebral cortex, most parts of the thalamus and hippocampus (Rogers et al. 1989). A single protein band immunoreactive for calbindin antiserum may thus mean lack of calretinin rather than no cross-immunoreactivity. Cross reactivity between calretinin antiserum and calbindin is easier to check because calbindin-like immunoreactivity is present nearly everywhere in the rat brain.

The purpose of this study is to compare the mapping of calbindin and calretinin. We shall show that both proteins could accurately be discriminated by appropriate immunolabelling. Auditory pathways, which contain both calbindin-like and calretinin-like immunoreactivities, will be detailed as an example. Our controls were analogous to those recommended by Larsson (1981) in order to discriminate the closely related peptides generated by the processing of pro-opiomelanocortin.

MATERIALS AND METHODS

Animals

The brains of 5 young adult Wistar rats were perfused with 0.9 % NaCl followed by 300 ml Helly's fluid consisting of 2 g $K_2Cr_2O_7$, 1 % Na_2SO_4, 5 % $HgCl_2$ and 4 % fresh formaldehyde (Langeron, 1942). After removal, the brains were cut transversally into 3 parts, post-fixed 6 hrs in Helly's fluid, rinsed twice in PBS pH 7.2, first overnight and then for 2h and finally paraffin embedded. Transverse sections (10 μm) were cut through the whole brains and at 200 μm intervals, three consecutive sections were stained with one of the following: a) luxol fast blue and cresyl violet (to identify nuclei and nerve tracts), b) calbindin antiserum and c) calretinin antiserum.

Immunohistochemistry

Deparaffinized sections were immunolabelled as described by Vacca et al. (1980). In principle the sections were incubated in a 1 % peroxide solution to inactivate endogenous peroxidases and then incubated at 4°C for 40 hrs in a humid atmosphere. Subsequently the sections were incubated with anti-rabbit gamma globulins, PAP complex and diaminobenzidine. The sections were lightly counterstained with haematoxylin to show the neurons no stained with the antisera. The calbindin and calretinin antisera were raised in rabbits using calbindin and the ß-galactosidase-calretinin fusion protein purified to homogeneity. Anti-calbindin was used at a dilution of 1:4000 and anti-calretinin at a dilution of 1:2000.

The specificity tests used replaced the two antisera with non-immune rabbit serum or by diluted antisera preabsorbed with one of the following: 1) purified calbindin,1 μg/ml; 2) purified calretinin, 1 μg/ml; 3) a mixture of both at the same dilution; 4) ß-galactosidase-calretinin fusion protein, 0.3 μg/ml; 5) E. coli ß-galactosidase, 3 μg/ml. These tests were done on consecutive sections, with sufficient frequency to ensure that all described structures were observed in the different experimental conditions.

Protein purification

Calretinin and calbindin were purified respectively from rat brain stem and rat cerebellum by affinity chromatography using anti-calbindin and anti-calretinin coupled to cyanogen bromide activated sepharose (Pasteels et al., 1987). The purified proteins were then electrophoresed on polyacrylamide gels, transferred to nitrocellulose sheets and immunoblotted. The molecular weights of the proteins were calculated using radioactive proteins of known molecular weights.

RESULTS AND DISCUSSION

Western blotting of purified calbindin and calretinin

Extracts of rat brain were applied either to a calbindin or calretinin affinity column and after washing off all unbound proteins the two calcium-binding proteins were released with glycine 0.2 M pH 2.0,NaCl 0.5 M and 10% redistilled dioxane. The two proteins were then separately electrophoresed, blotted and stained with each antiserum to test for cross-reactivity. Anti-calbindin gave some staining of calretinin (Pasteels et al. submitted) whereas anti-calretinin did not stain pure calbindin. The gel electrophoresis showed a 50 kDa band in the two affinity column eluates containing the proteins. This band was stained with all sera including non-immune serum.

It was concluded that the rabbit anti-calretinin does not cross-react with rat calbindin in this system but that anti-calbindin recognises calretinin.

Immunohistochemistry

Specificity of neuronal labelling : Sections of rat brain were stained separately with the two antisera and with the antisera treated with the pure proteins as described under the methods section. Five different types of neurons were observed in the rat auditory pathways (Table1). Type 1 neuron contains only calbindin-like immunoractivity as treating the antisera with calretinin, but not calbindin, had no effect on on the cells stained. Type 2 and 3 neurons stained with anti-calretinin but preincubating the antiserum with calbindin had no effect on staining. The difference between these two neuronal types is that the type 3 neurons stain non-speciffically with anti-calbindin whereas the type 2 neurons do not. The type 4 neurons stained specifically with both antisera i.e., staining with anti-calbindin could only be eliminated by treating the antisera with calbindin and calretinin staining was eliminated by treatment with calretinin. Type 5 neurons stained with neither antiserum.

No cells were stained with non-immune rabbit serum. Incubating either antiserum with bacterial ß-galactosidase did not modify the immunoreactivity observed using pure antiserum. Pre-incubation of calretinin with ß-galactosidase-calretinin fusion protein induced complete disappearance of the DAB reaction whereas, under the same conditions, the calbindin antiserum remained fully active.

Table 1. Specificity of immunoreactivity in rat auditory neurons stained with anti-calbindin and anti-calretinin

Neuronal Type	Calbindin antiserum				Calretinin antiserum Treatment				
	None	CaBP	CaR	Both	None	CaBP	CaR	Both	Content
1	+	-	+	-	-	-	-	-	Calbindin
2	-	-	-	-	+	+	-	-	Calretinin
3	+	-	-	-	+	+	-	-	Calretinin
4	+	-	+	-	+	+	-	-	Both
5	-	-	-	-	-	-	-	-	None

Clearly, in some neurons calbindin antiserum detected calretinin whereas in other neurons this cross-reactivity was not detected possibly because of the low amount of calretinin present. Calretinin antiserum however, showed no cross-reactivity with calbindin as observed by western blotting.

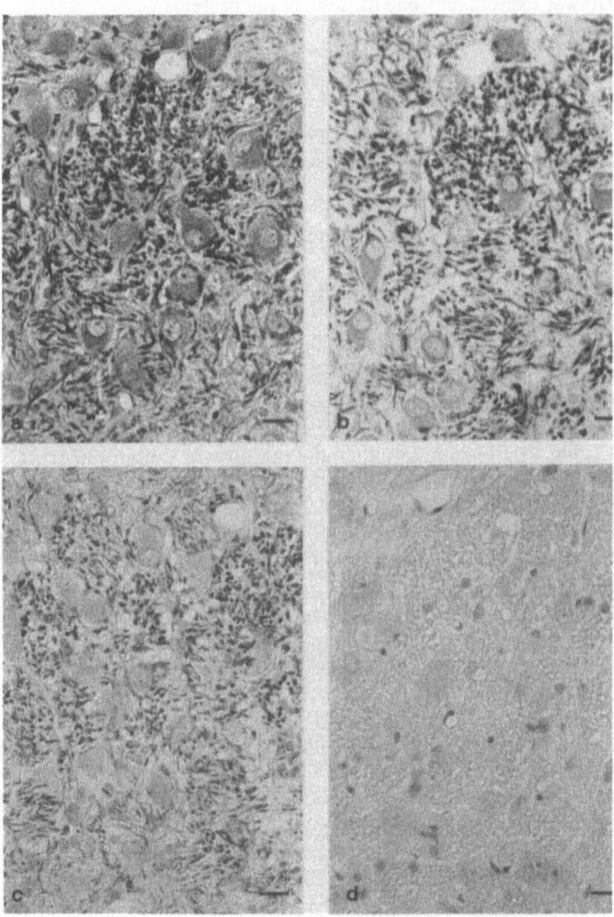

Fig. 1 Ventral cochlear nucleus. The neurons (type 3) are contrasted by calretinin (a) and calbindin (b) antisera but not by calbindin antiserum preincubated either with calretinin (c) or calbindin (d). The fibres of nerve VIII (type 4) are contrasted in a, b and c but not in d. Bar = 20 μm.

Staining pattern of rat auditory neurons with anti-calbindin and anti-calretinin : Neurons belonging to all groups were encountered along the auditory pathway. In the ventral cochlear nucleus, unlabelled neurons (type 5) and calretinin-positive neurons (types 2 or 3) were prominent (Fig.1) while neurons showing both immunoreactivities (type 4) were located in the medial third of the nucleus. Fibres of nerve eight also contained both proteins (Fig.1). Neurons containing calbindin alone were not present. Calretinin immunoreactivity was not encountered in neurons located elsewhere along the auditory pathway but thick calretinin-positive fibers arising from the ventral cochlear nucleus were observed up to the central nucleus of the inferior colliculus. True calbindin immunoreactivity (type 1) was observed in all the neurons of the medial nucleus of the trapezoid body, in very rare neurons of the lateral superior olive and in the ventral and dorsal nuclei lemnisci lateralis. As a consequence calbindin-immunoreactive fibres were far more numerous than calretinin-immunoreactive fibres in superior olive complex, lateral lemniscus and inferior colliculus. Their immuno-reactivity was unchanged if a calbindin antiserum pre-incubated with calretinin was used.

CONCLUSIONS

Immunohistochemical techniques exist to distinguish calbindin from calretinin containing cells. If the auditory pathway is typical of other neuronal paths, both proteins are rarely found in the same neurons. Calretinin is prominent in the auditory nerve and ventral cochlear nucleus while calbindin is more frequently encountered at more central levels. Unlabelled neurons are numerous in all nuclei except the medial nucleus of the trapezoid body . The specific expression of these two highly homologous proteins in neurons which are closely related raises two questions : a) does the two proteins have different functions ? and b) why do cells within the same pathway express only one or none of the proteins ?

ACKNOWLEDGEMENTS

This work was financially supported by the "Fonds de la Recherche Scientifique Médicale" (grant n° 3.4511.88), an EEC stimulation grant and a British Council grant in aid.

REFERENCES

Baimbridge, K.G., Miller, J.J., and Parkes, C.O., 1982, Calcium-binding protein distribution in the rat brain, Brain Res., 239:519-525.

Langeron, M., 1942, Précis de Microscopie, Collection de Précis Médicaux, Masson et Cie.

Larsson, L.-I., 1981, A novel immunocytochemical model system for specificity and sensitivity screening of antisera against multiple antigens, J. Histochem. Cytochem., 29:408-410.

Parmentier, M., 1989, Structure of the human cDNAs and genes coding for calbindin D28K and calretinin. in: Calcium Binding Proteins in Normal and Transformed Cells, R. Pochet, D.E.M. Lawson and C.W. Heizmann, Plenum, New York

Pasteels, B., Miki, N., Hatakenaka, S. and Pochet, R., 1987, Immunohistochemical cross-reactivity and electrophoretic comigration between calbindin D-27KDa and visinine, Brain Res., 412:107-113.

Pasteels, B., Rogers, J. Blachier, F., and Pochet, R. (1989), Calbindin and Calretinin localization in retina from different species, Visual Neuroscience, submitted

Pochet, R., Parmentier, M., Lawson, D.E.M., Pasteels, J.L., 1985, Rat brain synthesizes two vitamin D-dependent calcium-binding proteins, Brain Res., 345-251.

Rogers, J.H., 1987, Calretinin : a gene for a novel calcium-binding protein expressed principally in neurons, J. Cell Biol., 105:1343-1353.

Rogers, J.H., Khan, M., and Ellis, J., 1989, Calretinin and other CaBPs in the nervous system, in: Calcium Binding Proteins in Normal and Transformed Cells, R. Pochet, D.E.M. Lawson and C.W. Heizmann, eds., Plenum, New York

Taylor, A.N., 1974, Chick brain calcium-binding protein: comparison with intestinal vitamin D-induced calcium-binding protein. Arch. Biochem. Biophys., 161:100-108.

Vacca, L.L., Abrahams, S.J., Naftchi, N.E., 1980, A modified peroxidase-antiperoxidase procedure for improved localization of substance P in rat spinal cord. J. Histochem. Cytochem., 28:297-304.

Wilson, P.W., Rogers, J., Harding, M. Pohl V., Pattyn, G., and Lawson, D.E.M., 1988, Structure of chick chromosomal genes for calbindin and calretinin. J. Mol. Biol., 200:615-625.

CONTRIBUTORS

AUNIS Dominique
Groupe de Neurobiologie Structurale
et Fonctionelle
Unité INSERM U-44
Centre de Neurochimie du CNRS
5 rue Blaise Pascal
67084 Strasbourg Cedex France

BADER Marie-France
Groupe de Neurobiologie Structurale
et Fonctionelle
Unité INSERM U-44
Centre de Neurochimie du CNRS
5 rue Blaise Pascal
67084 Strasbourg Cedex France

BANVILLE Denis
BioTechnology Research Institute
National Research Council
Montreal H4P 2R2 Canada

BARGER Steven W.
Department of Cell Biology
Vanderbilt University
Nashville, Tn 37232 USA

BATTINI Renata
Istituto di Chimica Biologica
Universita di Modena
Via Campi 287
41100 Modena Italy

BAUDIER Jacques
Centre de Neurochimie du CNRS
INSERM U. 44
5 Rue Blaise Pascal
67084 Strasbourg France

BERCHTOLD Martin W.
Institute of Pharmacology and Biochemistry
University of Zürich-Irchel
CH-8057 Zürich Switzerland

BERGER Marianne C.
Institute of Pharmacology and Biochemistry
University of Zürich-Irchel
CH-8057 Zürich Switzerland

BLACHIER F.
Laboratory of Experimental Medicine
Brussels Free University
1000 Brussels Belgium

BLUM Janaki K.
Institute of Pharmacology and Biochemistry
University of Zürich-Irchel
CH-8057 Zürich Switzerland

BREHIER Arlette
INSERM, U 120
44 Chemin de Ronde
78110 Le Vésinet France

BREWER Linda M.
Division of Biological Sciences
National Research Council
Ottawa K1A OR6 Canada

BUFFA Roberto
Dipartimento di Patologia Umana
ed Ereditaria dell'Universita' di Pavia
Sezione di Anatomia Patologica 2
viale Borri 57
Varese Italy

BZDEGA T.
Department of Biological Chemistry
University of Maryland
School of Medicine
660 West Redwood Street
Baltimore, Maryland 21201 USA

CARAFOLI Ernesto
Laboratory of Biochemistry
Swiss Federal Institute of Technology (ETH)
Zurich Switzerland

CELIO M. R.
Institut für Anatomie
Christian Albrechts Universität zu Kiel
Olshausenstraße 40
D-2300 Kiel West-Germany

COX Jos A.
Department of Biochemistry

University of Geneva
30, Quai Ernest Ansermet
1211 Geneva 4   Switzerland

CRAIG T.A.
Department of Pharmacology
Vanderbilt University and
Laboratory of Cellular and Molecular Physiology
Howard Hughes Medical Institute
Nashville, Tennessee 37232   USA

CRUMPTON Michael J.
Cell Surface Biochemistry Laboratory
Imperial Cancer Research Fund
Lincoln's Inn Fields
London WC2A 3PX   UK

DELOULME Jean Christophe
Centre de Neurochimie du CNRS
INSERM U. 44
5 Rue Blaise Pascal
67084 Strasbourg   France

DONATO Rosario
Section of Anatomy
Department of Exper. Med. and Biochem. Sciences
Cas. Post. 81
06100 Perugia Succ. 3   Italy

DORIN Julia
MRC Human Genetics Unit
Western General Hospital
Edinburgh, EH4 2XU   UK

DRAKENBERG T.
Physical Chemistry 2
Chemical Centre
University of Lund
P.O. Box 124
S-221 00 Lund   Sweden

DUPRET Jean-Marc
INSERM, U 120
44 Chemin de Ronde
78110 Le Vésinet   France

EICHINGER Ludwig
Max-Planck-Institute for Biochemistry
8033 Martinsried   West Germany

ELLIS Jon
Physiological Laboratory
University of Cambridge
Downing street
Cambridge CB2 3EG   UK

FERRARI Stefano
Istituto di Chimica Biologica
Universita di Modena
Via Campi 287
41100 Modena   Italy

FILIPEK Anna
Nencki Institute of Experimental Biology

3 Pasteur street
02-093 Warsaw   Poland

FORSEN Sture
Physical Chemistry 2
Chemical Centre
University of Lund
P.O. Box 124
S-221 00 Lund   Sweden

GERGELY John
Department of Muscle Research
Boston Biomedical Research Institute
Department of Biological Chemistry
and Molecular Pharmacology
Harvard Medical School, Neurology Service
Massachusetts General Hospital
Boston MA   USA

GERKE Volker
Max-Planck-Institute for Biophysical Chemistry
Goettingen   West Germany

GINI Ambrogio
Dipartimento di Patologia Umana
ed Ereditaria dell'Universita' di Pavia
Sezione di Anatomia Patologica 2
viale Borri 57
Varese   Italia

GRABAREK Zenon
Department of Muscle Research
Boston Biomedical Research Institute
Department of Biological Chemistry
and Molecular Pharmacology
Harvard Medical School, Neurology Service
Massachusetts General Hospital
Boston MA
USA

GREGERSEN H.- J.
Institut für Anatomie
Christian Albrechts Universität zu Kiel
Olshausenstraße 40
D-2300 Kiel   West-Germany

GUERRA-SANTOS L.
Department of Pharmacology
Vanderbilt University and
Laboratory of Cellular and Molecular Physiology
Howard Huges Medical Institute
Nashville, Tennessee 37232   USA

HAGIWARA Masatoshi
Department of Pharmacology
Nagoya University, School of Medicine
Showa-ku, Nagoya 466   Japan

HAIECH Jacques
LCB, CNRS
31 Chemin Joseph Aiguier
13009 Marseille   France

HEIZMANN Claus W.
Universitäts-Kinderklinik

Abtlg für Klin. Chemie
Sternwiesstraße 75
CH-8032 Zürich   Switzerland

HIDAKA Hiroyoshi
Department of Pharmacology
Nagoya University, School of Medicine
Showa-ku, Nagoya 466   Japan

JAMES Peter
Laboratory of Biochemistry
Swiss Federal Institute of Technology (ETH)
Zurich   Switzerland

JOHANSSON C.
Physical Chemistry 2, Chemical Centre
University of Lund
P.O. Box 124
S-221 00 Lund   Sweden

JOHNSON J.D.
Department of Physiological Chemistry
The Ohio State University, Medical Center
Columbus, Ohio 43210   USA

JOHNSON Peter M.
Dept. of Immunology
University of Liverpool
PO Box 147
Liverpool   UK

KAEGI Urse
Universitäts-Kinderklinik
Abtlg für Klin. Chemie
Sternwiesstraße 75
CH-8032 Zürich   Switzerland

KAMPHUIS Willem
Department of General Zoology
University of Amsterdam
Amsterdam   The Netherlands

KAY Brian K.
Department of Biology
University of North Carolina at Chapel Hill
Chapel Hill NC 27599-3280   USA

KENTON Paul
Dept. of Immunology
University of Liverpool
PO Box 147
Liverpool   UK

KHAN Masood
Physiological Laboratory
University of Cambridge
Downing street
Cambridge CB2 3EG   UK

KILHOFFER M.-C.
Laboratoire de Biophysique
Université Louis Pasteur
Faculté de Pharmacie de Strasbourg

BL No.10
67048 Strasbourg Cedex   France

KOHAMA Kazuhiro
Department of Pharmacology
Gunma University School of Medicine
Maebashi 371   Japan

KORDEL J.
Physical Chemistry 2, Chemical Centre
University of Lund
P.O. Box 124
S-221 00 Lund   Sweden

KOSK-KOSICKA Danuta
Department of Biological Chemistry
University of Maryland
School of Medicine
660 West Redwood Street
Baltimore, Maryland 21201   USA

KREBS Joachim
Laboratory of Biochemistry
Swiss Federal Institute of Technology (ETH)
Zurich   Switzerland

KUZNICKI Jacek
Nencki Institute of Experimental Biology
3 Pasteur street
02-093 Warsaw   Poland

LAI F. Anthony
Departments of Biochemistry and Physiology
University of North Carolina
School of Medicine
Chapel Hill, NC 27599   USA

LANGLEY Keith
Groupe de Neurobiologie Structurale
et Fonctionelle
Unité INSERM U-44
Centre de Neurochimie du CNRS
5 rue Blaise Pascal
67084 Strasbourg Cedex   France

LAWSON D.Eric M.
Institute of Animal Physiology
and Genetics Research
Babraham Hall
Cambridge CB2 4AT   UK

LINSE S.
Physical Chemistry 2, Chemical Centre
University of Lund
P.O. Box 124
S-221 00 Lund   Sweden

LOMRI Nour-eddine
INSERM, U 120
44 Chemin de Ronde
78110 Le Vésinet   France

LUKAS T.J.
Department of Pharmacology,

217

Vanderbilt University
and Laboratory of Cellular
and Molecular Physiology
Howard Hughes Medical Institute
Nashville, Tennessee 37232   USA

MABUCHI Yasuko
Department of Muscle Research
Boston Biomedical Research Institute
Department of Biological Chemistry
and Molecular Pharmacology
Harvard Medical School, Neurology Service
Massachusetts General Hospital
Boston MA   USA

MAC MANUS John P.
Division of Biological Sciences
National Research Council
Ottawa K1A OR6   Canada

MALAISSE Willy J.
Laboratory of Experimental Medicine
Brussels Free University
1000 Brussels   Belgium

MANUEL Y KEENOY B.
Laboratory of Experimental Medicine
Brussels Free University
1000 Brussels   Belgium

MARE' Paolo
Dipartimento di Patologia Umana ed Ereditaria
dell'Universita' di Pavia
Sezione di Anatomia Patologica 2
viale Borri 57
Varese   Italy

MEISSNER Gerhard
Departments of Biochemistry and Physiology
University of North Carolina
School of Medicine
Chapel Hill, NC 27599   USA

MOSS Stephen E.
Cell Surface Biochemistry Laboratory
Imperial Cancer Research Fund
Lincoln's Inn Fields
London WC2A 3PX   UK

NEMCEK K.
Department of Physiological Chemistry
The Ohio State University
Medical Center
Columbus, Ohio 43210   USA

NOEGEL Angelika A.
Max-Planck-Institute for Biochemistry
8033 Martinsried   West Germany

OKAGAKI Tsuyoshi
The Physical Laboratories
Nihon University at Narashino
Funabashi 274   Japan

PARMENTIER Marc
I.R.I.B.H.N., ULB Campus ERASME
808 route de Lennik
1070 Bruxelles   Belgium

PERRET Christine
INSERM, U 120
44 Chemin de Ronde
78110 Le Vésinet   France

PIKE Wesley J.
Baylor College of Medicine
Department of Pediatrics
Endocrinology and Metabolism Section
8080 N. Stadium Drive
Houston, TX 77054   USA

POCHET Roland
Laboratoire d'Histologie
Faculté de Médecine
Université Libre de Bruxelles
2 rue Evers
1000 Bruxelles   Belgium

RESIBOIS Anne
Laboratoire d'Histologie
Faculté de Médecine
Université Libre de Bruxelles
2 rue Evers
1000 Bruxelles   Belgium

ROGERS John.H.
Physiological Lab
University of Cambridge
Downing street
Cambridge CB2 4EG   UK

ROHRENBECK Jürg
Max-Planck Institute for Brain Research
D-6000 Frankfurt 71   West Germany

ROUFOGALIS Basil D.
Laboratory of Molecular Pharmacology
Faculty of Pharmaceutical Sciences
University of British Columbia
Vancouver, B.C. V6T 1W5   Canada

SALVADORE Maurizio
Dipartimento di Patologia Umana ed Ereditaria
dell'Universita' di Pavia
Sezione di Anatomia Patologica 2
viale Borri 57
Varese   Italy

SCAILLET S.
Department of Biological Chemistry
University of Maryland
School of Medicine
660 West Redwood Street
Baltimore, Maryland 21201   USA

SCHLEICHER Michael
Max-Planck-Institute for Biochemistry
8033 Martinsried   West Germany

SENER A.
Laboratory of Experimental Medicine
Brussels Free University
1000 Brussels Belgium

SENSENBRENNER Monique
Centre de Neurochimie du CNRS
INSERM U. 44
5 Rue Blaise Pascal
67084 Strasbourg France

SIMON Jean-Pierre
Groupe de Neurobiologie Structurale
et Fonctionelle
Unité INSERM U-44
Centre de Neurochimie du CNRS
5 rue Blaise Pascal
67084 Strasbourg Cedex France

SOMMER Ernst W.
Institute of Pharmacology and Biochemistry
University of Zürich-Irchel
CH-8057 Zürich and Institute of Toxicology
ETH-Zürich
CH-8603 Schwerzenbach Switzerland

SONTAG Jean-Marie
Groupe de Neurobiologie Structurale
et Fonctionelle
Unité INSERM U-44
Centre de Neurochimie du CNRS
5 rue Blaise Pascal
67084 Strasbourg Cedex France

THOMASSET Monique
INSERM, U 120
44 Chemin de Ronde
78110 Le Vésinet France

THULIN E.
Physical Chemistry 2, Chemical Centre
University of Lund
P.O. Box 124
S-221 00 Lund Sweden

TOKUMITSU Hiroshi
Department of Pharmacology
Nagoya University
School of Medicine
Showa-ku, Nagoya 466 Japan

VAN ELDIK Linda J.
Department of Pharmacology
Vanderbilt University
Nashville, Tn 37232 USA

van HEYNINGEN Veronica
MRC Human Genetics Unit
Western General Hospital
Edinburgh, EH4 2XU UK

VILLALOBO Antonio
C.S.I.C., Instituto de Investigaciones Biomedicas
Madrid Spain

VORHERR Thomas
Laboratory of Biochemistry
Swiss Federal Institute of Technology (ETH)
Zurich Switzerland

WANG Kevin K.W.
Laboratory of Molecular Pharmacology
Faculty of Pharmaceutical Sciences
University of British Columbia
Vancouver, B.C. V6T 1W5 Canada

WATTERSON D.Martin
Department of Pharmacology,
Vanderbilt University and
Laboratory of Cellular and Molecular Physiology
Howard Hughes Medical Institute
Nashville, Tennessee 37232 USA

WAWRZYNOW A.
Department of Biological Chemistry
University of Maryland, School of Medicine
660 West Redwood Street
Baltimore, Maryland 21201 USA

WEBB Paul D.
Dept. of Immunology
University of Liverpool
PO Box 147
Liverpool UK

WILLIAMS R.J.P.
Inorganic Chemistry Laboratory
University of Oxford
South Parks Road
Oxford OX1 3QR UK

WILSON E.
Department of Pharmacology,
Vanderbilt University and
Laboratory of Cellular and Molecular Physiology
Howard Hughes Medical Institute
Nashville, Tennessee 37232 USA

WITKE Walter
Max-Planck-Institute for Biochemistry
8033 Martinsried West Germany

ZIMMER Warren E.
Department of Pharmacology
Vanderbilt University
Nashville, Tn 37232 USA

INDEX